The Frontiers Collection

The books in this collection are devoted to challenging and open problems at the forefront of modern science and scholarship, including related philosophical debates. In contrast to typical research monographs, however, they strive to present their topics in a manner accessible also to scientifically literate non-specialists wishing to gain insight into the deeper implications and fascinating questions involved. Taken as a whole, the series reflects the need for a fundamental and interdisciplinary approach to modern science and research. Furthermore, it is intended to encourage active academics in all fields to ponder over important and perhaps controversial issues beyond their own speciality. Extending from quantum physics and relativity to entropy, consciousness, language and complex systems—the Frontiers Collection will inspire readers to push back the frontiers of their own knowledge.

There are many modern definitions of consciousness stemming from the various philosophical, scientific, and even spiritual insights. It is generally accepted that consciousness is the condition of being aware of one's surroundings and one's own existence. In short, consciousness can be simply described as awareness of awareness. While this sounds trivial, it is an elusive and enigmatic problem for rigorous scientific inquiry, which remains hotly debated by neuroscientists and researchers in related fields. For most of the last century, the issue has been approached very hesitantly by the mainstream of the physical and life sciences. This has been in part due to the entrenched view that consciousness is non-material, and therefore outside the scope of science. This attitude has been gradually changing, and numerous studies on both conscious and unconscious processes have contributed to our current knowledge. Scientists generally agree that consciousness is a function of the brain, and since the brain is a material entity, consciousness should be a legitimate subject of scientific research. Therefore, information processing models such as neural networks could provide a useful framework for understanding the human mind. Are these models sufficient to grasp the foundational aspects of consciousness? Neuroscience investigates the mind by closely analyzing biological mechanisms underlying cognitive functions and focuses on the neuron-level processes and their organ- and organism-level manifestations. This addresses the questions of how cognitive functions are produced by the neural circuitry in a bottom-up hierarchy. The question still remains, however, whether we can associate the mind with the brain. Various theories propose conflicting answers to this question, and some of them lead to philosophical, spiritual and religious considerations, which scientists commonly distance themselves from. Staying within the realm of scientific inquiry, we encounter another burning question, which is whether we need to use the most advanced physical theories to understand consciousness or can we explain consciousness using deterministic classical physics concepts.

Greek philosopher Plato (429–347 BCE) set the stage for the millennia-long debate over consciousness when he discussed the differences between the material world (our body) and the immaterial world (the individual experience of the world created in our mind). He called this approach the "theory of forms." Plato believed that ideas were the basic building blocks of reality and the material world was just a shadow of the world of ideas. For Plato consciousness was a part of the ideal world creating imperfect mental images of the real world. His view that mind and brain are separate is called "mind-brain dualism," and it has been embraced by future generations of philosophers and scientists. Plato used the allegory of the cave to illustrate that what we do trying to understand the world through the use of our senses is metaphorically akin to watching shadows of the real world creating its image but not perceiving the actual external reality.

Fast forward 1600 years to Thomas Aquinas (1225–1274) and his discourse about mind and brain we find out that he did not view them as two separate entities interacting with each other, but instead as one and the same. However, he also argued for the existence of a separate, incorruptible soul. Thus, his philosophical position was somewhere between mind-brain dualism and the view that all we have is material existence. A half a millennium later, a modern-era debate about consciousness was

Series Editor's Foreword

Changes in scientific paradigms, also called scientific revolutions, typically occur suddenly and hence are sometimes called black swan events. However, these paradigm shifts sometimes are preceded by a gradual process of maturation of new ideas. This is how quantum mechanics burst into the scene of modern science exactly in 1900 ushered by Max Planck's discovery of photon quantization. While the emergence of quantum ideas was sudden and unexpected, it was only possible due to decades of painstaking experimentation with molecules and atoms as well as centuries of the development of sophisticated mathematical tools such as functional analysis and differential equations. Some parallels to these revolutionary changes in physics a 100 years ago can now be found in the life sciences where a wealth of data concomitant with the refinement of experimental techniques has prepared us for an outburst of creativity at the conceptual level. One such paradigm shift in the making is perhaps the case of quantum biology with specific inroads of quantum physics being made into neuroscience. This book introduces the reader to this nascent paradigm shift in our understanding of the brain and the mind.

The human brain is one of the most complex and fascinating structures in the world. The question about the molecular mechanisms involved in the human brain's functioning has been of great interest to not only neurophysiologists but also to information scientists, biophysicists and psychologists for decades. While much is known about the types, organizational structure of brain cells, and their electrical and biochemical activities, little is known about such enigmatic issues as where our memories are stored, or which molecular mechanisms are involved in information processing by brain cells. Speculations about these molecular mechanisms behind cognition abound and range from mundane to exotic. In the past few decades, heated debates centered on the possibility that at least some cognitive functions operate at a quantum level. Experimental demonstrations of such ideas are just starting to be tried owing to the recently acquired availability of modern experimental methods and instrumentation. What underlies all this is our difficulty to grasp not only the physical mechanism behind it, but even the definition of consciousness.

The word consciousness is derived from the Latin *conscientia*, meaning to know with, or to know together, and was originally used in the sense of moral conscience.

Josh Roeloffs
Westminster, CA, USA

ISSN 1612-3018 ISSN 2197-6619 (electronic)
The Frontiers Collection
ISBN 978-3-032-20067-9 ISBN 978-3-032-20068-6 (eBook)
https://doi.org/10.1007/978-3-032-20068-6

This Springer imprint is published by the registered company Springer Nature Switzerland AG
The registered company address is: Gewerbestrasse 11, 6330 Cham, Switzerland

Josh Roeloffs

The Classical Brain to the Quantum Mind

How Wavefunction Collapse Manifests Conscious Experience

Springer

ignited by philosophers such as Descartes in France and the Cambridge Platonists in England. Descartes considered consciousness to be a fundamental property of the mind and in fact defined thought as everything that is within us that we are aware of, which really means consciousness. For Descartes the form of thought represented an idea and his views can be briefly summarized as: (1) consciousness makes thoughts transparent to the mind, (2) consciousness involves reflection, and (3) conscious thought is intentional.

Spinoza envisaged consciousness to be mainly a moral issue, and he argued that "the order and connection of ideas is the same as the order and connection of things," which means that for each object, there is a corresponding idea, which is called mind-body parallelism. Hobbes understood the meaning of consciousness as shared knowledge, which is related to the function of conscience as a witness. John Locke, in his theory of personal identity, separated personal identity from the identity of matter and concluded that thinking is not the essence of the soul.

Leibniz was the first philosopher to propose a self-consistent theory of consciousness and described non-conscious mental states. Leibniz's theory of consciousness, being based on representations, allowed for transitions from conscious to non-conscious states. He regarded consciousness as a fully natural theory that opens up a new science of the mind, which he called pneumatology presaging the development of modern cognitive sciences. He further argued for a threshold theory of consciousness opening an avenue of investigations based on evolutionary origins of the emergence of consciousness. This brings us to the modern era when revolutionary advances in neuroscience and related fields have taken place with major consequences for the theory of consciousness.

The first connection between neuroscience and physics was made by Luigi Galvani who studied the effects of electricity on frogs in the late 1700s. He showed that animal tissues have intrinsic electric properties that enable physiological processes such as nerve conduction and muscle contraction. He, therefore, established a scientific foundation of electrophysiology which, after two centuries of studies led to the modern understanding of how nerves and muscles work. Ramón y Cajal pioneered investigations of the microscopic structure of the brain. More than a century since their creation, many of his illustrations of the arborizations of brain cells are still a useful educational tool. Among his numerous contributions to neuroscience, arguably the most important ones are the discovery of the axonal growth cone, and the experimental demonstration of the existence of gaps between nerve cells. These insights provided a solid basis for the so-called "neuron theory", which is the foundation of modern neuroscience. Cajal also hypothesized that neurons have an ability to grow in an adult and to create new connections, which explains how the brain makes new synaptic connections between neurons in the process of learning providing a basis for the synaptic theory of memory.

It took over 150 years from Galvani's experiments to discover how exactly neurons propagate electrical signals, which Sir Alan Hodgkin and Sir Andrew Huxley jointly uncovered while working with giant squid axons as model systems. Their work led to the discovery of the action potential whose mathematical description is now known as the Hodgkin–Huxley model of the neuron represented as an electrical circuit. Capacitors model the electrical charge storage of the cell membrane while resistors model the conductive properties of the membrane's ion channels and batteries represent the difference between the ion concentrations inside and outside the neuron, respectively. Hodgkin and Huxley shared their 1952 Nobel Prize in Medicine and Physiology with John Eccles who worked on the functions of synaptic connections between neurons. Eccles conceptualized the existence of three worlds encompassing both material and mental spheres. World 1 is the world of physical objects and states. It comprises matter and energy, including human brains, and man-made artifacts. World 2 forms our primary reality as the world of states of consciousness, knowledge and the sphere of our perceptions. Our conscious experiences are the basis of knowledge of World 1, which is thus a world of secondary reality. World 3 is the world of culture, and hence, it is restricted to humans, being completely unknown to animals. Eccles is perhaps the first neuroscientist to raise the hypothesis that mind-brain interactions are based on quantum physics. He envisaged these quantum effects to occur in the dendritic bundles of the cerebral cortex. As is well known, the release of vesicles from synaptic boutons induces a postsynaptic potential change and thus influences the electrical activity of the connected neurons. According to Eccles, the transfer of these vesicles between neurons is a quantum tunneling phenomenon, although there is no experimental evidence for this to be the case.

Subsequently, Nobel-Prize winning neuroscientist Eric Kandel discovered that as we learn, chemical signals change the structure of synaptic connections and showed that short-term and long-term memories are formed by electrical signals transmitted across synapses in a process called long-term potentiation (LTP). Specifically, repetitive pre-synaptic stimulation increases post-synaptic sensitivity and hence strengthens synapses. This is aptly expressed by saying that neurons that fire together, wire together. While LTP has provided a glimpse into the nature of learning and memory, the issue appears to be much more complicated. First, while long-term memory endures, LTP does not make permanent changes to synaptic strength but decays over hours to months. Second, LTP-based memory models suffer from the loss of signal fidelity. Third, memory requires the assimilation of information across different sensory inputs involving huge neural circuits that need to be integrated. Finally, external stimuli are associated within their context, so that new experiences are influenced by the current context, previous experiences, and even anticipation of the future. To explain this with only synaptic-based computing may easily exceed the capabilities of the brain. This indicates that another level of computation in the brain is needed for learning and memory functions, perhaps at deeper molecular levels in the interiors of neurons.

It follows from the above that higher cognitive functions, including learning, memory and consciousness, may represent emergent properties resulting from the ever-increasing computational complexity of the brain. This argument has been very eloquently presented by Alwyn Scott in his book: *Stairway to the Mind*. Within the current neuroscience paradigm, neurons are the basic computational units in the brain, and the connections between neurons, namely synapses, are seen as information relays. Hence, the brain can be viewed as a complex neural network. Within this network neurons work together transmitting electrical signals that follow the laws of classical physics to produce an output. Neurons are further assumed to be either "on" when sending an electrical signal, or "off" when at rest, much like switches in electrical networks. This viewpoint has led to the claim that consciousness arises from increasing computational complexity of the assemblies of neurons in the brain. When a critical level of complexity among neuronal assemblies is reached, interacting neurons create a conscious experience. Learning is then seen as causing modifications to the networks of neurons at the level of their synapses. Consequently, this effect is referred to as synaptic plasticity. Credit for this insight must be given to W. James who already in 1890 foresaw the ability of the brain to rewire itself, which he called neuroplasticity. It was not until 1949 that D. Hebb proposed that the plastic reorganization of synapses (synaptic plasticity) was the likely physical mechanism of learning and memory. However, molecular mechanisms of learning, long-term memory, and other cognitive processes including consciousness itself remain not only incomplete but very enigmatic to this day.

A consensus view about key characteristics of consciousness boils down to the following characteristics summarized succinctly by Donald Hoffman:

- Binding/unification: viewing an object as a composite whole with discernable features.
- Global accessibility and broadcast: back and forth interactions among specialized brain domains that inform decision-making.
- Selective attention and active exclusion: excluding or amplifying signals according to past and present context.
- Intentionality: the mapping of body, world, action, and their relations.
- Integration through time: holding on to incoming information long enough for it to be integrated and evaluated.
- Flexible evaluative system: evaluating perceptions and actions for rewards or penalties according to context.
- Agency and embodiment: inherent spontaneous activity and goal-directed behavior.
- A sense of self: recognition of self/other within a stable perspective.

Starting in 1994, annual conferences on the Science of Consciousness have been organized by Stuart Hameroff with venues alternating between Tucson, Arizona and overseas locations. These conferences have provided a broad forum for the development of concepts in the field of consciousness. Much of their focus has been on the Orchestrated Objective Reduction (Orch OR), a theory proposed jointly by Sir Roger Penrose and Stuart Hameroff. In addition to these conferences, the topic of consciousness was brought to the mainstream of scientific discourse through two books published by Sir Roger Penrose, namely: *Shadows of the Mind* and *The Emperor's New Mind*, which argued that the human brain cannot operate entirely in an algorithmic fashion. A decades-long collaboration between Penrose and Hameroff resulted in numerous papers wherein arguments were produced for the quantum nature of human consciousness with its substrate being the microtubule bundles in the brain's neurons. The key issue that emerged in this theory was the cause of the wave function collapse. Penrose and Hameroff argued that the collapse is spontaneous but orchestrated by gravitational forces acting between each tubulin dimer's quantum states in a superposition. This is a very unconventional hypothesis for a very "down-to-earth" field such as neuroscience. Orch OR claims that consciousness arises when gravitational instabilities in the fundamental structure of space-time cause the collapse of quantum wave functions in neuronal microtubules. The connection to quantum gravitational effects dates back to the late 1980s, when Penrose proposed that any observation of a quantum particle forces an interaction with the classical gravitational field associated with the apparatus, creating a conflict that drives the quantum particle to collapse into a definite state. Penrose went further, postulating that each time a quantum wave function collapses in the brain, and it gives rise to a moment of conscious experience. Hameroff connected Penrose's idea with the hypothesis that microtubules inside neurons are an ideal substrate for quantum effects due to their physical properties and biological functions. This has remained in the sphere of speculative hypotheses until 2023 when Aarat Kalra in the lab of Gregory Scholes demonstrated that quantum excitations triggered by ultraviolet quanta of light, photon, became absorbed by microtubules and created electronic quantum states in their constituent building blocks, tubulin dimers. This series of experiments provided evidence for the existence of quantum states in microtubules under laboratory conditions, and these states were adversely affected by the presence of anesthetic molecules in the solution. While providing support for the general idea of Orch OR, the observed quantum states extended only over distances comparable to the size of a tubulin dimer and existed only for a few nanoseconds. Although these length and time scales exceed those measured in another quantum biology effect, namely photosynthesis, they fall short of the distances and durations expected to play a role on neurophysiology. Moreover, an Italian team led by Câtâlina Curceanu tested the effects of gravity on quantum wavefunction collapse and did not observe any measurable evidence of gravity causing quantum collapse. It should be qualified though that Curceanu's team results do not rule out Orch OR theory in general, only variants of Orch OR based on the simplest version of the so-called Diósi-Penrose theory of gravity-related dynamical wavefunction collapse. This brings us to the current state of the art in the quantum consciousness field and the book by Josh Roeloffs *The Classical Brain to the Quantum Mind*.

This book provides a generally accessible overview of the field of consciousness studies ranging from psychological observations of human behavior to the quantum physics-based hypotheses of the mechanisms involved in the functioning of the brain. This is not a "tour de force" exposition of arcane ideas that only experts can grasp. The book explains difficult concepts in easy-to-understand terms and makes an effort to appeal to the lay reader. This doesn't mean that scientific rigor is sacrificed in exchange for simplicity. There is a clear message presented in this book and a build-up of evidence in support of the over-riding conceptual framework the author constructed. The message is that the brain operates on two interlocking levels: external and internal, slow and fast. These two levels can also be assigned classical and quantum characteristics, respectively. The conundrum of the wavefunction collapse may be resolved by the interactions between these two levels, and I find this new perspective both convincing and refreshing. This book is a must-read for anyone who is interested and intrigued by the enigma of consciousness, and especially by those of us who search for answers to such questions as: "how come autistic people and savants possess amazing cognitive abilities in some areas and lack them in others?", "where do sudden insights into complex problems come from?", "how can top athletes achieve their feats of physical performance which often defy the laws of physiology", "why can't subjective experiences be really shared with other people", "can free will be consistent with deterministic laws of physics?", etc. I invite the reader on a journey of exploration in search of answers to these questions with Josh Roeloffs as your tour guide.

Turin, Italy
October 2025

Jack A. Tuszynski

Preface

The author wants to acknowledge that the author does not feel like the "founder" of the ideas presented in this book. Instead, the author feels more like a translator who helped collect a few emerging phenomena together into one document so that the collective consciousness of humanity can explore its possibilities together.

This framework emerged from a simple conversation about relationship dynamics, miscommunication between varying degrees of consciousness, and underperformance from self-observation. Following the discussion, an insight emerged that it seems as though consciousness levels naturally manifest in human structures (business, governments, organizations) and then another insight arose that the ego observing one's performance which can result in the interruption of a flow state is similar in interesting ways to the observer's paradox in physics. This cascaded in a few minutes of thinking and note taking, where then, over the course of a few hours many of the core hypotheses of this book were present.

The author felt as though the book was writing itself, as the first draft was completed in less than 14 days of nearly continuous flow state following the initial insights. There was a noteworthy aspect of this process. The author, after the first insights and ideas arose, began researching current science to investigate if there was any supporting evidence in biology and physics for any of these hypotheses. While reading papers (especially in quantum biology, a field the author wasn't very familiar with), the author was shocked by the findings of these papers and their relevance to the hypotheses that were derived from these seemingly disconnected observations and random insights.

Wanting to embody the principles of the book itself, the author sought to get the first draft into the world so that, utilizing the potential of human consciousness collectively interfacing with the QID, the right feedback may have the chance to emerge to help the book reach its full potential. After finishing the first draft, the author sent it to a few of the most cited researchers in the book. Dr. Jack Tuszynski responded immediately and became critical in shaping the evidence and structure of the arguments. He helped to ensure that the work aligned with current physics and quantum biology research. Throughout the entire process it felt as if it was meant, or more precisely, as if it was probabilistically optimal, for Jack to be involved as he has been such an incredible partner in getting this book finished. He's been wonderful to work with and very humble. He's only helpfully suggested improvement and has never forcefully imposed upon the book, which has allowed it to maintain its superposition. As a result, he has helped shape and clarify many of the core proposals of the book, including the collapse mechanism, terminology around the QID, the detailed involvement of thermodynamics in information exchange, the biological mechanisms involved, the similarities and differences with other frameworks, the discussions on free will, and many other aspects throughout. Jack provided sources and research that has been integrated across every level of the book. The author has been blown away by his humility and willingness to explore new perspectives on long-held ideas in physics. His open mind and explorative nature enabled this work to become what it is today. The book wouldn't have been possible without Dr. Jack Tuszynski, and the author is deeply grateful to him and his life's work that was also so critical in the emergence of the ideas presented.

Also, this book and its content arise from a deep gratitude and respect for the contemplative and spiritual traditions across cultures that mapped much of this territory with language long before modern neuroscience, biology, and physics could attempt to explain the mechanisms of the experience. Their experiential wisdom provided a phenomenological foundation for this theoretical exploration. The book also has deep gratitude to the researchers whose work laid the foundation for these connections to be possible, particularly to those bold enough to study consciousness and quantum processes at the intersection of physics, neuroscience, psychology, and biology when it was professionally risky to do so.

The book builds from five observable premises to derive five theoretical proposals, which results in a testable framework. The early chapters will outline the five premises and will explore the proposals when they fit. Each chapter builds upon established science and explores the ideas with if-then logic derived from the premises to arrive at what may initially seem like extraordinary conclusions, yet the author hopes readers will maintain the superposition of these ideas and explore the chain of logic and their possibilities before collapsing them with judgment as all of us humble humans are in this together.

This work is intended for physicists, neuroscientists, consciousness researchers, cognitive scientists, psychologists, contemplative practitioners, philosophers of mind, quantum biologists, information theorists, educators, elite performers, those interested in personal development and consciousness studies, and, most importantly, any reader willing to maintain the superposition of the book's proposals and ideas. The ideas presented are offered as a framework to be tested, refined, challenged, and extended by the collective intelligence of humanity. The author hopes the book and its framework their own descriptive traits, and that they inspire others to pursue their most elaborate, farfetched insights without ego, maintaining the quantum coherence and potential that may reside within us all and the ideas that inhabit us.

Following the logic of the book, it seems it is likely that many others are working on extremely similar ideas as this is written, and the author looks forward to collaborating across the evolution of these ideas. The author avidly invites collaboration, good-faith critique, falsifiable evidence, experimental design and collaboration, and anything else that would help humanity center in towards truth. The author would also be happy to answer any questions or provide any clarification on ideas that were not clearly presented. If consciousness truly plays some sort of participatory role in the unfolding of reality, then all of humanity should be thought of as co-authors of this book to some degree.

The author is grateful for the warmth of the readers and is grateful to them all for taking the time to carefully consider everything presented. The author hopes that people from all over the world can take the tiny seeds presented here and uncover insights that flourish and allow humanity to thrive like never before.

Westminster, CA, USA Josh Roeloffs

Contents

Chapter 1
From Performance to Physics

The Performance Paradox

"Time slows down." Athletes report this moment when the basketball appears to float toward the rim, when a fastball's seams suddenly become visible, or when the game-winning catch unfolds in slow motion. Similarly, authors say entire chapters can arrive in their minds nearly complete. Musicians report they can watch their fingers navigate difficult passages without conscious control, and entrepreneurs have described how entire innovative system ideas can appear in one insight.

These states of focus and insight are referred to as flow states. Van der Linden et al. (2021) define flow as "a state of full task engagement that is accompanied with low-levels of self-referential thinking." Kounios and Beeman (2009) observe that "insights occur when a solution is computed unconsciously and later emerges into awareness suddenly" (p. 210), which distinguishes these moments from "incremental" analytic processing.

The conditions that generate these states appear consistent across contexts as Csikszentmihalyi (1990) identified four requirements for flow states to manifest. The first condition is that challenges must match one's skill level, then second, they must create balance at the edge of one's abilities, third, there must be clear goals with immediate feedback to maintain this balance, and fourth there must be intense concentration that naturally suppresses self-referential processing. He identified that when these conditions align, the phenomenological markers emerge where time dilates, self-consciousness vanishes, and action becomes effortless.

Vervaeke et al. (2018) developed this understanding by characterizing flow as a cascade of insights, "an extended 'aha!'" where "each instantiation of insightful restructuring procedurally primes the agent to further engage insightfully with his or her dynamic surrounding" (p. 13). They note that "this spontaneity is neither coincidental nor epiphenomenal. Instead, it is rooted in self-organizing cognitive processes that exhibit spontaneous structural and functional reorganization" (p. 2).

J. Roeloffs, *The Classical Brain to the Quantum Mind*, The Frontiers Collection,
https://doi.org/10.1007/978-3-032-20068-6_1

Most significantly, "flow is prevalent across differences in culture, gender, socioeconomic status, occupation, and age—and is described nearly interchangeably, irrespective of language" (Csikszentmihalyi, 1996, cited in Vervaeke et al., 2018, p. 2). This cross-cultural consistency suggests that something deeper than cultural conditioning or learned behavior may be at play.

The description of flow becomes more unusual when considered alongside stress research. Research by Jamieson et al. (2010) found that reframing stress as performance-enhancing improved GRE scores in both immediate laboratory tests and actual examinations months later. The intervention provided no additional knowledge, no new study techniques, and no extra preparation time for the participants. Students only received a different narrative about what stress meant. Yet this shift in perception alone created measurable, lasting performance improvements that persisted across months and different testing contexts.

How could a change in belief about stress alter test performance? Why would the way in which someone thinks about their physiological arousal influence their cognitive capacity to recall information and solve problems? If stress is a purely mechanical response that is independent of interpretation then reframing it should have no effect on outcomes. How could the way consciousness relates to experience appear to influence what becomes possible within that experience?

Consider what we know from direct experience, that we perform worse when we think about performing, that our best ideas arrive when we stop trying to have them, that time perception shifts dramatically in critical moments, and that groups can solve problems none of the individuals could have solved alone. We know that when performing our best, the moment we think "wow this is going great" often marks when things begin to take a turn for the worse. Cook-Greuter (2000) notes this effect as she observes that peak flow states end "as soon as one evaluates and judges the inner proceedings" (p. 11). In this, the very act of measurement hinders the performance.

Humanity has developed consistent language across cultures to describe these phenomena. Philosophers describe "wu wei" or "effortless action," athletes talk about being in "the zone," mystics speak of "no-mind," psychologists describe "flow," theologians discuss "sacred presence," and artists talk of "the muse."

Is it possible these descriptions document the same physical process using different language? If we examine the pattern across all these experiences, four consistent elements emerge. First, performance dramatically improves when self-conscious observation is decreased. Second, capabilities emerge from flow states that seemingly exceed normal function. Third, in flow states, there's a shift that happens suddenly, and fourth the insights received feel like they come from outside of oneself. Overall, across all cultures, there's one underlying theme which is that conscious self-monitoring hinders one's performance.

Therefore, premise 1 is that self-observation hinders performance. This is documented across human experience and validated through empirical research, and will be described in immense detail in further chapters. It forms the foundation from which the book will explore.

The Observer's Paradox

This pattern becomes particularly intriguing when considered alongside one of the most significant discoveries in physics, which is the role observation plays in determining reality. Since the 1920s, quantum mechanics has established that particles exist in superposition as they occupy multiple states simultaneously until observed or measured. This observation principle, which is fundamental to quantum mechanics, was first articulated in the Copenhagen interpretation (Bohr, 1928) and has been experimentally verified countless times, most famously in the double-slit experiment. When unobserved, particles pass through both slits simultaneously as waves. The act of observation forces them to choose one path, which causes them to collapse from probability into actuality (Feynman et al., 1965). As Toole et al. (2018) explicitly state, "The double-slit experiment informs us that a conscious observer can alter the outcome of a quantum measurement" (p. 117). At the quantum level, observation in some sense, actively participates in creating reality. While this is only one interpretation of the experiment's impact, it directly links consciousness to quantum collapse, and this validity will be explored extensively throughout the book.

Modern decoherence theory has refined the understanding of this process as Zurek (2003) demonstrates how classical reality emerges from quantum states through environment-induced superselection, or "einselection." Any interaction that gains information about a quantum system's state effectively "observes" it, as it selects certain stable pointer states while eliminating superposition. This points to another way in which observation, or measurement in Zurek's case, transforms quantum field states into classical experience.

Therefore, Premise 2 is that observation (measurement) collapses quantum states. This foundation from nearly a century of experimental physics provides the second building block of the framework.

While the mechanism of collapse, whether it occurs through consciousness, environmental interaction, or some deeper informational process, remains debated and will be explored in later chapters, the empirical fact remains that human observation alters quantum outcomes. At this point in the book, the goal is not to resolve *why* this happens but simply to recognize *that* it happens. In later chapters, this relationship will be revisited and the questions of *why* will be tied into proposals following the five premises that set them up.

Let's consider how this collapse of the quantum states parallels human performance. Just like quantum particles collapse from superposition when observed, human capabilities seem to collapse when self-observed. The basketball player who thinks "I can't miss" suddenly airballs, the writer who notices their flow loses it, and the test-taker worries about the time fails the test. If observation collapses quantum states in physics experiments, and if self-observation consistently collapses peak performance in humans, then is it possible that these phenomena share a similar underlying mechanism? Is there a tradeoff or a ratio between self-referential thinking and one's ability to harness quantum coherence in computation?

This reasoning leads to a testable hypothesis about the relationship between performance capability, quantum coherence in neural systems, and ego activation as the observing mechanism that induces decoherence in the quantum field.

Considering this, the book begins by proposing a theoretical relationship of which its plausibility will be explored in detail in coming chapters.

$$\mathbf{P = Q/E}$$

$$\textbf{Performance} = \textbf{Quantum computation/Ego interference}$$

The equation captures why self-observation destroys performance by proposing that ego activation (E) divides our access to quantum processing (Q). Throughout this framework, "ego"refers to the self-referential "I" which is the narrative center of conscious awareness continuously generated by the Default Mode Network. Through this proposal, this "I", that observes and comments on experience and provides this sense of self-referential thinking, serves as an internal quantum observer that collapses possibilities through the act of self-observation. This will be described in detail in coming chapters.

The book is aware that the evidence for quantum computation in biological systems has not been explored yet, but it seems valuable to have this framework in mind when considering the evidence later proposed, as it will be easier to understand how and why these mechanisms exist in this framework. As Einstein famously said to Heisenberg "It is the theory which decides what we can observe" (Heisenberg, 1971). At the very least a theory helps to effectively narrow our search for what to look for, so this is why this proposal takes place earlier in the book. The book hopes readers keep an open mind and maintains the superposition of the evidence presented following the proposal.

Considering this, the phenomenological markers of flow states do align with what this equation would predict. Van der Linden et al. (2021) observe that "people who experienced flow often, at least retrospectively, reported feeling in control, having a clear sense of direction (i.e., clear goals), and a condensed perception of time" (p. 2). Most significantly, they note that "time seems to fly when people are in a flow" and speculate that "such flow-related changes in time perception may be linked to the reduced sense of self, mediated by parts of the insular cortex" (p. 2). This connection between reduced self-awareness and altered time perception supports the framework's central claim that every moment of "I" thinking, every instance of self-monitoring, and every activation of one's ego may physically collapse quantum field states in the brain.

This mechanistically is describing Proposal 1 that P = Q/E. This equation, if valid, would provide an explanation for numerous phenomena including flow states, insights arriving whole, savant abilities, team synergy, test anxiety, beginner's luck, mystical experiences, and childhood learning advantages to name a few.

Humanity has seemingly catalogued these experiences for millennia. Warriors, artists, mystics, and athletes have mapped the experiential territory. What remains unclear is the mechanism. How do these experiences actually work? Why does self-observation consistently destroy performance?

If consciousness can collapse quantum states through internal measurement, then several questions emerge. Could the brain operate as a quantum processor? What aspect of consciousness would serve as the internal observer? How would such a system evolve, and what advantages would it confer? If the brain truly processes information through quantum mechanisms, what biological structures could possibly maintain quantum coherence in warm, wet neural tissue? How would evolution develop such sophisticated quantum machinery? If consciousness interfaces with quantum processing, what exactly is this quantum processing connecting to? Is reality real, or is reality only a relic of our brain's projections? Does something actually collapse the quantum field into a physical reality? Is the collapse mechanism physical at all?

These are questions this book attempts to explore while examining how the P = Q/E relationship might illuminate human performance and the fundamental nature of reality itself.

References

Bohr, N. (1928). The quantum postulate and the recent development of atomic theory. *Nature, 121*(3050), 580–590.

Cook-Greuter, S. R. (2000). Mature ego development: A gateway to ego transcendence? *Journal of Adult Development, 7*(4), 227–240. https://doi.org/10.1023/A:1009511411421

Csikszentmihalyi, M. (1990). *Flow: The psychology of optimal experience.* Harper and Row.

Feynman, R. P., Leighton, R. B., & Sands, M. (1965). *The Feynman lectures on physics. Vol. 3: Quantum mechanics.* Addison-Wesley.

Heisenberg, W. (1971). Physics and beyond: Encounters and conversations. [1st ed.] Harper & Row.

Jamieson, J. P., Mendes, W. B., Blackstock, E., & Schmader, T. (2010). Turning the knots in your stomach into bows: Reappraising arousal improves performance on the GRE. *Journal of Experimental Social Psychology, 46*(1), 208–212. https://doi.org/10.1016/j.jesp.2009.08.015

Kounios, J., & Beeman, M. (2009). The aha! Moment: The cognitive neuroscience of insight. *Current Directions in Psychological Science, 18*(4), 210–216. https://doi.org/10.1111/j.1467-8721.2009.01638.x

Toole, J. T., Kurian, P., & Craddock, T. J. A. (2018). Coherent energy transfer and the potential implications for consciousness. *Journal of Cognitive Science, 19*(2), 115–124. https://doi.org/10.17791/jcs.2018.19.2.115

van der Linden, D., Tops, M., & Bakker, A. B. (2021). The neuroscience of the flow state: Involvement of the locus coeruleus norepinephrine system. *Frontiers in Psychology, 12*, 645498. https://doi.org/10.3389/fpsyg.2021.645498

Vervaeke, J., Ferraro, L., & Herrera-Bennett, A. (2018). *Flow as spontaneous thought: Insight and implicit learning.* https://doi.org/10.1093/oxfordhb/9780190464745.013.8

Zurek, W. H. (2003). Decoherence, einselection, and the quantum origins of the classical. *Reviews of Modern Physics, 75*(3), 715–775. https://doi.org/10.1103/RevModPhys.75.715

Chapter 2
Quantum Infrastructure in Biology and the Brain

Having established premise 1 that self-observation destroys performance and premise 2 that observation (measurement) collapses quantum states, the book now turns to establishing premise 3 that the brain contains functional quantum infrastructure. For the proposal that self-observation collapses quantum states in the brain, as the $P = Q/E$ equation would predict, the framework must first demonstrate that the brain might actually maintain quantum coherence capable of computation.

Before exploring biological quantum coherence, it may be worth acknowledging what physics has already established. Wave functions are demonstrably real, as laboratories verify daily through interference experiments (Feynman et al., 1965). Superposition genuinely exists, which quantum computers rely upon for their operation (Nielsen & Chuang, 2000). Collapse observably happens with every measurement performed (von Neumann, 1932). And quantum effects provide undeniable computational advantages, as Shor's algorithm for factoring (Shor, 1997) and Grover's search algorithm (Grover, 1996) have mathematically demonstrated. Given these established facts, a natural question emerges. If quantum computation offers exponential advantages over classical processing, why would evolution ignore this resource?

The book proposes a fundamental reframe of what psychology calls System 1's intuitive processing (Kahneman, 2011). Consider the possibility that System 1 might access quantum computation instead of just representing an entirely primitive evolutionary system. Could it be possible that the speed of intuition and insight and the advanced pattern recognition that seems to violate classical processing constraints actually reflect quantum computational access?

For decades, many assumed quantum effects were too fragile to survive in warm, wet biological systems. As Craddock et al. (2014) note, "It has often been argued that living systems are too 'warm and wet'to support quantum effects; however, recent research suggests this is not necessarily always the case" (p. 1). As will be discussed in this chapter, breakthrough discoveries since Craddock wrote that have shattered this assumption.

J. Roeloffs, *The Classical Brain to the Quantum Mind*, The Frontiers Collection,
https://doi.org/10.1007/978-3-032-20068-6_2

Tegmark (2000) famously calculated that quantum decoherence in the brain would occur too rapidly for quantum computation. However, his calculations assumed isolated quantum states. The discovery of cooperative robustness by Babcock et al. (2024) fundamentally changes these calculations. As they demonstrate, "A counter-intuitive consequence of cooperativity is the fact that the robustness of a system to disorder can increase with the system size" (p. 4043). This means larger quantum networks in the brain would be more stable, not less. This directly contradicts the earlier skepticism. But considering the evidence presented from many sources in the last decade, the redundancy suggests evolution systematically selected for and incorporated multiple quantum mechanisms for robustness and functionality. Similarly Nishiyama et al. (2024b) argued against Tegmark, "It was claimed that quantum coherence in the brain cannot be maintained since the brain system is too warm, too wet and too noisy. The order of magnitude of the decoherence time is 10^20 s as estimated by Tegmark. However, his analysis has several major problems" (p. 3).

They identify specific flaws in Tegmark's calculations. First, he adopts the mass of a water molecule as approximately 18×940 MeV, when quantum brain dynamics should use the inverse of the moment of inertia (~ 4 meV) or the mass of polaritons formed from water fields and photon fields. Second, his procedure divides decoherence time scales by the number of relevant ions (10^6), which lacks physical justification. Most critically, Tegmark treats the brain as a closed system, while biological systems are open systems constantly supplied with metabolic energy (pp. 1–3). As they note, any proper analysis must "investigate the balance of decoherence and error corrections for quantum coherence in the flow of energy from the energy supply through the physical system to the heat bath" (p. 3). The discovery of cooperative robustness changes these calculations, considering it shows that larger quantum networks can actually become more stable.

This principle finds empirical support in avian magnetoreception, where Hiscock et al. (2016) showed that quantum coherence exceeding 1 microsecond creates dramatic performance improvements through what they term a 'spike' in directional precision. They note that "random mutations in the sequence of the protein… could have provided evolution with the scope to optimize the compass precision" (p. 4634), which demonstrates that evolution can actively refine quantum systems. Additional support for robustness comes from Pastawski et al. (2015) as this robustness has a natural analogue in holographic quantum error correction, that bulk information is redundantly encoded so that many local boundary perturbations do not corrupt protected content (Pastawski et al., 2015). By analogy, redundancy across microtubule networks could protect quantum computational content, which would help to explain why low-noise states such as flow yield reliable access despite biological imperfections. Additionally, quantum criticality can maximize information capacity per unit energy. Near critical points, quasidegenerate modes become nearly gapless, which enables large storage/processing with small energetic cost (Dvali & Gomez, 2014). Furthermore, this line of reasoning is supported by Kurian's (2025) model which suggests that some enzymatic complexes can actively create "decoherence-free subspaces," which maintain quantum correlations "not in spite but because of the environment" (p. 8). This suggests evolution did not just find ways to tolerate random noise, but it actually found ways to harness environmental interactions to protect and "steer" quantum states toward a more specific function for survival.

Biological Phenomena beyond Classical Explanation

Before further examining the evidence for quantum infrastructure in the brain, it may be valuable to establish that large biological systems already demonstrate phenomena that appear to violate classical physics. This recognition has emerged from over two decades of research into what are now called non-targeted effects (NTE) of radiation.

Mothersill and Seymour (2001) first comprehensively reviewed the phenomenon of radiation-induced bystander effects, noting that "the clastogenic effect of plasma from irradiated patients was well known" and that "the effect was known to persist for several years and to cause transgenerational effects" (p. 759). Even at that time, they recognized these effects challenged classical radiobiology, which assumed only directly irradiated cells would show radiation responses.

Over the following decades, their research group documented increasingly puzzling phenomena. By 2018, Mothersill et al. had designed experiments that found something that should not be possible under classical physics and biology, because fish appeared to respond to radiation events that had not yet occurred.

Their experimental design attempted to maintain rigor. Fish from two completely separate populations that had never met were allowed to swim together for two hours. They were then separated permanently. Only after separation was one group irradiated with 0.5 Gy X-rays. The non-irradiated fish, even though they had no further contact with the irradiated group, showed biological responses including proteomic changes with "six protein spots showed an increase in normalized volume only in fish which swam with the irradiated fish prior to radiation exposure" (Mothersill et al., 2018, p. 8). Additionally, calcium flux alterations occurred where "the group B fish which met before the group A fish were irradiated had calcium influxes that were significantly below those of the control fish" (p. 6). The fish also showed different protein expression patterns compared to traditional bystander effects, including "upregulation of proteins involved in neuroprotection" (p. 11). They emphasized the reach of previous studies across a wide variety of species and noted why this result was so surprising, saying,

"We have previously published extensively that postirradiation meeting causes effects in nonirradiated fish, rats, and mice, and other groups also have shown these organism level effects in frogs, mice, and water fleas, but this is the first report of preirradiation meeting, influencing outcome. At first sight, the results may seem beyond strange but the robust and well-tested assays require some consideration and explanation" (p. 7).

Importantly, their results showed that "it does not matter whether the fish met before or after 1 group was X-rayed" as there were bystander effects in both scenarios. The authors attempted classical explanations including social stress from meeting strangers, but their controls appeared to rule this out. They noted that "based on the clonogenic data, therefore, the conclusion is that whatever explanation there is for the results where 1 group were actually irradiated, they cannot be explained as solely due to the introduction of a new fish cohort into a tank already containing fish" (p. 10).

This is surprising from a classical perspective as the fish appeared to anticipate future radiation exposure of their former tank mates. They concluded that "some form of anticipation of a stressor may occur leading to a preconditioning effect or temporally displaced awareness after the fish become entangled" (Mothersill et al., 2018, Abstract).

Having documented phenomena that classical biology struggles to explain, Mothersill and colleagues have pursued quantum biological explanations. Matarèse et al. (2023), with Mothersill as co-author, propose specific quantum mechanisms that might underlie these anomalies. They document how "biophotons emitted by cells may encode quantum information" (p. 26) and suggest that biological systems might utilize "quantum tunneling, entanglement, voltage-gated ion channels, photon emissions, and calcium fluxes" (p. 22) for cellular communication. Their proposed mechanism involves cells potentially maintaining quantum entanglement after interaction, biophotons carrying quantum-encoded information between separated systems, ion channels operating through quantum tunneling to process this information, and biological responses manifesting from accessed quantum probability distributions. These are ideas that will be explored in coming pages.

Considering their research began as observations of "clastogenic effects" in irradiated patients' plasma, it has now evolved into recognition that biological systems might operate beyond classical constraints. If fish cells can coordinate responses to future events through quantum mechanisms, and if "respiration in mitochondria is thought to involve non-trivial quantum effects including entanglement, superposition, and tunneling" (Matarèse et al., 2023, p. 14), which are processes that occur in all cells including neurons, then a follow up question naturally emerges. Could the human brain utilize related quantum processing capabilities?

The brain contains cellular machinery similar to that documented to exhibit quantum behavior in fish. Every neuron has mitochondria, ion channels, and the capacity for biophoton emission. The difference might concern not whether quantum processes occur, but instead how evolution has developed these mechanisms. The following chapters examine evidence suggesting that the brain might maintain quantum coherence and may have evolved specialized structures to optimize it.

Development of Evidence for Quantum Infrastructure in Biology

The field of quantum biology has undergone a revolution in recent years, with each discovery, building upon the last, pointing towards the plausibility of quantum processes in the brain. The idea that the brain may utilize quantum processes traces back over five decades. Ricciardi and Umezawa (1967) first proposed treating the brain as a quantum many-body system, suggesting that long-range quantum correlations could underlie memory storage via spontaneously broken symmetry and Nambu–Goldstone modes. This framework was later extended into the Dissipative Quantum Model (Stuart et al., 1978, 1979), which treated the brain as an open quantum system that interacts continuously with its environment. Jibu and Yasue (1995) further grounded these ideas in biology as they proposed that microtubules and structured water dipoles might form coherent quantum states mediated by photon fields.

What remained speculative for decades has recently gained experimental support as modern findings demonstrate that the very substrates QBD theorists hypothesized, such as microtubules, exhibit quantum optical properties and information-processing capabilities.

The evidence began to mount in 2007 when Engel et al. demonstrated long-lived electronic coherence ($\approx$ 660 fs) in the FMO photosynthetic complex at 77 K, showing that coherent energy transfer could persist under cryogenic conditions. Then Panitchayangkoon et al. (2010) observed coherence at ~ 277 K (physiological conditions for the system) lasting ~300 fs, which directly tied coherence to biological function. As Craddock et al. (2014) summarize, "This limit was further pushed to 277 K, nearing physiological temperature, and effectively ruling out the 'warm and wet' limit for quantum phenomena in biology" (p. 1). Ball (2011) synthesized emerging evidence that evolution optimized biological systems to exploit quantum effects, with examples ranging from photosynthesis to magnetoreception in birds. This research began to point to ways in which biology appears to actively harness these effects for survival advantages.

Potentially most compelling is how Hiscock et al. (2016) theoretically demonstrated that if migratory birds achieve quantum coherence lasting over 1 microsecond in cryptochrome proteins, a "spike" would emerge in the quantum signal, which provides the angular precision necessary for transcontinental navigation. Their modeling suggests that "radical pair magnetoreception may be more of a quantum phenomenon than previously realized" (p. 4634). This theoretical work importantly showed that evolution could have optimized these quantum effects for biological function by selecting for survival advantages.

If photosynthesis and bird navigation already exploit quantum mechanics, and if quantum computation provides exponential speedup for certain problems, then examining whether the brain might have evolved similar capabilities becomes a logical next step.

Hameroff and Penrose (1996, 2014) proposed in their Orchestrated Objective Reduction (Orch OR) theory that microtubules could support quantum computation through objective reduction events. While controversial, their calculations suggested that microtubule quantum states could maintain coherence for 25 milliseconds at brain temperature, which they argued is sufficient for cognitive processing, as their theory posits that consciousness emerges from these orchestrated quantum computations in microtubules that are "tuned" by synaptic inputs and memory.

Craddock et al. (2014) further validated the quantum processes in microtubules, as they demonstrated that the tryptophan networks in microtubules show fluorescence quantum yields comparable to photosynthetic systems and can support "quantum coherent beating effects similar to those observed in the FMO photosynthetic complex" (p. 4). They conclude, saying, "Overall, we find that coherent energy transfer in tubulin and microtubules is biologically feasible" (p. 1). Nishiyama et al. (2024a) reported that these quantum states could support memory and consciousness, with timescales fast enough to prevent thermal decoherence.

Similarly, Toole et al. (2018) reinforced this shift, stating "The argument that biological systems are too 'warm and wet' to support quantum effects is becoming increasingly antiquated as research in the field of quantum biology progresses" (p. 115). They further noted that "not only is it becoming apparent that quantum processes may regularly take place in biological systems, but these processes may underlie the mechanisms of consciousness" (p. 115).

Direct experimental support for microtubules'role in consciousness comes from anesthesia research. Khan et al. (2024, cited in Wiest, 2025) found that rats administered brain-penetrant MT-binding drugs took significantly longer to fall unconscious under anesthesia, with a large effect size (Cohen's d = 1.9). This suggests consciousness depends on intact microtubule quantum processes, as disrupting them directly affects conscious states. Overall, this raises some interesting questions. If evolution harnessed quantum mechanics for energy transfer and navigation in animals, why not for computation in humans? Is it not plausible that evolution would select quantum processing capabilities for humanity as well?

The evolutionary pressure for such a development becomes clearer when considering the potential computational advantage, with recent models suggesting that biological quantum processing may be up to a billion times faster than its classical neural counterparts (Kurian, 2025). Considering this, the structural similarities are quite compelling as Craddock et al. (2014) found that "The fluorescence quantum yield for pure Trp is 0.14, and the experimentally observed yield for wild-type tubulin is 0.06 at room temperature. These are comparable to the room temperature fluorescence quantum yields of bacteriochlorophyll of 0.18, and the yield for LH1 of 0.08 and LH2 of 0.10" (p. 2). These findings, from 2014, suggested the brain may contain structures as quantum-capable as photosynthetic systems.

By 2018, already the field of quantum biology had strong evidence that quantum processes could happen in biological systems, including the brain, and significant discoveries since then suggest the brain may be much more quantum-capable than originally thought. Tan et al. (2019) experimentally demonstrated that quantum states could be protected from thermalization through a mechanism called confinement. To understand this, imagine quantum information as soap bubbles floating in air. Normally they pop quickly when they touch anything. But confinement is like creating an invisible container that keeps the bubbles from spreading out and popping. Using trapped-ion quantum simulators, they showed that "confinement can dramatically suppress information propagation and thermalization in such many-body systems" (p. 1). Their key finding was that long-range interactions could bind excitations together which prevents them from spreading.

As Tan et al. (2019) explained, "Confinement is a ubiquitous mechanism in nature, whereby particles feel an attractive force that increases without bound as they separate" (p. 1). Think of it like two magnets connected by a rubber band. The further you pull them apart, the stronger the force pulling them back together. They demonstrated this effect in spin chains, showing that "confinement, naturally induced by long-range interactions, may provide a novel mechanism for protecting quantum information without engineering disorder" (p. 4). Two years later, Konik (2021) analyzed the broader implications of this discovery in his commentary. He emphasized that "Confinement of excitations hinders this spread and can prevent the system from arriving at a thermal equilibrium" (p. 4) which highlights how this mechanism could preserve quantum coherence. In simple terms, by keeping quantum states bound together, they can't dissipate their special properties into the surrounding environment. It's like keeping ice cubes in a thermos instead of leaving them on a hot sidewalk.

Importantly Konik identified a fundamental gap in the field's understanding. While physics had proven confinement could preserve quantum states, no one knew how to harness this for practical quantum computation. He posed a challenge that would be very important, "Confinement thus is a promising generic mechanism for preserving quantum coherence through its ability to interrupt standard patterns of thermalization. How to convert this disruption into a means to store and manipulate quantum information remains, however, to be seen" (p. 4).

The field needed to discover systems that could both maintain quantum coherence, perform computations with it, and store data. Interestingly, this challenge would soon be answered. Building on earlier theoretical work by Toole et al. (2018), who found that "the electronic characteristics and spatial orientation of Trps within microtubules are comparable to light-harvesting structures in plants and bacteria, and that it is very feasible that the quantum processes experimentally observed to occur in these plant/bacteria structures may take place within the Trp network in tubulin" (pp. 119–120), the field was primed for experimental validation.

By 2022 and into 2023, the Overton window on the plausibility of quantum processes in biology and the brain had already begun to shift as Kauffman and Radin (2023) wrote, "with growing evidence for quantum biological effects, the plausibility that some aspects of brain function are quantum is increasing" (p. 17). This expanding sense of scientific plausibility provided the foundation for exploring how consciousness might interact with these quantum processes through specific mechanisms. Yet some of the most revealing studies were just about to be published.

In 2023, Kalra and colleagues conducted an experiment that measured electronic energy migration within microtubules. The researchers used tryptophan autofluorescence and an external fluorescence quencher (AMCA) to track energy migration, allowing them to measure how far quantum excitation could travel before being absorbed or dissipating. Using time-resolved fluorescence methods, they found that "a diffusion length of 6.64 nm suggests that energy transfer between tryptophan residues across a single tubulin dimer can occur in a microtubule" (Kalra et al., 2023, p. 353). They emphasized how extraordinary this behavior is for a biological structure as they noted that while proteins generally perform poorly as light-harvesting systems, "the lattice-type structure of microtubules that incorporates periodic arrays of aromatic amino acids… is qualitatively similar to the structure of molecular aggregates and light-harvesting complexes like phycobilisomes, potentially enabling excitation energy transport over long distances" (p. 352). They describe how they found that the "photoexcitation diffusion length in microtubules (6.6 ± 0.1 nm) is substantially higher than that predicted by conventional Förster theory," which is a level of efficiency "comparable to that reported in some photosynthetic complexes" (p. 352). The authors concluded that "considering the factors… that make tryptophan a far-from-optimal chromophore, our results demonstrate that electronic energy migration is surprisingly efficient in microtubules" (p. 352).

Their analysis showed "a 2D diffusion coefficient of $3.15 \pm 0.1 \times 10^{-5}$ cm^2 s^{-1} and a corresponding diffusion length of 6.64 ± 0.1 nm for microtubules" (p. 353). The authors described this value as "surprisingly high in microtubules, given that they are optimized to play structural roles in the cell" (p. 358). This observation points toward an unexpected efficiency in energy transfer that operates at the molecular scale of the tubulin lattice.

Kalra's team also discovered that this coherence can be influenced by pharmacological compounds known to affect consciousness. They reported that "the presence of etomidate and isoflurane decreased the 2D diffusion lengths from 6.6 ± 0.1 nm to 5.6 ± 0.1 nm and 5.8 ± 0.2 nm, respectively" (p. 356). These results imply that anesthetics directly alter the efficiency of excitonic transport within microtubules, an observation that aligns with long-standing theories proposing that anesthesia interacts with quantum processes in the brain. When paired with Khan et al. (2024) finding that protecting microtubules delays unconsciousness, a pattern emerges where consciousness appears linked to intact quantum processing in microtubular structures.

When describing the broader implications of these measurements, the authors wrote, "Our studies indicate that microtubules are, unexpectedly, effective light harvesters" (p. 352). In ordinary biological terms, this means that microtubules display photonic properties more commonly associated with photosynthetic systems. Their capacity to conduct excitation energy over nanometer scales and to respond sensitively to environmental changes invited new questions about whether these properties could serve computational or informational functions inside living cells.

The researchers concluded that "photoexcitation diffusion cannot be explained via conventional Förster theory" (p. 358), which means that the classical dipole-dipole model of energy transfer cannot account for their observations. They noted that a "satisfactory explanation of our observations may include solvent screening by the protein matrix, electronic coupling beyond the point dipole approximation in Förster theory, and accommodation for molecular exciton states in tubulin" (p. 358). They also observed that "the tuning of photoexcitation diffusion length through biochemical parameters of microtubules (tubulin polymerization, introduction of anesthetics) was a further intriguing observation" (p. 358), adding that "protein polymers may be suitable for active materials in biologically sourced electronic devices where UV photoexcitation is desired" (p. 358).

The implication is that the energy migration must involve collective or quantum-mechanical effects. Architecture like this would enable energy to move coherently across organized protein structures, which aligns with a growing body of work showing that biological order can sustain quantum phenomena at physiological conditions. In practical terms, this means the excitation energy within a microtubule does not behave like isolated particles passing information one step at a time. They found that it behaves more like a shared field that spreads across several molecules at once, which allows them to act in unison. This delocalized behavior would be the same principle that gives photosynthetic systems their extreme energy efficiency. Seeing it in microtubules means that the building blocks of neurons already possess a structure capable of quantum-level coordination inside living cells. For decades, this kind of coherence was assumed to be impossible in warm biological environments, yet these findings show that nature has already solved the problem at the molecular level.

Taken together, these findings frame microtubules as organized molecular networks capable of energy migration and modulation through environmental and biochemical factors. From the framework's perspective, these observations provide the kind of empirical bridge that early quantum-mind theorists could only hypothesize. If quantum coherence can occur across the structural framework of the cell, and if anesthetic molecules can influence this coherence, then the biological infrastructure for quantum information processing may already exist within the body's cellular architecture.

The next big breakthrough came from Babcock et al., as their initial work (Babcock et al., 2023) showed that microtubules contain vast networks of tryptophan residues that function as quantum optical systems, essentially acting like billions of tiny antennas that can transmit and receive quantum information. By 2024, it was peer-reviewed, experimental evidence, demonstrating that "networks of tryptophan (Trp) form diverse architectures in transmembrane proteins, cytoskeletal filaments, subneuronal elements" containing "up to >10^5 Trp UV-excited transition dipoles in microtubule architectures, which leads to an enhancement of the fluorescence quantum yield (QY) that is confirmed by our experiments" (Babcock et al., 2024, p. 4035).

The study provided important experimental measurements showing quantum yields increasing from 12.4% for isolated tryptophan to between 15.7 and 19.5% for microtubules, with an average of 17.6%, representing up to a 70% enhancement. This constitutes direct bench-top evidence that cytoskeletal assemblies support robust quantum effects at physiological conditions. As Babcock et al. (2024) conclude, these networks exhibit "collective and cooperative quantum effects" that "support robust quantum states in protein aggregates, with observed consequences even under thermal equilibrium conditions" (p. 4044).

The significance of superradiance cannot be overstated. In conventional quantum systems, decoherence typically increases with system size, as more components mean more opportunities for environmental interference to destroy quantum states. Superradiance flips this relationship. As Babcock et al. (2024) explain, "A counterintuitive consequence of cooperativity is the fact that the robustness of a system to disorder can increase with the system size" (p. 4043). Their experimental findings showed that "the QY is almost unaffected when a disorder strength equal to room-temperature energy (~ 200 cm^−1) is considered, and a QY enhancement is still observable even at 1000 cm^−1 disorder" (p. 4041).

This cooperative robustness means quantum effects don't just happen to survive at room temperature, instead, they actually become stronger as networks grow larger. This aligns with the kind of long-range interaction that Tan et al. (2019) had shown could create confinement. Then, the brain may be able to use these quantum states for active processing and transmission of quantum information. As Babcock et al. (2024) note, "photons from brain metabolic activity could be absorbed rapidly via superradiant states for ultrafast information transfer" (p. 4043).

Potentially most interesting, Babcock et al. (2024) propose that axons themselves may serve as quantum waveguides. They say, "Confining a superradiant optical mode to one dimension in a waveguide has been proposed to extend emitter interactions to extremely long range, raising the tantalizing possibility that axons might serve as such waveguides between giant superradiant emitters in the brain" (p. 4043). Their calculations show superradiance enhancements "approaching ~7000" (p. 4043) for neuronal MT bundles, with effects spanning "many orders of magnitude across which the brightest (hundreds of femtoseconds) and darkest (tens of seconds) states can coexist in these Trp lattices" (p. 4035).

This builds directly on earlier work by Craddock et al. (2012), who discovered that CaMKII phosphorylation patterns can encode stable "bytes" of memory directly into microtubule lattices. They showed that each activated CaMKII holoenzyme, which is a hexagonal molecular complex triggered by calcium influx, can write up to six phosphorylation "bits" into hexagonal patches of tubulin dimers, creating molecular-scale memory units within the cytoskeleton. In this model, microtubules act as a solid-state memory substrate capable of storing highly stable information that persists even as synaptic proteins turn over. This provided the line of theory needed for recent exploration.

Nishiyama et al. (2024a) recent findings suggest that these classical molecular memory codes are embedded within a fully quantum-capable infrastructure as they provided mathematical verification that these systems can actually function as biological quantum systems. Their calculations demonstrate that superradiant states in microtubules could operate on sub-picosecond timescales, less than a trillionth of a second, fast enough to complete quantum computations before heat and noise can destroy them. They calculate that these systems could function using biologically available energy (~1 meV), transmit information at rates up to 10^4 bits per second, and could even create holographic interference patterns in cellular water for distributed information storage.

As Nishiyama et al. (2024a) conclude, "Since the time scales of superradiant emission of light and sound less than picosecond are much smaller than those of thermal effects (~10 ps), super-radiance is achieved without thermal loss via a microtubule" (p. 302). This is the prominent statement. The speed of the superradiant emissions allow for burst of computation faster than the noise that causes decoherence, which enables this quantum system to function.

The latest breakthrough came from Nishiyama et al. (2025), who demonstrated that this quantum infrastructure could actually store and process information which directly answers Konik's 2021 challenge about how to "store and manipulate quantum information" (p. 4). Using waveguide quantum electrodynamics (wQED), they showed that tryptophans can function as "data qubits" which are the quantum equivalent of computer bits that can exist in multiple states simultaneously when entangled with water molecules acting as "Josephson quantum filters." These filters help maintain the delicate quantum states by screening out environmental noise.

Their discovery provided what Konik said was missing, as they found that sub-radiant states in tryptophans can create stable quantum memory for storage, that water molecules can function as Josephson filters to enable active control of quantum states for manipulation, and that the combined system can perform actual quantum computations at body temperature for processing.

Their calculations show that superradiance and sub-radiance work as complementary mechanisms, with superradiance enabling rapid information transfer while sub-radiant states provide stable information storage at body temperature (310 K).

It seems as though evolution's solution involves tryptophan networks whose long-range interactions create confined quantum states resistant to thermalization. While Tan et al. (2019) used long-range interactions in engineered ion traps, evolution seemingly discovered, and selected for, water-mediated tryptophan networks that create similar confinement effects. These biological systems also incorporated the storage and manipulation capabilities that Konik (2021) identified as essential. Considering this, Nishiyama et al. (2025) state, "Using the robustness of data qubits, we demonstrated the feasibility of quantum information processing that can be performed by tryptophans in a microtubule as local events" (p. 21).

Additional support comes from Patwa et al. (2024), who analyzed tryptophan chromophore networks across multiple neuroprotein structures including microtubules, actin filaments, and amyloid fibrils. Their findings reveal significant quantum coherence stability is possible even while under thermal stress. When testing microtubules containing 20,800 tryptophan molecules under static disorder five times greater than room temperature conditions, they observed only a 3.08% reduction in quantum yield. As they note, "The quantum yield dampening by only 3.08% when the static disorder is five times that of room temperature demonstrates its robustness" (p. 8).

Also important, their analysis shows that quantum enhancement actually increases with system size. Comparing isolated tryptophan molecules to large microtubule assemblies showed quantum yield improvements of 15.76% under ideal conditions and 12.31% even under extreme static disorder. As Patwa et al. explain, "even with extremely large static disorder strengths, at thermal equilibrium, the quantum yield for microtubules is enhanced as the system size grows" (p. 10). They emphasize that "Such quantum yield robustness to static disorder has recently been experimentally confirmed for microtubules at room temperature" with additional theoretical predictions showing "enormous superradiant enhancements for axonal microtubule bundles" (p. 8).

Perhaps most significantly for the framework, they discovered that quantum coherent effects in these neuroproteins reveal "an ultrafast mechanism that our brains may use to process information, which is paired with an extremely long-lived mechanism for coordinating biological function" (p. 21). This dual-timescale architecture suggests quantum processing operates at speeds far beyond classical neural transmission while maintaining stable coordination across extended biological timescales.

The progression from physics principles to the biological implementation now appears to be manifesting across multiple research groups as Tan et al. (2019) demonstrated confinement could preserve quantum coherence through long-range interactions. Then Konik (2021) posed the challenge of achieving storage and manipulation. Next, Babcock et al. (2024) showed microtubules form superradiant networks with the necessary long-range interactions. Then Nishiyama et al. (2024a) proved they can transmit information holographically, and finally, Nishiyama (2025) demonstrated they can store and process quantum information using biological confinement, while Patwa et al. (2024) confirmed these quantum optical properties remain robust even under extreme thermal conditions and actually strengthen with system size.

The theoretical analysis by Nishiyama et al. (2024a, 2024b, 2025) combined with the experimental support by Kalra et al. (2023), Patwa et al. (2024) and Babcock et al. (2024) suggest that microtubule tryptophans, coupled through water-mediated interactions, can function as data qubits at 310 K as superradiant channels enable ultrafast transfer, subradiant modes provide the memory, and waveguide QED affords control. This dual-timescale architecture may provide the computational substrate necessary for consciousness to interface between quantum possibilities and classical reality, this potential relationship will be explored in detail in coming chapters. As Patwa et al. (2024) observe, these discoveries call for fundamental reconceptualization. They write,

"Such a paradigm shift can greatly enhance our understanding of nature, to visualize biological architectures as chromophore lattices synchronized by long-range interactions, and imbued with unique and specific photophysical properties that are enhanced by collective light-matter interactions governed by the equations of quantum optics. Such a shift reflects a return to understanding, in the (paraphrased) words of Richard Feynman and with the ancients, how external light from the fiery sun causes trees and plants to grow from the carboniferous air; and in parallel symbiosis how oxygen-metabolizing organisms may have evolved their protein architectures to exploit 'internal' photonic emissions for information processing and to mitigate potentially damaging wavelengths in the cell" (Patwa et al. 2024, p. 21).

From the framework's perspective, these findings outlined in this section suggest that biology's, and the brain's, quantum architecture may be far more robust and effective than previously thought.

Neurofilamentary Pre-firing Mechanism

Throughout the last few years as well, discoveries in neurofilamentary dynamics may provide insight into how quantum computations could be filtered before classical neural processing. Neurofilaments are neuron-specific proteins that provide structural support to the neuronal cytoskeleton and appear to be implicated in both axonal and synaptic signaling through their dynamical behavior (Bragina & Conti, 2018). In synaptic signaling, neurofilamentary proteins show differential expression in the presynaptic and postsynaptic compartments of glutamatergic (excitatory) and GABAergic (inhibitory) synapses. These proteins also seem to contribute to axonal signaling processes.

Research suggests that axons may process information at multiple timescales (Singh et al. 2021, 2021) in patterns that have been described as vortex-like electromagnetic signals, which may be detectable through quantum optical methods (Singh et al., 2021). Each axon contains thousands of densely packed neurofilaments beneath its membrane. The data points to the possibility of an intricate mechanism in which electromagnetic resonance in the neurofilaments might rapidly identify relevant neural paths or circuits, termed "branch selection," at microsecond speeds (10^{-6} s), which could then serve as inputs to ion-channel-mediated transmission occurring on millisecond timescales (10^{-3} s) (Singh et al., 2021).

Particularly intriguing, Singh et al. (2021) observed that four ordered structures within the cytoskeletal filaments appear to exchange energy approximately 250 microseconds before a neuron fires. If confirmed, this would suggest a pre-firing stage of computation occurring roughly a thousand times faster than classical membrane-level spike initiation. Such a mechanism could potentially integrate multiple temporal domains into a unified architecture, possibly extending the traditional Hodgkin-Huxley model to include branch selection, spike time-gap regulation, and synaptic plasticity (Singh et al., 2021).

This pre-firing mechanism finds additional support from Bandyopadhyay and colleagues who "stimulated MT resonances in cultured neurons and observed the MT resonance state 'spanning across multiple neurons and controlling membrane voltage'" (Saxena et al., 2020; Singh et al., 2021, 2021; cited in Wiest, 2025, p. 3). This demonstrates that microtubule resonances can directly influence and coordinate neural firing patterns across multiple cells, which suggests there may be a quantum-to-classical control mechanism operating faster than membrane-level processes.

Within the framework, this pre-firing layer offers a plausible physical substrate for switching and filtering that will be explored in later chapters. The thousand-fold speed difference between electromagnetic branch selection and neural firing could provide a temporal window in which the brain surveys candidate outcomes and commits only the most relevant possibilities to classical action. This remains an open hypothesis, yet it fits naturally with the LC-NE to DMN to calcium to microtubule pathway developed later, where network-level state sets the boundary conditions for subcellular coherence and collapse.

Considering this, the brain's multiscale organization may admit a compact description via entanglement-renormalization. The Multiscale Entanglement Renormalization Ansatz (MERA) organizes entanglement across scales and induces network geometries with hyperbolic/AdS-like structure (Vidal, 2007; Swingle, 2012). If microtubule tryptophan arrays act as "boundary tensors" with mesoscale assemblies forming deeper layers, biology could realize a MERA-like architecture. Architecture like this would enable quantum information to flow efficiently from molecular to network scales, which aligns with the rapid, scale-bridging computation attributed to System 1. The framework offers this as a guiding analogy while acknowledging uncertainty about the direct relational identity between neural tissue and MERA networks.

Experimental Evidence of Quantum Processes

While these discoveries establish the brain's quantum infrastructure, the important question remains, which is "does consciousness actually utilize these quantum capabilities?" This question received experimental evidence from Kerskens and López Pérez (2022). Kerskens and López Pérez (2022) detected zero-quantum-coherence signals in living human brains that tracked conscious state (present during wakefulness, absent during sleep), synchronized with heartbeat-evoked potentials, and correlated with short-term memory performance. These findings point to non-classical spin correlations in the awake human brain; no known classical mechanism reproduces all reported constraints simultaneously, although independent replication remains needed. The data constitute direct spectroscopic evidence that consciousness-related brain functions operate non-classically, making classical-only accounts of consciousness increasingly difficult to sustain.

To put this in perspective, Kerskens and López Pérez essentially used a degraded detection system and found quantum signals hundreds of times stronger than the best equipment should theoretically detect under classical physics. This would only appear possible if consciousness somehow actively amplifies or mediates quantum processes. As they conclude, "Our findings suggest that we may have witnessed entanglement mediated by consciousness-related brain functions. Those brain functions must then operate non-classically, which would mean that consciousness is non-classical" (p. 2). This provides the first direct experimental evidence that consciousness itself is in some way quantum mechanical. The requirement for conscious awareness while ruling out all classical mechanisms suggests consciousness could play a part in actively enabling the brain's quantum functionality. At the very least, they are deeply intertwined. Findings like these may complement the rapid pre-firing dynamics described earlier, together implying that quantum processes remain active across multiple temporal and spatial scales. Overall, this aligns well with this chapter's core premise that the brain possesses functional quantum architecture, and of course will be explored further in later chapters.

Energy Release and Insight

If the brain computes in quantum-enabled ways, it would be expected to see measurable energy signatures when quantum states collapse into classical outcomes before conscious awareness of the result. Interestingly, this seemingly has been documented in neuroscience.

Kounios and Beeman (2009) found that "insight solutions were associated with a burst of high-frequency (i.e., 40-Hertz gamma-band) activity starting about 300 milliseconds before the button-press" (p. 212). This sudden gamma burst, which is a discrete energy release, matches what one would expect from quantum collapse. The burst occurs after unconscious processing but before conscious awareness, right where quantum-to-classical transition would happen. The 300-millisecond timing provides support, because this is sufficient for quantum collapse in microtubules to propagate through classical neural channels to motor response. While one cannot currently directly observe quantum processes in the brain, this gamma signature represents the kind of classical "aftershock" that quantum collapse would likely produce.

In holographic quantum systems, information can reorganize across the entire network much faster than classical processing would allow. The time required for this reorganization grows logarithmically with system size, which means reorganization can happen nearly instantaneously even in large networks. The ~40 Hz gamma burst preceding insight appears consistent with this kind of rapid information reorganization that is followed by a classical readout (Sekino & Susskind, 2008; Swingle et al., 2016).

Toole et al. (2018) provide additional support in connecting these phenomena, as they note that "coherent energy transfer across multiple microtubules in a neuron and across many brain cells via synchronous neuron firing may have implications in brain activity, behavior, and even consciousness" (p. 121). These results suggest that suitable biological infrastructure for quantum computing may already exists in the brain.

Theoretical Validation of Bose-Einstein Condensates at Body Temperature

A further line of theoretical support for neural quantum processing comes from recent calculations by Cavaglià et al. (2023), who demonstrate that the energetic conditions within neuronal membranes are sufficient to sustain Bose–Einstein condensates at physiological temperature.

Ordinarily, condensate formation is associated with near-absolute-zero conditions, yet their analysis shows that neural membranes meet all thermodynamic criteria for biological quantum coherence. They note that "these assumptions are satisfied by the neuronal membrane phosphate groups' dipoles provided both AP energy and the rate of AP generation is sufficiently high," with "externally supplied energy by APs channeled into a single collective dynamical mode" (p. 6). Their calculations shows that maintaining weak condensates requires "the pumping rate for biochemical energy above the threshold of (100 K) k_B/s… [with] the rate of energy loss due to thermal interactions… less than 20 K kB/s, i.e., 5 times lower than the rate of energy supply" (p. 5). When applied to physiological conditions at 310 K, they calculate that neural action potentials provide energy supply rates of "$5\ k_BT/s < S = E_{ap-d}/dt < 50\ k_BT/s$" which "clearly exceeds the minimum requirement of $0.3\ k_BT/s$" (p. 6).

Based on these calculations, the authors conclude that "a weak Bose-Einstein condensate can be generated in the dipolar system of neuronal membrane's head groups" (p. 6). While experimental confirmation is still needed, the analysis supports the energetic plausibility of condensate formation in living neural tissue. When placed alongside the experimental evidence for quantum coherence in microtubules (Babcock et al., 2024; Nishiyama et al., 2025) and the consciousness-linked coherence effects reported by Kerskens and López Pérez (2022), it further strengthens the foundation for the framework's premise that neural tissue may possess the quantum computational infrastructure necessary for the $P = Q/E$ mechanism.

Processing Speeds

A common objection to quantum consciousness theories concerns scale. How could seemingly fragile quantum effects matter in the warm, noisy environment of the brain? One might explore how exponential computational advantages could emerge from even minimal quantum coherence. As a thought experiment, consider that the human brain contains approximately 7×10^{27} atoms. If a very small fraction of these atoms maintained quantum superposition, the number would still be astronomical, potentially creating massive computational advantages through quantum parallelism even if these moments of quantum processing were short lived.

Recent research, however, provides a more direct and concrete picture of the computational scale involved. Instead of estimating based on a percentage of atoms, Kurian (2025) estimates that the maximum computational speed of the standard classical model for a neuron, the Hodgkin-Huxley mechanism, is approximately 10^3 operations per second. By comparison, the superradiant protein fibers found in eukaryotic cells are calculated to have an upper bound of approximately 10^{13} operations per second. This suggests that the quantum mechanisms available to biology may be up to a billion times faster than the conventional model of neural communication.

Furthermore, this research indicates that biological quantum systems are extremely efficient, operating within two orders of magnitude of the Margolus-Levitin bound, which is the absolute physical limit for computation. When extrapolated over the history of life on Earth, this distinction becomes even more significant. The total computational capacity of life using classical neural processing is estimated to be around 10^{40} operations. The capacity of "superradiant life" utilizing these quantum processes is estimated to be around 10^{60} operations, which is a difference of twenty orders of magnitude (Kurian, 2025).

From this perspective, even subtle improvements in coherence could yield exponential computational gains, which would help to explain why small changes in consciousness, through meditation, flow, psychedelics, or other mechanisms, could produce measurable shifts in perception and performance. Evolution had billions of years to optimize even tiny quantum advantages into the sophisticated consciousness humanity experiences today.

Many researchers now argue that classical models alone cannot explain the brain's extraordinary efficiency. Patwa et al. (2024) conclude,

> "Our results pose opportunities for a paradigm shift in the theory of neuronal information processing and signaling ... classical energy scalings cannot account for the sub-neuronal information processing capacity of the brain, given its extremely low input power of around 20 W. There must be another physical mechanism that enables the human brain to achieve the computational efficiency that it does... A tantalizing possibility is that extended protein architectures... including axons in the brain, may form a highly interconnected, ultrafast quantum-optical network... This mechanism would be much faster than chemical Hodgkin-Huxley-type transport... which is currently used as a standard paradigm in neuroscience" (p. 20).

This quantum infrastructure, which operates at speeds and scales far beyond classical neural processes, provides the physical substrate necessary for phenomena that have puzzled neuroscience for quite a while. If the brain truly maintains quantum coherence in warm, wet biological tissue, processing information through superradiant networks that operate billions of times faster than classical signaling, then perhaps this provides reason for capabilities that seem to transcend normal computational limits.

If the brain indeed holds this capability, a new set of questions naturally follows. Why can humanity access it only intermittently? What controls this access? How does the switching mechanism work, and why would it have evolved? If the infrastructure is so deeply embedded in biology, why has it remained hidden in plain sight? Wouldn't some humans have figured out ways to use it more effectively by now? These questions guide the exploration that unfolds in the coming chapters.

References

Babcock, N. S., Montes-Cabrera, G., Oberhofer, K. E., Chergui, M., Celardo, G. L., & Kurian, P. (2024). Ultraviolet superradiance from mega-networks of tryptophan in biological architectures. *The Journal of Physical Chemistry B, 128*(17), 4035–4046. https://doi.org/10.1021/acs.jpcb.3c07936

Babcock, N. S., Kattnig, D. R., Oncul, M., Lewis, A. M., Zdrazil, L., Strizhak, V., et al. (2023). Ultraviolet superradiance from mega-networks of tryptophan in biological architectures. *Journal of Physical Chemistry Letters, 14*(10), 2158–2165. https://doi.org/10.48550/arXiv.2302.01469

Ball, P. (2011). Physics of life: The dawn of quantum biology. *Nature, 474*(7351), 272–274. https://doi.org/10.1038/474272a

Bragina, L., & Conti, F. (2018). Expression of neurofilament subunits at neocortical glutamatergic and GABAergic synapses. *Frontiers in Neuroanatomy, 12*, Article 74. https://doi.org/10.3389/fnana.2018.00074

Cavaglià, M., Deriu, M. A., & Tuszynski, J. A. (2023). Toward a holographic brain paradigm: A lipid-centric model of brain functioning. *Frontiers in Neuroscience, 17*, 1302519. https://doi.org/10.3389/fnins.2023.1302519

Craddock, T. J. A., Friesen, D., Mane, J., Hameroff, S., & Tuszynski, J. A. (2014). The feasibility of coherent energy transfer in microtubules. *Journal of the Royal Society Interface, 11*(100), 20140677. https://doi.org/10.1098/rsif.2014.0677

Craddock, T. J. A., Tuszynski, J. A., & Hameroff, S. (2012). Cytoskeletal signaling: Is memory encoded in microtubule lattices by CaMKII phosphorylation? *PLoS Computational Biology, 8*(3), e1002421. https://doi.org/10.1371/journal.pcbi.1002421

Dvali, G., & Gomez, C. (2014). Black holes as critical point of quantum phase transition. *European Physical Journal C, 74*, 2752. https://doi.org/10.1140/epjc/s10052-014-2752-3

Engel, G. S., Calhoun, T. R., Read, E. L., Ahn, T. K., Mančal, T., Cheng, Y. C., et al. (2007). Evidence for wavelike energy transfer through quantum coherence in photosynthetic systems. *Nature, 446*(7137), 782–786. https://doi.org/10.1038/nature05678

Feynman, R. P., Leighton, R. B., & Sands, M. (1965). *The Feynman lectures on physics. Vol. 3: Quantum mechanics.* Addison-Wesley.

Grover, L. K. (1996). A fast quantum mechanical algorithm for database search. In *Proceedings of the Twenty-Eighth Annual ACM Symposium on Theory of Computing* (pp. 212–219). Association for Computing Machinery. https://doi.org/10.1145/237814.237866

Hameroff, S., & Penrose, R. (1996). Orchestrated reduction of quantum coherence in brain microtubules: A model for consciousness. *Mathematics and Computers in Simulation, 40*(3–4), 453–480. https://doi.org/10.1016/0378-4754(96)80476-9

Hameroff, S., & Penrose, R. (2014). Consciousness in the universe: A review of the 'Orch OR' theory. *Physics of Life Reviews, 11*(1), 39–78. https://doi.org/10.1016/j.plrev.2013.08.002

Hiscock, H. G., Worster, S., Kattnig, D. R., Steers, C., Jin, Y., Manolopoulos, D. E., Mouritsen, H., & Hore, P. J. (2016). The quantum needle of the avian magnetic compass. *Proceedings of the National Academy of Sciences, 113*(17), 4634–4639. https://doi.org/10.1073/pnas.1600341113

Jibu, M., & Yasue, K. (1995). *Quantum brain dynamics and consciousness: An introduction.* John Benjamins Publishing.

Khan, S., Huang, Y., Timuçin, D., Bailey, S., Lee, S., Lopes, J., Gaunce, E., Mosberger, J., Zhan, M., Abdelrahman, B., Zeng, X., & Wiest, M. C. (2024). Microtubule-stabilizer epothilone B delays anesthetic-induced unconsciousness in rats. *eNeuro, 11*(8), ENEURO.0291-24.2024. https://doi.org/10.1523/ENEURO.0291-24.2024

Kahneman, D. (2011). *Thinking, fast and slow.* Farrar, Straus and Giroux.

Kalra, A. P., Benny, A., Travis, S. M., Zizzi, E. A., Morales-Sanchez, A., Oblinsky, D. G., Craddock, T. J. A., Hameroff, S. R., MacIver, M. B., Tuszyński, J. A., Petry, S., Penrose, R., & Scholes, G. D. (2023). Electronic energy migration in microtubules. *ACS Central Science, 9*(3), 352–361. https://doi.org/10.1021/acscentsci.2c01114

Kauffman, S. A., & Radin, D. (2023). Quantum aspects of the brain-mind relationship: A hypothesis with supporting evidence. *Biosystems, 223*, 104820. https://arxiv.org/abs/2101.01538. https://doi.org/10.1016/j.biosystems.2022.104820

Kerskens, C. M., & López Pérez, D. (2022). Experimental indications of non-classical brain functions. *Journal of Physics Communications, 6*(10), 105001. https://doi.org/10.1088/2399-6528/ac94be

Konik, R. (2021). Quantum coherence confined. *Nature Physics, 17*(6), 669–670. https://doi.org/10.1038/s41567-021-01211-5

Kounios, J., & Beeman, M. (2009). The aha! Moment: The cognitive neuroscience of insight. *Current Directions in Psychological Science, 18*(4), 210–216. https://doi.org/10.1111/j.1467-8721.2009.01638.x

Kurian, P. (2025). Computational capacity of life in relation to the universe. *Science Advances, 11*, eadt4623. https://doi.org/10.1126/sciadv.adt4623

Matarèse, B. F. E., Rusin, A., Seymour, C., & Mothersill, C. (2023). Quantum biology and the potential role of entanglement and tunneling in non-targeted effects of ionizing radiation: A review and proposed model. *International Journal of Molecular Sciences, 24*(22), 16464. https://doi.org/10.3390/ijms242216464

Mothersill, C., & Seymour, C. (2001). Radiation-induced bystander effects: Past history and future directions. *Radiation Research, 155*(6), 759–767. https://doi.org/10.1667/0033-7587(2001)155[0759:RIBEPH]2.0.CO;2

Mothersill, C., Smith, R., Wang, J., Rusin, A., Fernandez-Palomo, C., Fazzari, J., & Seymour, C. (2018). Biological entanglement-like effect after communication of fish prior to X-ray exposure. *Dose-Response, 16*(1). https://doi.org/10.1177/1559325817750067

Nielsen, M. A., & Chuang, I. L. (2000). *Quantum computation and quantum information*. Cambridge University Press.

Nishiyama, A., Tanaka, S., & Tuszynski, J. A. (2024a). Quantum brain dynamics: Optical and acoustic super radiance via a microtubule. *Biosystems, 234*, 105118. https://doi.org/10.3390/foundations4020019

Nishiyama, A., Tanaka, S., Tuszynski, J. A., & Tsenkova, R. (2024b). Holographic brain theory: Super-radiance, memory capacity and control theory. *International Journal of Molecular Sciences, 25*(4), 2399. https://doi.org/10.3390/ijms25042399

Nishiyama, A., Tanaka, S., & Tuszynski, J. A. (2025). Waveguide quantum electrodynamics: Tryptophans entangled with water as data qubits in a microtubule. *Dynamics, 5*, 7. https://doi.org/10.3390/dynamics5010007

Panitchayangkoon, G., Hayes, D., Fransted, K. A., Caram, J. R., Harel, E., Wen, J., Blankenship, R. E., & Engel, G. S. (2010). Long-lived quantum coherence in photosynthetic complexes at physiological temperature. *Proceedings of the National Academy of Sciences, 107*(29), 12766–12770. https://doi.org/10.1073/pnas.1005484107

Patwa, H., Babcock, N. S., & Kurian, P. (2024). Quantum-enhanced photoprotection in neuroprotein architectures emerges from collective light-matter interactions. *Frontiers in Physics, 12*, 1387271. https://doi.org/10.3389/fphy.2024.1387271

Pastawski, F., Yoshida, B., Harlow, D., & Preskill, J. (2015). Holographic quantum error-correcting codes: Toy models for the bulk/boundary correspondence. *Journal of High Energy Physics, 2015*(6), 149. https://doi.org/10.1007/JHEP06(2015)149

Ricciardi, L. M., & Umezawa, H. (1967). Brain and physics of many-body problems. *Kybernetik, 4*(2), 44–48. https://doi.org/10.1007/BF00290174

Saxena, K., Singh, P., Sahoo, P., Sahu, S., Ghosh, S., Ray, K., Fujita, D., & Bandyopadhyay, A. (2020). Fractal, scale free electromagnetic resonance of a single brain extracted microtubule nanowire, a single tubulin protein and a single neuron. *Fractal and Fractional, 4*(2), 11. https://doi.org/10.3390/fractalfract4020011

Sekino, Y., & Susskind, L. (2008). Fast scramblers. *Journal of High Energy Physics, 2008*(10), 065. https://doi.org/10.1088/1126-6708/2008/10/065

Shor, P. W. (1997). Polynomial-time algorithms for prime factorization and discrete logarithms on a quantum computer. *SIAM Journal on Computing, 26*(5), 1484–1509. https://doi.org/10.1137/S0097539795293172

Singh, P., Sahoo, P., Ghosh, S., Saxena, K., Manna, J. S., Ray, K., Krishnananda, S. D., Poznanski, R. R., & Bandyopadhyay, A. (2021). Filaments and four ordered structures inside a neuron fire a thousand times faster than the membrane: Theory and experiment. *Journal of Integrative Neuroscience, 20*(4), 777–790. 10.31083/j.jin2004082.

Singh, P., Sahoo, P., Saxena, K., Manna, J. S., Ray, K., Ghosh, S., & Bandyopadhyay, A. (2021). Cytoskeletal filaments deep inside a neuron are not silent: They regulate the precise timing of nerve spikes using a pair of vortices. *Symmetry, 13*(5), 821. https://doi.org/10.3390/sym13050821

Singh, P., Saxena, K., Sahoo, P., Ghosh, S., & Bandyopadhyay, A. (2021). Electrophysiology using coaxial atom probe array: Live imaging reveals hidden circuits of a hippocampal neural network. *Journal of Neurophysiology, 125*(6), 2107–2116. https://doi.org/10.1152/jn.00478.2020

Stuart, C. I. J. M., Takahashi, Y., & Umezawa, H. (1978). On the stability and non-local properties of memory. *Journal of Theoretical Biology, 71*(4), 605–618.

Stuart, C. I. J. M., Takahashi, Y., & Umezawa, H. (1979). Mixed-system brain dynamics: Neural memory as a macroscopic ordered state. *Foundations of Physics, 9*(3–4), 301–327.

Swingle, B. (2012). Entanglement renormalization and holography. *Physical Review D, 86*(6), 065007. https://doi.org/10.1103/PhysRevD.86.065007

Swingle, B., Bentsen, G., Schleier-Smith, M., & Hayden, P. (2016). Measuring the scrambling of quantum information. *Physical Review A, 94*(4), 040302. https://doi.org/10.48550/arXiv.1602.06271

Tan, W. L., Becker, P., Liu, F., Pagano, G., Collins, K. S., De, A., Feng, L., Kaplan, H. B., Kyprianidis, A., Lundgren, R., Morong, W., Whitsitt, S., Gorshkov, A. V., & Monroe, C. (2019). *Observation of domain wall confinement and dynamics in a quantum simulator*. arXiv preprint arXiv:1912.11117. https://doi.org/10.1038/s41567-021-01194-3

Tegmark, M. (2000). Importance of quantum decoherence in brain processes. *Physical Review E, 61*(4), 4194–4206. https://doi.org/10.1103/physreve.61.4194

Toole, J. T., Kurian, P., & Craddock, T. J. A. (2018). Coherent energy transfer and the potential implications for consciousness. *Journal of Cognitive Science, 19*(2), 115–124. https://doi.org/10.17791/jcs.2018.19.2.115

Vidal, G. (2007). Entanglement renormalization. *Physical Review Letters, 99*(22), 220405. https://doi.org/10.1103/PhysRevLett.99.220405

von Neumann, J. (1932). *Mathematische Grundlagen der Quantenmechanik*. Springer. https://doi.org/10.1007/978-3-642-61409-5

Wiest, M. C. (2025). A quantum microtubule substrate of consciousness is experimentally supported and solves the binding and epiphenomenalism problems. *Neuroscience of Consciousness, 2025*(1), niaf011. https://doi.org/10.1093/nc/niaf011

Chapter 3
The Mathematical Forcing Function

Chapters 1 and 2 have established three of the five premises. The book has examined how self-observation destroys performance, drawn from established physics showing that observation collapses quantum states, and presented evidence that the brain contains quantum infrastructure. The framework now turns to **premise 4 that at least one human has accessed information beyond personal experience.**

While this premise may appear extraordinary, readers are invited to hold it in superposition, consider the probability, and examine the evidence. If biology supports quantum-capable processing, it becomes reasonable to ask whether humans have ever accessed information beyond classical expectation, even inadvertently.

These are not new questions. They have been explored for more than a century by scientists who tried to understand the connections between mind and matter. Carl Jung was one of the first in the modern era to take these ideas seriously. He began to discuss the theory of relativity with Albert Einstein while Einstein was in Zürich, and Jung began to worked closely with Nobel-winning physicist Wolfgang Pauli for decades to develop the concept of synchronicity which they defined as meaningful connections that appear to transcend space and time. Jung was committed to grounding his psychology in physics and thermodynamics, and he searched for language that could unite the psyche and nature within one continuous framework (Jung, 1963; Hürter, 2023).

Near the end of his life, Jung described an experience that captures the kind of phenomenon under discussion in this chapter. In his autobiography, *Memories, Dreams, Reflections* (1963b), he wrote,

> "At that time I had to deliver a lecture in B. I returned to my hotel around midnight. I sat with some friends for a while after the lecture, then went to bed, but I lay awake for a long time. At about two o'clock—I must have just fallen asleep—I awoke with a start, and had the feeling that someone had come into the room; I even had the impression that the door had been hastily opened. I instantly turned on the light, but there was nothing. Someone might have mistaken the door, I thought, and I looked into the corridor. But it was still as death. 'Odd,' I thought, 'someone did come into the room!' Then I tried to recall exactly what had happened, and it occurred to me that I had been awakened by a feeling of dull pain, as

J. Roeloffs, *The Classical Brain to the Quantum Mind*, The Frontiers Collection,
https://doi.org/10.1007/978-3-032-20068-6_3

> though something had struck my forehead and then the back of my skull. The following day I received a telegram saying that my patient had committed suicide. He had shot himself. Later, I learned that the bullet had come to rest in the back wall of the skull. This experience was a genuine synchronistic phenomenon such as is quite often observed in connection with an archetypal situation—in this case, death. By means of a relativization of time and space in the unconscious it could well be that I had perceived something which in reality was taking place elsewhere. The collective unconscious is common to all; it is the foundation of what the ancients called the 'sympathy of all things.' In this case the unconscious had knowledge of my patient's condition. All that evening, in fact, I had felt curiously restive and nervous, very much in contrast to my usual mood" (p. 136).

For Jung, this was a real observation of how mind and matter might briefly intersect. He interpreted it as evidence that the unconscious could relativize time and space and access shared information through the collective field as he describes. Including Jung's account here serves as historical grounding for the ideas explored in this chapter. It aims to reinforce to that questions about non-local information and the link between consciousness and physics are not new conjectures of the quantum age as they have been pursued for over a century by researchers trying to bridge psychology and the physical sciences.

This chapter does not claim that all reported phenomena are genuine, far from this actually. Its purpose is to treat the cumulative evidence as a probability problem instead of a belief problem. Even a *single* verified instance of non-local access to information in human history would flip the burden of proof and require a re-examination of current models of cognition, but what is the probability that all reported instances of these types of phenomena are false? This is what this chapter explores.

The Data and Pillars

The scientific literature contains over 100 documented cases of savant syndrome. These are individuals who demonstrate abilities that appear to violate humanity's understanding of how knowledge and skills are acquired. These are peer-reviewed, carefully documented cases published in respected journals.

As Treffert (2014) notes after decades of studying savants, "Some savants… clearly know things they never learned. The only possible way to know things one never learned… is for that knowledge to be factory installed, genetically transmitted" (p. 566). Yet even this explanation fails for cases involving original creative works, novel mathematical insights, or languages with no genetic connection to the individual. Treffert is understandably operating from a classical perspective of the brain's function. Considering the evidence presented in previous chapters of the plausibility of quantum infrastructure in the brain, this book will explore another option that may shed more light on where these savants' knowledge could come from.

The First Pillar: Documented Savant Cases

For over a century, science has ignored savants and other outlier gifts as noisy signals, which may have been reasonable at the time. However, if in the last 10–15 years evidence has emerged that biology and the brain could have quantum functions, then this creates mathematical realities that must be faced. The odds present a direct challenge to dismissal through skepticism. What are the odds, when accounting for the savant cases alone, that all 100+ documented cases are entirely false? Even under extreme skepticism, assuming each case has a 99% chance of being explained away by deception, misdiagnosis, or error, the probability that all 100 cases are false is:

$$0.99^{100} \approx 36.6\%$$

That means, even under the most skeptical conditions, there's a 63.4% chance that at least one savant case represents genuine access to knowledge never learned. Even if these probabilities are not fully independent or are affected by selection bias in some way, the disjunctive conclusion still holds that multiple heterogeneous data streams each increase the posterior probability that at least one instance likely represents genuine access to non-local information.

The Second Pillar: Contemporary Scientific and Phenomenological Reports

The mathematical forcing function becomes more compelling when considering contemporary research on non-local information access. Wahbeh et al. (2018) found that 93.2% of scientists and engineers ($n = 175$) reported at least one exceptional experience transcending conventional space-time boundaries. This prevalence, essentially equal to the general population (94.0%), suggests that even those professionally committed to materialist paradigms regularly experience phenomena consistent with quantum information access.

Furthermore, Wahbeh et al. (2022b) collected detailed phenomenological accounts from 521 individuals describing how they access information "not limited by conventional notions of time and space." Of these, 219 specifically reported precognitive experiences which was defined as knowing future events before they occurred. Even considering only these precognition cases and assuming a 99% independent probability that each represents delusion, misinterpretation, or deception, the chance that all are false is about 10.8%. In other words, even under extreme skepticism there approximately remains an 89% probability that at least one report reflects genuine non-local information access.

The Third Pillar: Laboratory Research under Controlled Conditions

Controlled laboratory experiments provide more perspective as Beischel and Schwartz (2007) conducted a triple-blind study where mediums provided accurate information about deceased individuals ($p = 0.007$, effect size $= 0.5$) despite having no contact with or information about the living relatives, using a proxy sitter who knew nothing about the deceased, and having sitters blind to which readings were theirs when scoring.

They concluded, "The results suggest that certain mediums can anomalously receive accurate information about deceased individuals" (Beischel & Schwartz, 2007). The study's well-grounded design systematically eliminated most conventional explanations, including cold reading, sensory cues, and even telepathy from living persons which leaves non-local information access as the most viable explanation. If information about deceased individuals can remain accessible with no physical or mental pathway, then where is it coming from?

The Fourth Pillar: Historical and Contemplative Testimony

Throughout history, millions across cultures have reported experiencing consciousness beyond ego identification and accessing information beyond personal experience. The framework focuses on a specific, testable claim that through certain contemplative or altered states, the sense of "I" can temporarily dissolve while awareness persists, and this state is sometimes accompanied by insights or knowledge not previously possessed. This phenomenon appears to be consistent across contemplative traditions as Buddhist monks, Christian scholars, Sufis, secular meditators and many other religious groups may disagree on theology but they seem to converge on this experiential report. Through certain practices, the ordinary sense of self can reportedly dissolve while consciousness remains present, and this is often accompanied by direct knowing or receiving revelation.

While anecdotal, the global and historical prevalence of these descriptions invites some statistical considerations. If, say hypothetically, one million individuals throughout recorded history reported experiencing ego dissolution that was accompanied by access to information beyond prior knowledge, and if 99.9% of those reports were attributable to delusion, misinterpretation, or cultural framing, the probability that all remaining cases are false would still approach zero. Although this reasoning is clearly not a formal quantitative analysis, it seems worth considering these known reports as additional context as it similarly suggests that at least some individuals may have experienced phenomena consistent with non-local information access.

Additional Unexplained Experiments and Phenomena

The convergence of these types of experiences still extends beyond the previous four pillars as Wahbeh et al. (2022a) document a variety of additional categories of phenomena that suggest non-local information access is plausible, as each is supported by controlled laboratory studies that span multiple decades and institutions.

Mind-to-mind information transfer has been tested through the Ganzfeld protocol and replicated in over 120 published experiments comprising approximately 4000 individual trials across multiple laboratories. Where chance expectation would be 25% accuracy as a baseline, the overall hit rate for those in the experiments reached over 30%. These results have been discussed in mainstream academic psychology journals and subjected to multiple meta-analyses (Storm et al., 2010, as cited in Wahbeh et al., 2022a, p. 9). Related studies using physiological measures like electrodermal activity and EEG have shown that one person's focused attention can measurably influence another isolated person's unconscious responses to a degree.

The modest but consistent deviation from chance in these experiments suggests that some form of non-local correlation may be possible under specific psychological conditions. Decades before these protocols were developed, Carl Jung had already explored similar anomalies while studying J. B. Rhine's early ESP experiments. Jung noted that statistical improbability alone made "mere chance" explanations insufficient, and he admitted that conventional causality could not account for these experimental results that appeared to link distant minds or events. He explicitly rejected mystical or transcendental explanations for these surprising experimental results, as he argued that new physical principles might eventually explain these effects. He wrote, "Here, for want of a demonstrable cause, we are all too likely to fall into the temptation of positing a transcendental one. But a 'cause' can only be a demonstrable quantity. A 'transcendental cause' is a contradiction in terms, because anything transcendental cannot by definition be demonstrated" (1969c, pp. 39–40).

He proposed what he termed an acausal connecting principle, as he described meaningful coincidences that occur "when an event is observed without experimental restrictions" and the observer's emotional state "alters space and time by contraction" (p. 39). Jung's language was tentative at the time to maintain scientific rigor, but he seemed to anticipate a framework in which consciousness and physical reality could interact through non-local means as he concluded that "we are finally compelled to assume that there is in the unconscious something like an a priori knowledge or an immediacy of events which lacks any causal basis" (p. 40). Jung's work provides historical and theoretical grounding for the idea that non-local information access might be both psychologically and physically lawful as an emergent property of consciousness as it may interact with a shared informational field. His studies and ideas around that topic will be examined in detail later in the book.

In other areas of research and experimentation, remote perception involves individuals accessing information about distant locations under rigorous experimental conditions. A classified US government program conducted over 500 operational missions from 1972 to 1995, with results evaluated by both a professor of statistics and a skeptical psychology professional. Both evaluators concluded the evidence was "statistically significant and could not be attributed to methodological flaws" (Mumford et al., 1995; Utts, 2016, as cited in Wahbeh et al., 2022a, p. 8). Multiple meta-analyses of public domain experiments continue showing highly positive evidence, with practical applications now including stock market prediction and archaeological site location.

Precognitive responses demonstrate physiological reactions to randomly selected future events occurring 1–10 s before stimulus presentation. The body appears to react to emotionally charged images before they are randomly selected and displayed. Mossbridge et al. (2012, as cited in Wahbeh et al., 2022a, p. 9) analyzed 26 studies and found significant pre-stimulus effects with an overall effect size of 0.21 and statistical significance of $p < 2.71 \times 10^{-12}$. Reverse priming experiments, where emotional words appear after target images instead of appearing before, which continues to show the expected priming effects across 90 independent replications.

Terminal lucidity describes patients with severe neurodegenerative conditions who surprisingly display normal cognitive function immediately before death, despite objective medical evidence that indicates that biological function like this should be impossible for them. Batthyány and Greyson (2021, as cited in Wahbeh et al., 2022b) reviewed 124 cases in dementia patients, finding that "more than 80% showed complete remission with the return of memory, orientation, and responsive verbal ability" (p. 10). One documented case involved a cancer patient with brain metastases and minimal functional brain tissue remaining who regained awareness and conversed coherently with family for five minutes before death.

While this book entirely understands, and even shares, the skepticism of any given individual report of the types of phenomena described in this chapter, it also has to acknowledge the significant amount of evidence for non-local information. When these extensively documented phenomena are added to the savant cases, contemporary reports, laboratory validation, and historical testimony, the mathematical improbability that all categories represent error, fraud, or misinterpretation becomes overwhelmingly unlikely. These patterns suggest that there is some sort of systematic access to information beyond classical learning mechanisms that is documented across multiple independent research programs using different methodologies, populations, and institutional settings.

The Combined Forcing Function

When combining all lines of evidence, even when only from the first four pillars, the statistical argument becomes more evident within the scientific documentation. Considering there are 100+ savant cases, this means there is a 63.4% probability of at least one being real even when applying 99% skepticism. Contemporary prevalence shows 93.2% of scientists report exceptional experiences, and the phenomenological evidence includes 219 precognition reports with an 89.2% probability of at least one being genuine even under extreme skepticism. Then laboratory evidence comes through the triple-blind studies with $p < 0.01$, all the while historical testimony includes millions of contemplative reports across cultures with near-zero probability of all being false.

The convergence creates what might be called a mathematical forcing function, as multiple independent categories of evidence, each using different methodologies and populations, arrive at similar conclusions. This suggests that at least one human at some point has demonstrably accessed non-local information in some way. If this is the case, this shifts the burden of proof as it changes the important question from whether this type of access is possible to exploring why it wouldn't be. Naturally, this raises new questions about where this non-local information resides, what physical and biological processes allow for this access, and why this ability remains mostly inaccessible for the vast majority of people.

Conventional Explanations

As Treffert (2014) acknowledges, existing theories fail to explain the savant phenomenon:

> "There have been many theories put forth to try to explain savant syndrome ranging from early heredity theories to present day Quantum theory. Some of the neuropsychological theories such as weak central coherence, mind blindedness are interesting as they apply to the autistic savant. But 50% of savants are not autistic" (p. 555).

Treffert is showing how conventional explanations have consistently fallen short as he described that genetic theories yield inconsistent results as "the search for a savant 'gene' is underway, one study finding such a gene, but another not confirming that finding" (Treffert, p. 556), while learning theories face the challenge that "compensatory learning, reinforcement and repetition-compulsion may also play a role, but then, if those dynamics producc savant syndrome, why wouldn't that apply to all persons with autism or other CNS limitations?" (p. 556), and inherited memory cannot explain original compositions or novel mathematical insights.

Contemplative experiences of accessing knowledge beyond personal experience similarly resists the conventional explanation through hallucination, wishful thinking, or cultural conditioning, especially because of their consistency across unconnected traditions, and Treffert notes that quantum theories are already being proposed as current neuroscience has no satisfactory mechanism for how someone could access complex information they never encountered.

If the brain has the quantum infrastructure as described in the previous chapter, it opens the door for these ideas to be plausible. The framework's proposal that consciousness can access non-local information through quantum mechanisms finds interesting support from recent theoretical work. Nishiyama et al. (2024) suggest that "manipulating holograms by external electric fields, we might be able to affect our memory and subjective experiences" (p. 10). If consciousness interfaces with quantum information holographically, then both savant abilities and contemplative insights might represent naturally occurring variations in this interface. Some individuals might possess neural architectures that maintain stronger coherence with the quantum information field, which could potentially enable access to mathematical constants, musical compositions, linguistic structures, or complex patterns without conventional learning.

This holographic model could explain why savant abilities often emerge suddenly after brain injury and why contemplative practices that reduce ego activity can produce similar access. Damage to ego-generating regions or systematic ego suppression through meditation might not create these abilities but rather remove the barriers that normally restrict access to quantum information already present in the holographic field.

A Savant Case Study

Even though there are hundreds of savant cases, Jason Padgett represents a fascinating case study. Following the logical conclusions of the framework, it would predict that if someone had suppressed ego and was utilizing the quantum capabilities of their brain, they could potentially observe some aspects of the quantum process in real time, and if so, it's entirely possible they would see geometric shapes like those who takes psychedelics do and would have incredible intuitive skills to visualize math, especially geometry. As will be explored in later chapters, the language of quantum would inherently be interpreted as geometric and archetypal in nature. Padgett provides some evidence for these ideas.

Padgett, who at the time had never studied math or physics in his life, was assaulted and received damage to ego-generating regions; he reported instant mathematical visualization which closely matched quantum interference patterns (Padgett & Seaberg, 2014).

Post-injury, he reports, "When things moved, they moved in these bizarre stop-action frames, like individual discrete picture frames with a line connecting them" (Padgett, TEDxTacoma, 2016a). Notably, he was compelled to draw what he saw. His geometric drawings from that time are public and match quantum physics he'd never studied. He describes it saying, "When he showed me these [interference patterns], a feather could have knocked me over because the interference patterns were identical to many of my drawings" (Padgett, TED Archive, 2016b). He was drawing quantum mechanical phenomena without knowing what they were. Padgett's story is worth investigating and many of the details he reports and conclusions he has come to are extremely relevant to the topics of this book. Overall, Padgett's case is hypothesis-consistent and motivates testable predictions from the text; it does not, on its own, establish any sort of mechanism or provide the tangible evidence necessary, but it seemed valuable to acknowledge the similarities in his case.

The Next Question

The mathematical forcing function created by savant documentation, contemporary scientific reports, phenomenological testimony, and controlled laboratory studies establishes the highly probable odds that at least some humans can and do access non-local information. A few more logical questions arise. If consciousness can indeed transcend personal knowledge boundaries and access expanded information states, what determines differential access? How does the brain navigate between normal and expanded modes? What mechanisms filter the vast quantum field configurations into useful insights?

Treffert proposes a mechanism he calls "paradoxical functional facilitation,"

> "What I have come to call the 'three R's' reflects the process Kapur termed 'paradoxical functional facilitation' in 1996 in which one area of the brain is released from the inhibiting influence of some other brain area. In the case of savants, both congenital and acquired, there is brain damage in one area, frequently the left hemisphere, with recruitment of still intact brain tissue in another area of the brain, rewiring of circuitry to that new area, and release of dormant capacity, through a disinhibiting process, of information and skills already stored in that newly recruited area" (p. 556).

The answer may lie in understanding what damaged regions in savants and what contemplative practices both affect which is the ego-generating, self-referential areas of the brain, and there's been detailed research on this. The left hemisphere damage is particularly significant. Neuroscience has established that these areas maintain large portions of humanity's sense of self, our narrative identity, and our continuous self-observation. Also interesting is Treffert's observation about human potential. He writes, "These instances raise many interesting questions about dormant capacity within us all, and raise the even more challenging question of how to tap those buried abilities without enduring some CNS catastrophe" (p. 566).

This points towards the possibility that these abilities are buried and then released once the injury occurs. Treffert notes that developing savant abilities doesn't impair other functions, "Such a 'dreaded trade-off' or loss of skills does not occur... To the contrary, vigorously training the talent leads... to increased language, social, and daily living skills" (p. 567). This suggests that the capacity may exist in all brains, something may be actively suppressing it, and if executed correctly, strategic modulation could access it safely.

There are three big questions that emerge from the research and statistical likelihood of non-local information. Where does this information come from? Why would something evolve to inhibit humanity's access to this ability? What mechanism would modulate humanity's access to it?

First, the book will explore where the information may come from, then it will go on to explore why evolution developed in this way, and why humanity has had limited access to it.

References

Beischel, J., & Schwartz, G. E. (2007). Anomalous information reception by research mediums demonstrated using a novel triple-blind protocol. *EXPLORE, 3*(1), 23–27. https://doi.org/10.1016/j.explore.2006.10.004

Hürter, T. (2023). *Too big for a single mind: How the greatest generation of physicists uncovered the quantum world* (D. Shaw, Trans.). The Experiment.

Jung, C. G. (1963). Memories, dreams, reflections (R. Winston & C. Winston, Trans.; A. Jaffé, Ed.). Collins and Routledge & Kegan Paul. (Original work published 1961).

Nishiyama, A., Tanaka, S., Tuszynski, J. A., & Tsenkova, R. (2024). Holographic brain theory: Super-radiance, memory capacity and control theory. *International Journal of Molecular Sciences, 25*(4), 2399. https://doi.org/10.3390/ijms25042399

Padgett, J. (2016a, May 12). *Alternate realities from relativity* [video]. TEDxTacoma. https://www.youtube.com/watch?v=CyYQIcZacvA

Padgett, J. (2016b, November 1). *How math saved my life* [video]. TED Archive. https://www.youtube.com/watch?v=GDU7lEmiiD8 (Filmed 2015).

Padgett, J., & Seaberg, M. (2014). *Struck by genius: How a brain injury made me a mathematical marvel*. Houghton Mifflin Harcourt.

Treffert, D. A. (2014). Savant syndrome: Realities, myths and misconceptions. *Journal of Autism and Developmental Disorders, 44*, 564–571. https://doi.org/10.1007/s10803-013-1906-8

Wahbeh, H., Radin, D., Mossbridge, J., Vieten, C., & Delorme, A. (2018). Exceptional experiences reported by scientists and engineers. *EXPLORE, 14*(5), 329–341. https://doi.org/10.1016/j.explore.2018.05.002

Wahbeh, H., Speirn, P., Pederzoli, L., & Tressoldi, P. (2022a). What do we know about the phenomenology of exceptional human experiences? A thematic analysis of 20 years of research. *F1000Research, 10*, 497. https://doi.org/10.12688/f1000research.51746.3

Wahbeh, H., Radin, D., Cannard, C., & Delorme, A. (2022b). What if consciousness is not an emergent property of the brain? Observational and empirical challenges to materialistic models. *Frontiers in Psychology, 13*, 955594. https://doi.org/10.3389/fpsyg.2022.955594

Chapter 4
The Quantum Information Dimension (QID)

Information Conservation

Before introducing the second proposal, the framework must establish the final premise. **Premise 5 is that information cannot be destroyed,** which represents one of the most rigorously tested principles in physics.

Modern physics has increasingly recognized information as fundamental as Landauer's principle establishes that information is physical, as it requires a minimum amount of energy to erase (Landauer, 1961). Quantum information theory demonstrates that quantum states contain and process information in ways classical systems cannot replicate (Bennett & Shor, 1998). The holographic principle proposes that all information within a volume can be represented on its boundary (Susskind, 1995). Wheeler's "it from bit" hypothesis even frames physical reality as emerging from information (Wheeler, 1990). These developments point toward information as a conserved quantity as fundamental as energy or momentum.

Hawking (1976) initially proposed that information falling into black holes was permanently destroyed, it created a crisis in physics by violating quantum mechanics' core tenet of unitarity. The decades-long debate that followed, involving physicists like Susskind (1993) and ‘t Hooft (1993), ultimately resolved in favor of information preservation. Even Hawking himself eventually conceded (Hawking, 2005), acknowledging that black holes, often seen as the universe's ultimate destroyers, cannot erase information. It may be scrambled beyond recognition, encoded holographically on the event horizon, or slowly released through Hawking radiation, but it persists.

If information cannot be destroyed, then every pattern, every relationship, every possible configuration that has ever existed or could exist must be preserved somewhere in the universe.

J. Roeloffs, *The Classical Brain to the Quantum Mind*, The Frontiers Collection,
https://doi.org/10.1007/978-3-032-20068-6_4

The Logical Chain

Let us now follow the evidence to its logical implications. The book has established that the brain maintains quantum coherence in microtubules with 40–70% efficiency (Babcock et al., 2023). It has demonstrated it's plausible that at least one human has accessed non-local information which may include mathematical constants, musical compositions, or linguistic structures, that cannot be explained by genetics or unconscious observation (Treffert, 2014). It has established that it is known from physics that information cannot be destroyed and must exist somewhere.

If it's known that the world operates from quantum principles, and if the brain has quantum processing capabilities, and if at least one human has accessed information never learned, and if that information must exist somewhere due to conservation laws, then two questions arise. First, where does this information reside? Second, how does consciousness access it?

Proposing the Quantum Information Dimension

Following this logic, the book arrives at what appears to be a necessary theoretical construct. If information cannot be destroyed and if biological systems exhibit access to structured information that isn't learned through classical means, then this information must exist within a structure that transcends ordinary spacetime.

Therefore, proposal 2 introduces the Quantum Information Dimension (QID) as a coherent, pre-temporal substrate of quantum field configurations. This is akin to what quantum field theory already describes mathematically as configuration space which is the infinite-dimensional space of all possible field values at all spacetime points (Peskin & Schroeder, 1995; Zee, 2010). The QID represents the ontological reality of what Feynman's path integral formalism explores as the complete possibility space that quantum fields can occupy before any particular outcome is realized (Feynman & Hibbs, 1965; Schulman, 1981).

The QID is termed "pre-temporal" not in the sense of existing before time began, but as logically prior to the emergence of classical, experienced spacetime through collapse. In quantum field theory, spacetime coordinates simply parameterize field configurations, while the field configurations themselves exist as lawful structures independent of any particular moment in experiential time. The QID represents this configuration space in its uncollapsed totality as the complete Hilbert space of quantum configurations before measurement selects specific spacetime trajectories as physical.

This resembles how a Feynman path integral sums over all possible histories before identifying the trajectory that becomes observed, the QID would act as "the thing" that Feynman path integral sums over as the QID contains all lawful field configurations as potential. The way in which reality emerges from this space will be explored in the coming chapter. The QID can therefore be viewed as a continuous field of lawful potential that contains every configuration permitted by quantum mechanics. Within it, distinctions that could exist are encoded as amplitude and phase relations that form a landscape of lawful possibility. Quantum systems evolve through this space by exploring those relations, and the physical processes we observe are the stabilized expressions of this deeper, uncollapsed field.

When the framework speaks of "information," it does not refer to symbols, human knowledge, or fully intact works of art. In it, information represents the set of lawful distinctions that could, in principle, be realized. Within the QID, those distinctions exist as uncommitted relational patterns that define the full landscape of lawful possibilities. One could think of these as the potential answers to every yes or no question the universe can pose where each relation represents a binary outcome that remains uncommitted until interaction occurs. Collapse would then be the act that selects one of those relations, the yes, and commits it into the "classical record." The unrealized alternatives, the no, are not destroyed but become dispersed through entanglement with the environment. In this proposal, this dispersion appears locally as entropy, while globally the total informational state remains conserved within the QID. Each commit adds to the classical record and diffuses unchosen potential into the wider informational field, but the unitarity of the total state remains intact because the information associated with both outcomes continues to exist within the broader QID. In this way, the universe's history unfolds through a continuous sequence of binary selections where every commit adds to the classical record while the unrealized branches persist as diffuse informational potential within the field. Every photon, atom, and biological decision can be viewed as the realization of one among many possible paths contained within the QID's configuration space. The totality of those alternatives constitutes what Wheeler (1990) described through the phrase "It from Bit," where reality appears to emerge from a sequence of informational choices. The framework extends this view into "Bit from It," where each bit of realized information arises from the deeper informational substrate that physics already describes.

If information is conserved (Susskind, 1995), if non-local patterns can be accessed through quantum coherence (Tegmark, 2008; Dvali and Gomez, 2014), and if biological systems make use of this coherence (Cao et al., 2020), then a pre-temporal informational substrate appears necessary. Through this lens, the QID is less a speculative addition to physics than a synthesis of what is already implied by unitarity, quantum field theory, and holographic encoding, but the difference is mostly interpretive as the framework treats the wavefunction as ontologically real instead of just being a description of probability. Therefore, the QID, which is this continuous informational field or evolving Hilbert space of lawful configurations, may appear holographic when viewed from within spacetime, as the boundaries that can be observed, from atomic, to biological, to cosmological, can then be interpreted as holographic projections of information written into this field (t'Hooft, 1993; Susskind, 1995).

Recent work in physics and quantum critical phenomena provides potential physical mechanisms for the QID. At quantum critical points, collective Bogoliubov modes become nearly gapless, with energy gaps scaling as 1/N, where N represents occupation number. This creates quasi-degenerate manifolds capable of storing vast amounts of structured information at minimal energetic cost (Dvali and Gomez,

2014). Findings like these resonate with the holographic principle (t'Hooft, 1993; Susskind, 1995), which shows that information can be encoded non-locally across spacetime boundaries, which suggests that physics itself may emerge from these conserved informational structures.

In an analysis of the universe's computational capacity, Kurian (2025) calculates that the "total operations that can have been performed in the history of the entire matter-dominated universe" is approximately 10^120. He notes that this value, when expressed in fundamental constants, "hints at a holographic principle… and per the Bekenstein bound… attained by black holes and other objects with event horizons" (p. 2). The Bekenstein bound (Bekenstein, 1973) indicates that the maximum information a region can contain is determined by its surface area, which links information to fundamentally geometric structure. This relationship supports the view that the universe's information content is finite, conserved, and tied to boundary geometry, which aligns naturally with the QID's conception as a holographically encoded informational substrate.

Through this framing, the QID becomes a necessary and coherent element of a unified ontology as it functions as the informational foundation from which quantum and classical realities emerge, while holography, field theory, and thermodynamic principles describe the conservation and translation of that information across scales. The QID may therefore represent the field through which structure, symmetry, and pattern, what later chapters describe as archetypal families of form, exist as lawful potentials that biological and physical systems can access through coherence. From the paper's perspective, this interpretation organizes around the idea that information, energy, and consciousness are interwoven expressions of a single, conserved process.

Building on Established Theoretical Frameworks

This possibility aligns with several established theoretical frameworks in physics, though it extends them in new directions. Wheeler's (1990) "it from bit"hypothesis proposed that physical reality emerges from information. The holographic principle (t'Hooft, 1993; Susskind, 1995) demonstrates that information can be encoded in ways that transcend ordinary spatial relationships. Tegmark's (2008) Mathematical Universe Hypothesis argues that reality *is* mathematical structure and is not just being described by it.

These pioneering models converge on recognizing an underlying layer of possibility beneath physical reality. Although, they leave a few important questions unaddressed. What exactly is this substrate made of? How does consciousness interface with it? Did this substrate drive the evolution of consciousness itself?

The framework proposes an inversion of Wheeler's formulation. Instead of thinking about it like "it from bit"(physical reality emerging from information), the mechanism may be much closer to "bit from it" as information emerges from collapsed quantum field configurations. As the fundamental substrate would be the

Quantum Information Dimension which exists in superposition, it wouldn't be information itself. What we call information may only manifest when something collapses these field states into discrete, distinguishable patterns.

This reframes the central question from "How does physics emerge from information?" to "What selects which quantum field configurations become classical information?" And to "What role does consciousness play in this collapse?" Wheeler focused on physics emerging from information but didn't address the role consciousness would play in this emergence. While Tegmark (2008) discusses how self-aware substructures would experience themselves within the mathematical universe, he treats consciousness as an emergent property and does not explore whether it might be evolution's answer to navigating quantum field possibilities. None of these frameworks adequately address why biological systems would develop the specific capacity to interface with fundamental quantum fields.

Reframing Quantum Information Theory

Traditional Quantum Information Theory (QIT) treats information as an abstract measure of distinguishable states, essentially a bookkeeping tool for quantum systems. While Timpson (2004) rigorously critiques treating information as a substance with physical properties, both traditional and critical approaches may miss an important distinction to this discussion. The framework proposes that the quantum field is the fundamental substrate as the QID contains all possible quantum field configurations in superposition, but these are not yet "information" in any meaningful sense. Information would emerge when wave functions collapse these field states into specific, measurable configurations.

The phenomena documented earlier reveals patterns that require explanation because savants who access mathematical constants they never learned, consciousness that appears to affect quantum systems (Radin et al., 2012), and humans who demonstrate knowledge beyond personal experience all suggest some type of non-classical information access. These phenomena become comprehensible if consciousness can interface with a substrate of quantum configurations.

For example, when a savant accesses π to thousands of digits, they demonstrate knowledge that cannot be explained through classical learning or genetic transmission, so the framework proposes that such individuals may be accessing quantum field configurations that, upon collapse, manifest as mathematical information. The pattern that we recognize as π exists within the infinite configurations of the quantum field. Through mechanisms to be explored in following chapters, certain individuals appear capable of receiving information or collapsing these specific configurations into classical information. This perspective suggests that this unique access to information emerges from the interaction between consciousness and quantum field states. The quantum field contains all possible configurations in superposition, but these only become measurable, distinguishable information through the process of collapse.

Table 4.1 Quantum information dimension versus classical information theory versus quantum information theory

Property	Classical info theory	Quantum info theory (QIT)	Quantum information dimension (QID)
Nature of reality	Statistical/abstract measure	Quantum state descriptions	Quantum field as substrate
Role of observer	External measurement device	Collapses wave function	Gradient collapse
Location	Locally encoded in matter	Non-local via entanglement	Pre-spacetime dimension
Relationship to physics	Describes physical states	Mathematical formalism	Generates physical states
Consciousness role	Not addressed	Measurement problem (unresolved)	Evolved biological interface
Temporal status	Time-bound	Time-evolution equations	Pre-temporal, eternal
Accessibility	Via sensory/learned channels	Via quantum measurement	Via quantum coherence in brain
Conservation	Shannon entropy	Unitarity preservation	Eternal informational patterns

Each collapse event transforms quantum potential into classical bits, which then creates the information we observe and measure.

As Chap. 2 described, the discovery of quantum coherence in brain microtubules (Babcock et al., 2024; Nishiyama et al., 2025) provides potential biological infrastructure for this interface. As Kurian's (2025) model extends this principle to all eukaryotic life, including "aneural organisms," this indicates that quantum information processing is a widespread biological strategy. If brains maintain quantum coherence, the brain could theoretically access quantum field configurations before collapse. The specific mechanism by which consciousness directs this collapse, and why most individuals have limited access to this capability, will be addressed in the following chapters (Table 4.1).

Why the QID Remains Uncollapsed

Proposing the Quantum Information Dimension as a pre-temporal quantum substrate, from which all classical states are selected, raises a fundamental question. The framework has established that observation (measurement) collapses quantum states, so one may wonder what prevents this ultimate source field from collapsing entirely? Answering this question offers insight into the QID's relationship with time and the nature of collapse itself.

Collapse requires three conditions which include an agent capable of measurement (from thermal noise to consciousness), temporal sequence (the collapse creates time instead of requiring it), and interaction capability (information exchange must be possible). The QID remains uncollapsed because it exists as the source field from which collapsed states emerge.

Following the framework's logic, the QID exists before time itself emerges. The QID itself exists prior to these collapse mechanisms. It is the infinite possibility space from which all collapse agents select specific actualities. There is no "when" for it to collapse because "when" only exists after collapse begins (this will be described in detail in further chapters). There is another important factor which is that each collapse event creates new possibilities.

Like a reservoir that refills faster than it drains, collapses draw from the QID while simultaneously generating new potential. Every collapsed moment in history creates more possible futures. Consider how a conscious decision would branch into more possibilities than any simple particle interaction. From this, the expanding complexity continuously "refills" the QID. This creates a dual protection. First, the QID exists in a different domain that is upstream from spacetime where collapse occurs. Second, the very act of collapse generates new superposition possibilities faster than they can be exhausted. The bootstrap protection remains as observation requires collapse agents, collapse agents require time to operate, time emerges from collapse events, and collapse events require the QID.

Convergence of Independent Evidence for the Quantum Information Dimension

The framework's proposal of a Quantum Information Dimension does not arise in a vacuum, as it represents a formal synthesis of a concept that has been independently discovered and documented through radically different methodologies across science, art, and contemplative practice. The consistency of these descriptions, arrived at without collaboration, suggests they may be observing and describing a genuine feature of reality. The following table highlights some of these convergent findings (Table 4.2).

Now considering everything so far, if consciousness can access information beyond personal experience, if quantum coherence in biology serves a bigger role, and if reality may fundamentally be informational, then understanding these mechanisms could transform the conception of human potential and the nature of existence itself. The savants, mystics, and flow states might not be anomalies but glimpses of consciousness operating closer to its fundamental capacity as reality's interface.

Then considering the strong scientific evidence that the brain possesses quantum computational abilities and considering the plausibility of the Quantum Information Dimension, fundamental questions emerge about the mechanisms that enable consciousness to interface with quantum processing. If the $P = Q/E$ relationship

Table 4.2 A wide variety of converging ideas about the existence of a substrate

Source	Methodology	Description of field	Key insight
Wheeler (1990)	Theoretical physics	"It from bit"—Physical reality emerges from binary information	Information is more fundamental than matter
Tegmark (2008)	Mathematical physics	Mathematical universe—Reality IS mathematical structure	Universe is literally made of math/ information
Jung (1969b, 1969a)	Depth psychology, clinical practice	Collective unconscious—"sheer objectivity, as wide as the world" containing "pre-existent forms"	A universal substrate that connects all human cognition
Hawkins (2020)	Contemplative practice	"Infinite, nonlinear field of information and consciousness"	Direct observation through ego transcendence
Rick Rubin (2023)	Creative practice	"The source… an inexhaustible offering always available"	Accessed through creative surrender
Lex Fridman (2024)	Ayahuasca experience	Universal "glow" of "whatever makes humans special" throughout the cosmos	Psychedelic ego dissolution reveals field
Jason Padgett	Savant syndrome	"Every moment has potential to be anything"—Sees reality's quantum unfolding	Brain injury reveals pre-collapse structure
Biblical (John 1:1)	Mystical revelation	"In the beginning was the word (logos)"	Information/pattern preceded physical creation
Buddhist (Yogācāra)	Meditative insight	Ālaya-vijñāna—Storehouse consciousness containing all karmic imprints	Universal repository of all experience/ potential

describes how consciousness accesses quantum capabilities, then something must function as the observer that collapses quantum field states into classical experience.

The following chapters will explore what collapses quantum configurations into reality, evolution's role in consciousness, how consciousness even exists in the first place, what consciousness is, and how the brain is even able to navigate this quantum universe at all.

References

Babcock, N. S., Montes-Cabrera, G., Oberhofer, K. E., Chergui, M., Celardo, G. L., & Kurian, P. (2024). Ultraviolet superradiance from mega-networks of tryptophan in biological architectures. *The Journal of Physical Chemistry B, 128*(17), 4035–4046. https://doi.org/10.1021/acs.jpcb.3c07936

Babcock, N. S., Kattnig, D. R., Oncul, M., Lewis, A. M., Zdrazil, L., Strizhak, V., et al. (2023). Ultraviolet superradiance from mega-networks of tryptophan in biological architectures. *Journal of Physical Chemistry Letters, 14*(10), 2158–2165. https://doi.org/10.48550/arXiv.2302.01469

Bekenstein, J. D. (1973). Black holes and entropy. *Physical Review D, 7*(8), 2333–2346.

Bennett, C. H., & Shor, P. W. (1998). Quantum information theory. *IEEE Transactions on Information Theory, 44*(6), 2724–2742. https://doi.org/10.1109/18.720553

Cao, J., Cogdell, R. J., Coker, D. F., Duan, H. G., Hauer, J., Kleinekathöfer, U., Jansen, T. L. C., Mančal, T., Miller, R. J. D., Ogilvie, J. P., Prokhorenko, V. I., Renger, T., Tan, H. S., Tempelaar, R., Thorwart, M., Thyrhaug, E., Westenhoff, S., & Zigmantas, D. (2020). Quantum biology revisited. *Science Advances, 6*(14), eaaz4888. https://doi.org/10.1126/sciadv.aaz4888

Dvali, G., & Gomez, C. (2014). Black holes as critical point of quantum phase transition. *European Physical Journal C, 74*, 2752. https://doi.org/10.1140/epjc/s10052-014-2752-3

Feynman, R. P., & Hibbs, A. R. (1965). *Quantum mechanics and path integrals*. McGraw-Hill.

Fridman, L. (2024, August 2). *Lex Fridman reflects on Ayahuasca trip with Elon Musk* [video]. Lex Clips. YouTube. https://youtu.be/8wvyBN-fEsE?si=F1v7b7knGrQrGOMb

Hawking, S. W. (1976). Breakdown of predictability in gravitational collapse. *Physical Review D, 14*(10), 2460–2473. https://doi.org/10.1103/PhysRevD.14.2460

Hawking, S. W. (2005). Information loss in black holes. *Physical Review D, 72*(8), 084013. https://doi.org/10.1103/PhysRevD.72.084013

Hawkins, D. R. (2020). *The map of consciousness explained: A proven energy scale to actualize your ultimate potential*. Hay House LLC.

Jung, C. G. (1969a). Synchronicity: An acausal connecting principle (R. F. C. Hull, Trans.). In H. Read, M. Fordham, & G. Adler (Eds.), *The collected works of C. G. Jung: Vol. 8. The structure and dynamics of the psyche* (2nd ed.). Princeton University Press. (Original work published 1952).

Jung, C. G. (1969b). *The archetypes and the collective unconscious* (R. F. C. Hull, Ed. & Trans.; 2nd ed., Vol. 9, part 1). Princeton University Press. (Original work published 1959).

Kurian, P. (2025). Computational capacity of life in relation to the universe. *Science Advances, 11*, eadt4623. https://doi.org/10.1126/sciadv.adt4623

Landauer, R. (1961). Irreversibility and heat generation in the computing process. *IBM Journal of Research and Development, 5*(3), 183–191. https://doi.org/10.1147/rd.53.0183

Nishiyama, A., Tanaka, S., & Tuszynski, J. A. (2025). Waveguide quantum electrodynamics: Tryptophans entangled with water as data qubits in a microtubule. *Dynamics, 5*, 7. https://doi.org/10.3390/dynamics5010007

Peskin, M. E., & Schroeder, D. V. (1995). *An introduction to quantum field theory*. Westview Press.

Radin, D., Michel, L., Galdamez, K., Wendland, P., Rickenbach, R., & Delorme, A. (2012). Consciousness and the double-slit interference pattern: Six experiments. *Physics Essays, 25*(2), 157–171. https://doi.org/10.4006/0836-1398-25.2.157

Rubin, R. (2023). *The creative act: A way of being*. Penguin Press.

Schulman, L. S. (1981). *Techniques and applications of path integration*. Wiley.

Susskind, L. (1995). The world as a hologram. *Journal of Mathematical Physics, 36*(11), 6377–6396. https://doi.org/10.1063/1.531249

Susskind, L., Thorlacius, L., & Uglum, J. (1993). The stretched horizon and black hole complementarity. *Physical Review D, 48*(8), 3743–3761. https://doi.org/10.1103/PhysRevD.48.3743

't Hooft, G. (1993). *Dimensional reduction in quantum gravity*. arXiv preprint gr-qc/9310026. https://arxiv.org/abs/gr-qc/9310026

Tegmark, M. (2008). The mathematical universe. *Foundations of Physics, 38*(2), 101–150. https://doi.org/10.1007/s10701-007-9186-9
Timpson, C. G. (2004). *Quantum information theory and the foundations of quantum mechanics* [Doctoral dissertation]. University of Oxford. https://doi.org/10.48550/arXiv.quant-ph/0412063
Treffert, D. A. (2014). Savant syndrome: Realities, myths and misconceptions. *Journal of Autism and Developmental Disorders, 44*, 564–571. https://doi.org/10.1007/s10803-013-1906-8
Wheeler, J. A. (1990). Information, physics, quantum: The search for links. In W. H. Zurek (Ed.), *Complexity, entropy, and the physics of information* (pp. 3–28). Westview Press.
Zee, A. (2010). *Quantum field theory in a nutshell* (2nd ed.). Princeton University Press.

Chapter 5
Collapse and Evolution's Solutions

Physics continues to wrestle with a central question about how the quantum world gives rise to classical reality. Donadi et al. (2021) summarize it directly, writing that "one question is left open: what triggers the collapse of the wave function" (p. 3). The answer to this question is critical for understanding physics and the biological evolution of consciousness within a quantum universe.

For decades, John Wheeler argued that resolving this mystery would require rethinking reality in informational terms. He suggested that the physical world might ultimately emerge from binary "yes-or-no" outcomes which is an intuition later expressed in his *it from bit* hypothesis (1990). Wheeler believed that at some fundamental level, nature "decides" whether a possibility becomes real or not. At the point the wave function collapses, it's a binary "yes or no," but he didn't offer a mechanism for how or when this occurs. His framing sharpens the puzzling questions still faced today. What physical interaction produces an irreversible answer? Where is the threshold between information flow that can be undone and information that is permanently committed? How is this elementary step counted in nature, and how does it relate to Planck's constant? And if the universe is constantly committing answers, why would biological systems evolve architectures designed to steer some of those commits toward survival-relevant outcomes?

There are many theories that try to provide answers to many of these questions, and especially some that provide answers to what collapses the wavefunction. Some theories propose that consciousness itself causes collapse. These perspectives, often recently associated with thinkers like Federico Faggin (2021, 2024) and Bernardo Kastrup (2019), treat consciousness as ontologically fundamental and suggest that physical reality arises from it. The framework diverges from this view. From the book's perspective, conscious awareness participates in reality but does not exclusively or fundamentally cause the collapse of quantum possibilities into classical outcomes as will be explored in the coming pages.

On this point of collapse, the framework aligns more closely with Roger Penrose, who picked up Wheeler's intuition that quantum possibilities must physically resolve

J. Roeloffs, *The Classical Brain to the Quantum Mind*, The Frontiers Collection,
https://doi.org/10.1007/978-3-032-20068-6_5

into actualities, and agrees that collapse is objective and physical and not dependent on a conscious observer. However, Penrose proposed that gravitational thresholds trigger this collapse, which is an idea that is central to his Orch OR theory with Stuart Hameroff (Hameroff & Penrose, 1996, 2014). For years this theory was at the forefront of consciousness research, but recent experimental results call the gravitational hypothesis into question. Donadi et al. (2021), after testing gravity-related collapse in highly controlled conditions, concluded that "Penrose's proposal for a gravity-related collapse of the wave function, in the present formulation, is ruled out" (p. 17). While they note that the idea remains appealing, they conclude it likely requires "a radically new approach" (p. 17).

As Penrose (1990) recognized, "an important gap exists at the point where quantum and classical physics meet" and speculated that "the conscious brain might be taking advantage of whatever new physics is needed to fill this gap to achieve its non-algorithmic effects" (p. 1). This continues to keep the door open to explore new ways to think about physics including its definitions and traditional views on how collapse may occur as well as the roles that consciousness and the brain itself play in these processes. This is exactly what this book aims to do.

Overall, the central problem is still unresolved. If collapse does not depend on conscious awareness and gravitational thresholds cannot account for it, then how does quantum potential resolve into classical outcomes? What triggers the shift from possibility to actuality?

The rest of this chapter explores these questions and considers how evolution may have navigated them to shape the architecture of consciousness.

Orch OR

The Orchestrated Objective Reduction theory (Hameroff & Penrose, 1996, 2014) deserves significant recognition for establishing the foundational framework to connect consciousness with quantum physics. By proposing that consciousness arises from quantum processes in neural microtubules, Orch OR demonstrated the quantum computational potential of these cellular structures and established the scientific credibility of quantum processes operating in warm, wet biological environments. The theory's emphasis on microtubules as quantum information processors has catalyzed decades of research into quantum biology in neural systems and has played an important role in upending the assumption that quantum effects decohere too rapidly in biological tissues to be functionally relevant. Orch OR's rigorous mathematical treatment of quantum coherence in biological systems opened essential scientific dialogue about the intersection of quantum mechanics and neuroscience.

As described, this book's framework aligns most with Orch OR as it agrees that collapse is a physical process. The divergence between the ideas is in what exactly is causing the collapse. What is central to Orch OR's mechanism is Penrose's hypothesis that gravitational effects trigger wave function collapse when quantum superpositions reach a sufficient mass threshold. From that perspective, the gravitational collapse would convert coherent quantum states in microtubules into classical outcomes that help to create moments of conscious experience. The theory fundamentally depends on gravity's ability to collapse quantum superpositions at the cellular scale within biologically relevant timescales.

However, recent experimental evidence points to limitations in the gravitational collapse mechanism. The underground experiment conducted 3100 meters beneath a mountain by Donadi et al. (2021) directly tested gravity-related wave function collapse and "rules out the natural parameter-free version of the Diósi-Penrose model" (p. 2), concluding that "Penrose's proposal for a gravity-related collapse of the wave function, in the present formulation, is ruled out" (p. 17). This experimental falsification undermines Orch OR's core mechanism for generating conscious moments.

Derakhshani et al. (2022) conducted a comprehensive mathematical analysis of Orch OR under these experimental constraints. Using the spatial resolution limits established by the underground experiments, they found that the gravitational collapse mechanism cannot support the timescales and coherence requirements that Orch OR needs for biological consciousness. Their calculations show that achieving realistic conscious moments would require coherent superpositions that "vastly exceed any of the coherent superposition states achieved with state-of-the-art optomechanics or macromolecular interference experiments" (p. 10).

Derakhshani et al. (2022) acknowledge this limitation and consider what could work, noting that their results "do not rule out Orch OR theory in general" but rather "rule out variants of Orch OR based on the simplest version of the DP theory of gravity-related dynamical wavefunction collapse" (p. 13). They suggest that future theories might emerge that avoid energy conservation violations, yet acknowledge that "objections to the Orch OR theory based on environmental decoherence considerations will likely still remain" (p. 13) and finally they state, "We conclude that Orch OR theory, when based on the simplest version of gravity-related dynamical collapse, is highly implausible in all the cases analyzed" (p. 8).

These findings point to both mathematical and conceptual challenges for Orch OR. Even with these challenges with the gravitational collapse mechanism, recent experimental evidence validates Orch OR's central substrate hypothesis that quantum processes in microtubules play a fundamental role in consciousness. Kalra et al. (2023) demonstrated that anesthetics directly interfere with quantum energy transfer in microtubules, which reduces diffusion lengths from 6.6 nm to 5.6–5.8 nm. Given that these same substances reliably eliminate consciousness in clinical settings, this finding provides empirical support for a causal relationship between microtubule quantum processing and conscious states. As Kalra's team notes, their observations "suggest that etomidate and isoflurane 'dampen' energy transfer between tryptophan and AMCA due to changes in the dielectric screening of electronic couplings" (p. 356).

When combined with Khan et al. (2024) finding that protecting microtubules with binding drugs delays the onset of unconsciousness, and considering all the evidence presented in Chap. 2 for the viability of quantum processes in microtubules, it seems to suggest Orch OR correctly identified microtubules as a critical piece of the consciousness puzzle, even though the gravitational collapse mechanism itself cannot account for the required timescales and coherence.

Beyond the specific gravitational mechanism being ruled out, the theory does not provide an evolutionary explanation for why consciousness would have developed as a quantum phenomenon in the first place. If quantum processes operate throughout matter, what evolutionary pressures would have shaped consciousness to interface with them in specialized ways? And if collapse occurs through mechanisms other than gravity, how might biological systems have evolved to participate in or modulate those processes?

Overall, this experimental falsification of Orch OR's gravitational collapse creates space to consider an alternative approach that can address both the mathematical constraints and the evolutionary questions that gravitational collapse theories cannot resolve.

What Collapses the Wavefunction?

Physics faces a fundamental challenge in understanding quantum collapse mechanisms. As Donadi et al. (2021) observe, "There is no reason to believe that measurements are so special to temporarily suspend the quantum dynamics given by the Schrödinger equation and replace it with a completely different one" (p. 2). They note that "the postulate of wave function collapse introduced by von Neumann, and now part of the standard mathematical formulation of the theory… clearly has a phenomenological flavour" (p. 2). Similarly, Melkikh (2015), writing on quantum field theory's approach to the collapse of the wave function, argues that quantum field theory already contains the collapse mechanism without requiring additional postulates. He emphasizes that "there is another possibility to explain the causes of the wave function collapse, which is not associated with any additional assumptions or creation of a new theory" (p. 47).

This creates an interesting dichotomy with Penrose's previous statement. Donadi et al. and Melkikh describe that the current equations should be enough to derive a hypothesis of collapse. If they are correct that new equations are not needed, then is it possible the interpretation of these equations is the problem?

Modern quantum mechanics already provides mathematical tools for describing how environments drive collapse. Donadi et al. (2021) note that collapse dynamics, if they exist, should vary in strength. They write, "More realistically, if collapses occur at all, they are part of the dynamics. In some cases, they are weak and can be neglected. In some other cases, such as in measurements, they become strong and rapidly change the state of a system" (p. 3). This seemingly points toward a gradient perspective that aligns with evidence showing how decoherence rates depend on environmental coupling. Lindblad superoperators describe how quantum systems interact with environments and generate open-system dynamics that lead to decoherence (Breuer & Petruccione, 2002; Schlosshauer, 2007). Melkikh (2015) gives a field-level account of how measurement involves particle creation and annihilation, and these processes introduce nonlinearity into quantum field theory equations, which breaks the superposition principle during measurement. As Melkikh explains, "Superposition principle is not valid for the system of equations of quantum field theory for particles and fields, because this system is a non-linear" (p. 47). He concludes that from this perspective, "In this case, the term 'collapse of the wave function' becomes a consequence of the other statements of quantum mechanics instead of a separate postulate of quantum mechanics" (p. 47). He goes on to write, "Every measurement process is accompanied by either the creation or annihilation of particles (quasi-particles)… the creation and annihilation of particles plays a key role in solving the problem of the wave function collapse" (p. 48).

Earlier work by Bonilla and Guinea (1992) showed that strong coupling between quantum and classical variables drives the system into a nonlinear regime where "an initial superposition of states evolves toward a definite final state" (p. 7726). These perspectives reflect a coherent picture where collapse emerges from information exchange within quantum fields, with Melkikh's (2015) analysis demonstrating that when particle creation and annihilation processes are properly included, the nonlinearity of quantum field theory equations naturally produces what appears as collapse. Does an explanation for collapse already exist within standard quantum field theory as an environmentally modulated gradient process? If so, then the remaining questions concern the quantitative structure of this process like what sets the minimum size of each information exchange, what determines the thermodynamic cost, and how do these limits combine to govern collapse rates?

Therefore, proposal 3 is that collapse occurs through a gradient process where information exchange between quantum systems and their environment progressively destroys interference patterns. When particle creation and annihilation introduce sufficient uncertainty, quantum coherence degrades proportionally until classical behavior emerges. Each collapse event commits a discrete quantum of action on the order of h into the classical record, with the logically irreversible part bounded by Landauer's limit ($k_B T \ln 2$). The interplay of these constants sets the informational exchange rate of the universe. A true, irreversible collapse requires both a commit event (a selection) and an environmental recording of the discarded alternatives (entropy confirmation). This entropy manifests physically as the dispersion of phase information into photons, phonons, and molecular vibrations that now encode fragments of the system's history. In this view, decoherence is the continuous gradient that links coherent dynamics to classical outcomes, and collapse is the completion of that process once the environment confirms the record.

Landauer (1961) established that erasing one bit of information has an unavoidable thermodynamic cost of $k_B T \ln 2$ which showed that information is physically consequential. This cost emerges specifically from logically irreversible operations. As Bérut et al. (2012) explain, "An operation is said to be logically irreversible if its input cannot be uniquely determined from its output… In particular, the erasure of information, the RESET TO ZERO operation, is logically irreversible and leads to an entropy increase of at least $k_B \ln 2$ per erased bit" (p. 2). The key distinction is that reversible operations, where all information is preserved, can occur without energy cost, while any operation that discards information about previous states necessarily dissipates heat.

In direct tests, Bérut et al. (2012) report that "the mean dissipated heat saturates at the Landauer bound in the limit of long erasure cycles" (p. 187) and that their result "demonstrates the intimate link between information theory and thermodynamics" (p. 187). A second high-precision experiment by Jun, Gavrilov, and Bechhoefer (2014) strengthens this connection. They write,

> "Thus, we have shown experimentally that the full-erasure protocol, which involves the compression of phase space from two macroscopic states to one, asymptotically requires $k_B T \ln 2$ of work, while the very similar no-erasure protocol, which has no such phase-space compression, is reversible" (p. 4).

Their results are fascinating as two nearly identical protocols differ in one respect which is whether information about the initial state is preserved or erased. When preserved, the process costs nothing and remains reversible, but when erased, the process requires $k_B T \ln 2$ of work and becomes irreversible. This distinction is important as the thermodynamic cost emerges only when the number of accessible states is reduced. When no information is erased, no entropy is produced, and the process becomes thermodynamically reversible.

Bérut et al. (2012) further demonstrated that this bound is approached gradually as they found that "the irreversible heat dissipation associated with a finite-time procedure should decrease as $1/\tau$" and that "the mean dissipated heat decreases with the duration τ and approaches the Landauer's bound $k_B T \ln 2$ for long times" (p. 13). This experimental result establishes that information erasure operates as a continuous, rate-dependent process where faster operations dissipate more energy while slower, quasi-static operations approach the theoretical minimum as the erasure unfolds progressively, with thermodynamic cost determined by the rate and completeness of the process.

From the framework's perspective, this points to irreversibility and energetic cost arising right when possibilities are discarded. But this raises a question about the nature of information processing in physical reality. If erasing information costs energy, and quantum mechanics tells us energy comes in discrete packets defined by Planck's constant ($E = hf$), how do these two fundamental limits connect? When a quantum system collapses from many possibilities to one, the quantum correlations that allowed interference between these possibilities are broken, and which-path information becomes irreversibly transferred to environmental degrees of freedom. Does this transfer follow Landauer's bound? And if so, what sets the minimum size of the quantum information being transferred?

When extending Landauer's principle to its quantum limit, Planck's constant defines the smallest quantum of action through which information can transition from potential to physical expression, while Landauer's limit quantifies the minimal energy that must be dissipated for that transition to become irreversible. This creates a unified information-energy framework where h represents the quantum information exchange limit while $k_B T \ln 2$ represents the classical thermodynamic limit.

Together, they describe the complete cost structure of reality's information processing architecture.

From within the system, we can only detect the informational rules by which events unfold. Whether those rules are closer to "code" or to the lawful structure of reality itself may be hard to separate in practice. Because these irreversible compressions discard the distinguishability of prior states, their physical footprint behaves mathematically like a one-way hash. To illustrate, a hash function may map both "hello" and "world" to the output digest "a3c5d7." When only the digest is known, there is no way to determine which input produced it or whether it came from many other possible inputs. This reflects what Bérut et al. (2012) described when defining logical irreversibility, writing that "An operation is said to be logically irreversible if its input cannot be uniquely determined from its output" (p. 2). A hash function is logically irreversible by design, while quantum collapse is logically irreversible by physics. When a quantum system collapses, many potential microstates are mapped into a single recorded macrostate, and the details of the alternatives become physically unrecoverable. The environment retains only a digest of that event, which is its thermodynamic footprint, while the specific pathways that led there are lost, in the way that the pre-image of a cryptographic hash cannot be reconstructed from its output.

In this process, quantum collapse shares an interesting resemblance to collision resistance in information theory. In a well-formed hash function, it becomes vanishingly unlikely that two different inputs yield the same digest. Collapse events behave in a similar way through what may be called environmental uniqueness. Each collapse disperses phase information into countless environmental degrees of freedom that include photons, phonons, and molecular vibrations. Because these interactions unfold across enormous numbers of particles and modes, no two collapses leave identical thermodynamic footprints. Each one carries its own signature of energy, phase, and configuration that becomes written into the environment's record.

In this way, entropy becomes the cost of certainty, of final collapse, as it would be the energy the universe spends to turn quantum potential into classical reality. In that sense, collapse represents nature's own form of irreversible encryption as decoherence proceeds until a phase transition occurs at which enough information has been committed to the physical "record" that the prior superposed configurations can no longer be retrieved.

Interestingly, Wheeler's (1992) "it from bit" hypothesis seemingly anticipated this connection:

> "Every 'It' — every particle, every field of force, even the spacetime continuum itself — derives its way of action, its very existence entirely, even if in some context indirectly, from the detector-elicited answers to yes-or-no questions, binary choices, bits. In another way of wording the idea, all things physical, all its, must in the end submit to an information-theoretic description" (p. 282).

Wheeler proposed that physical reality emerges from binary information, from yes/no answers to fundamental questions. But he didn't specify the mechanism by which these binary choices become physical facts. The framework provides a mechanism for Wheeler's intuition as information exchange triggers quantum collapse, and each commit is quantized by h. In this view, collapse represents the same physical act as Landauer erasure but generalized to the quantum domain. When a system transitions from superposed possibilities to a single classical outcome, one Planck-sized unit of action is committed, and the discarded alternatives are expressed as entropy. Each quantum collapse represents one of Wheeler's fundamental yes/no decisions, because among all possible QID configurations, information exchange above the h-threshold forces a binary choice, committing one possibility while discarding alternatives.

Within this framework, entropy can be pictured as an environmental ledger that accumulates discarded alternatives. Formally, which-path information flows into environmental degrees of freedom and raises the relevant von Neumann entropies, which makes revival extremely unlikely. Like confirmations in a blockchain, each discarded quantum possibility gets written into environmental modes such as phonons, photons, and thermal fluctuations, where it remains permanently recorded. The more commits that have been written, the deeper the entropy becomes, making reversal increasingly difficult. Interestingly enough, this physical mechanism mirrors the same problem blockchain technology faced which was how to make irreversible commitments in a distributed network. Both came to the resolution that statistical improbability accumulated to sufficient degrees creates irreversibility without requiring any central authority. In blockchain, multiple independent validators make transaction reversal exponentially unlikely. In physics, environmental recording through countless molecular interactions makes quantum commit reversal thermodynamically prohibitive. In this blockchain analogy, significant decoherence exists in a state analogous to unconfirmed blockchain transactions, still potentially reversible, while high-entropy states represent decisions confirmed by countless environmental "validators" and as enough "confirmations" occur the "transaction" becomes practically irreversible, resulting in "collapse."

This reframing helps to clarify why entropy appears to increase universally as each quantum collapse generates new environmental records, and the universe continuously processes vast numbers of these information commits. Temperature emerges as the measure of how readily the environment can serve as a validation network for these quantum commits, since higher temperatures provide more active environmental modes that can rapidly record and confirm discarded quantum possibilities, while lower temperatures reduce this recording capacity, which allows quantum coherence to persist longer. Each quantum collapse commits one possibility while discarding all alternatives, but the discarded possibilities do not disappear or get destroyed. They become entropy which is written permanently into the environmental record where they contribute to the universe's growing informational history, or one may say, ledger.

This connection is further grounded mathematically by the equivalence of Shannon's information entropy (Shannon, 1948) and Boltzmann's thermodynamic entropy (Boltzmann, 1877), which share an identical functional form and differ only by the constant $k_B \ln 2$. The historical roots of this unification trace back to Kelvin's absolute temperature scale (Thomson, 1848) and Clausius's kinetic theory (1857), which established that temperature measures particle collision rates.

Within the framework, each collision represents an opportunity for h-sized information exchange between quantum system and environment. Higher temperatures mean more collisions per second which creates more opportunities for quantum correlations to be broken and which-path information to be irreversibly transferred to environmental modes. Therefore, temperature does not directly cause collapse, but it governs the rate at which environmental interactions that are capable of recording quantum information occur. At absolute zero, collision rates drop to zero and quantum coherence can persist indefinitely, but at room temperature, collisions occur at incredibly high rates which create continuous statistical pressure toward environmental information transfer, especially for massive particles.

This makes E = hf a measure of update frequency, which measures how many times per second the system commits new positional information to the environmental "record." The similarity between Shannon and Boltzmann's formulations suggests that information loss and thermodynamic entropy are different aspects of the same process. Von Neumann (1932) introduced quantum entropy for quantum systems, and Shannon (1948) later developed classical information entropy. The close mathematical relationship between these formulations created the foundation for modern quantum information theory. Modern developments support this view as experimental confirmations of Landauer's principle in quantum regimes (Bérut et al., 2012; Jun et al., 2014) demonstrate that information processing costs apply across both classical and quantum domains.

In this framework, decoherence is the gradual loss of phase coherence as information leaks into environmental modes, while collapse is the point at which the environment retains a durable record that makes reversal infeasible. In other words, decoherence marks the opening of recording channels, and collapse completes when the environment itself confirms the outcome by maintaining a stable record.

When this is considered, the picture becomes a bit more clear as Planck's constant defines the minimum commit size, energy becomes the rate of commits, entropy represents the permanent record of committed actions, and temperature governs how frequently environmental interactions capable of recording quantum information occur. Donadi et al. (2021) explored whether collapse occurs as an ongoing physical process rather than only during measurements. While their experiment ruled out the specific parameter-free Diósi-Penrose model they tested, their framing supports the view that collapse, if it occurs, operates as a continuous gradient process instead of being a binary measurement-triggered event. This aligns with the framework's proposal that environmental recording of quantum information occurs at varying rates depending on temperature, material properties, and isolation from recording channels.

As will be explored in this chapter, BEC experiments demonstrate that when environmental validation networks are suppressed through cooling, commits slow dramatically, coherence persists, and classical behavior fails to emerge (Dutton et al., 2004; Hau, 2016).

The thermodynamic framework supports this gradient perspective through the generalized Landauer bound as Bérut et al. (2012) showed experimentally that information erasure is not an all-or-nothing event, as it actually scales continuously with the success probability of reset. They found "for a reset operation with efficiency $p < 1$, the Landauer bound is generalized: $\langle Q \rangle \geq k_B T\,[\ln 2 + p \ln p + (1 - p) \ln(1 - p)]$" (p. 3). When erasure is incomplete ($p < 100\%$), the thermodynamic cost is lower than the full $k_B T \ln 2$, which demonstrates that information processing can operate at intermediate levels of commitment. This supports the framework's proposal that collapse exists on a continuous spectrum where environmental recording can be partial as thermodynamic cost scales with the degree of irreversibility achieved.

Melkikh provides a concrete example of gradient collapse through fullerene interference experiments as he writes, "For the fullerene, one could definitely say that collapse to a point will never happen. With increasing temperature of the fullerene, the interference pattern gradually disappears when the fullerene emits photons with increasingly shorter wavelength… there is no such wavelength at which the collapse definitely occurs" (p. 50). This gradient collapse model finds support from studies that show decoherence as a concrete, environment-dependent process governed by the availability of environmental "recording channels" that can absorb quantum phase information. Uhlemann et al. (2013) identified a source of decoherence in high-resolution electron microscopy through "magnetic field noise from thermally driven currents" in conductive materials (p. 1). By cooling a metal tube in the beam path from room temperature to liquid nitrogen conditions, they demonstrated that coherence loss scaled directly with temperature. As they concluded, the experiment "clearly proves the thermodynamic nature of the observed noise effect and shows that the phenomenon is related to the material of the test tubes." (p. 2). Their quantitative analysis shows a linear relationship where contrast degradation followed $(\xi \operatorname{var} B)(T) = C\backslash\mu_0 k_B\, T/R^2$, with the variance of magnetic field fluctuations rising proportionally with temperature (p. 4). From the framework's perspective, higher temperatures activate more thermally driven current loops throughout the conductive material, which create additional environmental "microphones" that record phase information through accumulated tiny magnetic kicks. Cooling appears to quiet these recording channels, which allows quantum coherence to persist for longer durations.

Building on these effects, Beierle, Zhang, and Batelaan (2018) also showed that environmental recording capacity depends strongly on material properties. Using diffracted electron matter waves passing near surfaces, they found clear coherence loss when electrons approached a silicon surface but "no reduction in contrast measured for an electron passing close to the gold surface" under otherwise identical conditions (p. 4). Their results ruled out several established theoretical models and favored explanations based on plasmonic excitation, concluding that "the experiment is consistent with decoherence theory and rules out established Coulomb interaction models in favor of plasmonic excitation models" while noting that "the observed effect is strongly material dependent" (p. 5). Within this framework's interpretation, silicon's semiconductor properties provided active dielectric and plasmonic modes that could serve as recording channels for quantum phase information, while gold lacked accessible channels in the relevant frequency range under these experimental conditions.

The theoretical foundation for these observations emerges from Howie's work (2011, 2014) on aloof-beam interactions and thermal effects. Howie demonstrates that decoherence often results from accumulated scattering events involving energy losses "too small to measure" individually but whose collective effect destroys interference patterns (2011, p. 1). He derives an explicit thermal enhancement factor FT $= \coth(\hbar\omega/kT)$ that demonstrates how thermal occupation of environmental modes amplifies interaction probabilities, with the enhancement becoming linear in temperature FT $\approx 2kT/\hbar\omega$ for low-frequency oscillations where thermal energy exceeds the quantum energy scale (2014, p. 5). This enhancement occurs because thermal energy populates environmental oscillator modes, which creates more opportunities for both energy loss and energy gain interactions that can record quantum phase information. In experiments involving electron beams in narrow tubes, Howie, aligning with the findings from Uhleman et al., reported decoherence "proportional to temperature" and attributed the effect to "very low frequency magnetic fields emitted by current loops in Johnson noise fluctuations" (2014, p. 5). This analysis shows how rising temperature systematically opens additional low-frequency environmental channels that can absorb which-path information, even when individual interactions remain spectroscopically invisible.

Further experimental evidence provides a valuable demonstration of this gradient-driven view at its limit as research on Bose-Einstein condensates (BECs) also points to what occurs when environmental "recording channels" for collapse are systematically eliminated. Early experiments by Dutton, Hau, and colleagues (2004) showed that light pulses can be slowed by seven orders of magnitude, spatially compressed by factors of 20,000, and even "completely stopped, stored, and subsequently revived in an atomic medium, with millisecond storage times" (Dutton et al., 2004, p. 33). While halted, the light's phase and amplitude information becomes imprinted holographically into the condensate as "the pulse makes an imprint—a hologram, really—in the atom cloud" (Dutton et al., 2004, p. 33). When the control field is restored, the pulse resumes seamlessly, reviving its full informational structure.

Subsequent work extended these results dramatically. Hau's group (2016) achieved storage times up to 0.5 minutes, demonstrating that single photons can be mapped onto matter-wave imprints inside a condensate, stored with coherence preserved, and later revived with greater than 50% efficiency. As they describe, "Optical information imprinted in the light pulse is stored and manipulated by converting the light pulse into a small atom pulse that is a perfect matter copy of the extinguished light pulse" that "can be moved around and reshaped" (p. 1). During storage, the imprints can be sculpted and repositioned, and the revived photons reflect those edits. Multiple pulses can be stored simultaneously, reordered deterministically, and retrieved on demand.

Within the framework, these BEC results provide direct empirical support for two key claims. First, decoherence appears to be a physical process where photons, thermal noise, and other environmental interactions actively maintain classicality through irreversible information recording. Second, collapse may exist on a spectrum where protected regimes allow quantum states to persist when the environmental recording channels are systematically suppressed. The BECs achieve this by approaching absolute zero, where thermal populations of environmental modes drop to near zero, which effectively silences the "microphones" that would otherwise record and commit quantum phase information.

The BEC findings complement the temperature-dependent decoherence studies in revealing the bidirectional nature of quantum-classical transitions. Where Uhlemann et al. (2013) and Howie (2011, 2014) demonstrate how warming environments accelerate collapse by activating more recording channels, BEC research shows how environmental cooling preserves coherence by suppressing these recording mechanisms. Together, they suggest that quantum states respond proportionally to their level of environmental protection and do not collapse through mysterious or instantaneous mechanisms. The temperature dependence observed across both extremes supports the view that environmental recording capacity may be the fundamental parameter governing the quantum-to-classical transition. The BEC findings also complement Melkikh's (2015) particle creation mechanism as it shows the bidirectional nature of quantum-classical transitions. While Melkikh demonstrates how environmental interference gradually destroys quantum coherence through particle creation and annihilation, BEC research shows how environmental isolation preserves coherence by suppressing these similar processes.

Melkikh (2015) identifies three reasons that collapse can be understood as a fully physical process and describes the mechanism explicitly saying, "As a result of the creation (annihilation) of a particle, an additional uncertainty arises, which 'smears' the interference pattern. The imposition of such a large number of uncertainties in the repetitive measurements leads to the classical behavior of particles" (p. 47). His three key aspects of this process include the nonlinearity of quantum mechanics when particles and fields interact introduces natural instability in superposed states, the particle creation and annihilation during measurement events add positional uncertainty which progressively destroys interference patterns, and "when many such events occur, the rapid accumulation of uncertainty drives a faster loss of coherence" (p. 50).

Considering Melkikh's (2015) analysis within the framework, this suggests that biological systems may modulate environmental factors internally that BEC researchers control artificially. The temperature differential between BECs (near absolute zero) and biological systems (310 K) remains significant, yet the underlying principle appears consistent. Environmental isolation preserves quantum states, while environmental interference destroys them, with a gradient scale of collapse happening between them, regardless of the specific mechanisms involved.

Thermo Field Dynamics (Takahashi & Umezawa, 1975; Umezawa, 1993) provides a formal setting where the universe's global information remains pure even as local systems appear mixed. In this doubled-Hilbert-space representation, temperature corresponds to entanglement between physical and mirror degrees of freedom, which makes entropy an explicit measure of information exchange. From the framework's perspective, this offers a mathematical parallel to process described so far in this chapter where total information is conserved in the Quantum Information Dimension, and gradient collapse and environmental recording create the appearance of irreversibility. TFD doesn't specify collapse thresholds, but it situates the framework's thermodynamic interpretation of quantum processes within established quantum field theory.

These convergent results provide a range of physical support for the framework's proposal that collapse operates as a gradient process governed by environmental recording capacity. Instead of absolutely requiring conscious observation or some mysterious mechanisms, quantum decoherence appears to reflect a temperature-dependent density of environmental modes available to irreversibly record phase information. The findings suggest that collapse rates scale with this recording capacity, which varies according to both thermal activation of environmental channels and material-specific availability of appropriate modes to serve as quantum information receptors. Overall, the framework posits collapse as a quantifiable, environmentally modulated process where temperature serves as the primary variable determining how readily quantum information gets committed into classical records.

The Evolutionary Solution

From the framework's perspective, let's consider the points that have been explored and proposed so far. It has been explored and proposed that the brain maintains quantum coherence in microtubules, that reducing ego activity enhances performance dramatically, that this mechanism appears to be universal across human consciousness, and that collapse occurs in a gradient manner.

From the evolutionary perspective, the survival and computational challenges humanity's ancestors faced would have included pattern recognition for predator detection, rapid assessment of complex social dynamics, navigation through uncertain terrains, and creative problem-solving for tool use, all of which would benefit from the significant parallel processing that quantum computation could provide, but at the exact same time, these early humans would have needed a consistent identity for social cooperation, the ability to take decisive action without overthinking, and then all of this would need to occur through metabolically sustainable, fast processes.

This raises a few fundamental questions. First of all, why would evolution develop such an elaborate quantum biological mechanism in the human brain? Wouldn't the metabolic cost of maintaining quantum coherence in warm, wet neural tissue be enormous? How could this be possible? From the initial perspective, the benefit seems insufficient to justify the biological investment.

The necessity of this evolutionary development finds deeper grounding in computational theory as Kurian (2025) demonstrates that while "all computations can be embedded in a logically reversible framework" theoretically requiring no energy dissipation, "in practice, any computationally restricted agent will dissipate energy, because the unobservable degrees of freedom cannot be tracked to reversible precision" (p. 7). This fundamental limitation becomes even more severe for quantum systems, where "this requirement for reversible precision at the thermodynamic limit is impossible for quantum systems obeying Heisenberg uncertainty relations" (p. 7).

This computational restriction points to why biological systems cannot maintain perfect quantum coherence indefinitely as even with the quantum infrastructure documented in previous chapters, the brain faces an insurmountable tracking problem, because maintaining complete reversibility would require simultaneously knowing both the position and momentum of every particle with perfect precision. Heisenberg uncertainty makes this physically impossible as the more precisely the brain tracks where particles are, the less it can know about where they're going, and vice versa. This trade-off means biological systems must constantly choose what information to preserve and what to sacrifice in a cost effective manner. This is a difficult problem for biological systems but evolution's solution appears to be very elegant.

Kauffman and Radin (2023) independently recognize this evolutionary paradox, noting that "if mind is epiphenomenal, then the proposal that mind evolved from simple to complex is incoherent without an evolutionary selective advantage" (p. 3). What would this evolutionary selective advantage be? Toole et al. (2018) help to explain some of the benefits of biological quantum mechanics, saying, computational powers such as these "afford a biological system like the brain a degree of non-deterministic (i.e., non-algorithmic) behavior" (pp. 117–118), which provides the evolutionary advantage of non-algorithmic processing that is essential for human understanding and consciousness.

Through this lens, evolutionary logic suggests a possibility that brains that could temporarily access quantum processing for critical decisions, such as predator detection, pattern recognition, and invention or innovation would have been selected for as it would have provided significant survival advantages. However, wouldn't continuous quantum computation be metabolically costly, prevent quick decision making, and be destabilizing for the identity maintenance that is required for social cohesion? Could it feasibly have the best of both worlds? How would evolution find a way to balance these needs and benefits in biology?

Vervaeke and Ferraro (2013) describe that evolution has indeed been selecting for these constraints and show that there is a need for a switching mechanism between two types of systems saying, "There are constraints of efficiency that put selective pressure on processing while there are constraints of resiliency that enable new possibilities of processing. Cognition evolves its fitness to its environment in a dynamic self-organizing manner" (pp. 8–9). Simply put, evolution faced a conundrum. If a being had too little contemplation, it would die; but if it had too much contemplation, it would die. Is it possible evolution would select for the mechanism that best modulates its access between modes of computation? How would evolution select to modulate this spectrum of contemplation? What would this mechanism be?

The conventional objection to quantum consciousness is that environmental decoherence destroys quantum effects, and from the perspective of the observer effect, consciousness seems to get in the way of learning more about quantum phenomena. It's interpreted to be a problem biology must overcome. But what if consciousness' "interference" with quantum states isn't an unfortunate side effect?

What if its Interference IS the Point?

As Chap. 2 described, recent advances in quantum biology provide important evidence. Comprehensive reviews (Cao et al., 2020) have cemented quantum biology as a legitimate field, with applications ranging from vision to avian navigation, which challenges outdated assumptions about coherence at biological temperatures. Most significantly, these studies reveal a counterintuitive principle as they show that nature doesn't avoid decoherence, instead, it actively exploits it. As Cao et al. (2020, p. 8) demonstrate, "Nature does not engineer the bath to avoid decoherence to direct functional processes… Nature, rather than trying to avoid dissipation, specifically exploits it together with the engineering of site energies and excitonic coupling to direct energy transport." This principle extends beyond energy transfer, as they state, "The basic physics behind thermalization is used to impose direction" (p. 8).

The conventional objection to quantum consciousness is that decoherence ruins quantum effects in warm biological systems. But what if nature uses decoherence functionally as a design principle? What if decoherence is a feature not a bug?

Furthermore, research shows that without proper quantum-classical interfaces, biological systems fail to function. Cao et al. (2020) show that in purely classical descriptions, "in a world where the behavior of electrons is governed by the fundamental equations of quantum mechanics and that of the nuclei by classical physics, there would be no preferential 'downhill' energy flow, strongly impairing the macroscopic function of these complexes" (p. 5).

This line of reasoning could offer a missing evolutionary link. Just like photosynthetic systems evolved to exploit collapse for directed energy transport, consciousness might have evolved to exploit decoherence.

Therefore Proposal 4 is that biology utilizes decoherence in the brain to optimize survival. It uses decoherence to find the optimal fit between System 1 (quantum) and System 2 (classical) processing that is most effective for survival.

But if evolution uses decoherence as a tool, what specific mechanism would implement this tool in complex nervous systems like humans possess? What would collapse quantum function in the brain?

The answer may require examining what uniquely characterizes human consciousness compared to simpler organisms. While bacteria use decoherence for photosynthesis and birds for navigation, these appear as fixed, automatic processes. Humans demonstrate something different, which includes the ability to consciously modulate between focused attention and open awareness, between rigid thinking and creative flow, and between self-consciousness and self-forgetting. This modulable quality suggests humans evolved a controllable mechanism. But what aspect of human consciousness could provide this control? The framework proposes that the answer may lie in the very characteristic that makes humans uniquely self-aware, namely the continuous self-referential processing that creates the sense of being a persistent "I" across time. This sense of "I" is often referred to as the ego.

Consider how self-awareness correlates with performance. When athletes think about their movements, they fumble. When speakers focus on their words, they stutter. When test-takers monitor their performance, they choke. Conversely, when self-awareness drops away in flow states, performance dramatically increases. This consistent pattern across all domains suggests that self-referential processing may directly interfere with the brain's quantum access.

The Default Mode Network (DMN), the brain's self-referential processing center, must be suppressed for peak performance to emerge. Menon (2023) provides definitive evidence saying, "Suppression of the DMN and network switching to engage the cognitive control networks are implemented by a common mechanism… These task-induced dynamical processes contribute to the ubiquitous process of deactivation of the DMN and its disengagement from cognitive control systems" (p. 2474). There appears to be correlation here as the narrative-generating self must go offline for flow states to emerge. As Menon further notes, "DMN activation is observed immediately preceding mind wandering… associated with performance errors in cognitive tasks" (p. 2478). When the ego reasserts itself through DMN activation, performance degrades.

This aligns with findings on insight preparation as Kounios and Beeman (2009) showed that "preparation for solving an upcoming problem with insight involved directing attention inwardly" while analytical solving involved "directing attention outwardly" (p. 212). Importantly, their "inward attention" refers to withdrawing from external focus instead of increasing self-reflection. They found reduced visual cortex activity and enhanced readiness to detect "weakly activated, subconscious solutions." This represents a reduction in self-referential processing which creates space for flow states and insights to emerge. As they note, insight solutions arrive suddenly, which is consistent with accessing parallel quantum computation.

This interference could make mechanistic sense. When humans think about themselves, including their performance, image, or goals, the DMN generates a lot of neural activity. This activity produces electromagnetic fields that would inevitably interact with any quantum states in nearby microtubules. Evolution would have likely selected for this kind of system, as a modulable collapse mechanism based on self-referential processing would provide a wide variety of survival advantages.

Therefore, Proposal 5 is that the ego, operating through the Default Mode Network, functions as the biological switch between quantum processing (System 1) and classical processing (System 2), with varying degrees of activation creating a spectrum of hybrid states which are commonly referred to as the varied levels of consciousness.

This self-referential processing system requires precise definition, as the framework's use of "ego"differs from colloquial usage where "having a big ego" implies arrogance or self-importance. Here, ego refers to the comprehensive self-referential processing system that creates and maintains the sense of being a distinct, continuous self. This definition aligns with Gallagher's (2000) philosophical framework, which distinguishes between the "minimal self" as "a consciousness of oneself as an immediate subject of experience" and the "narrative self" as "a more or less coherent self that is constituted with a past and a future in the various stories we tell about ourselves" (p. 15). The ego, as the framework uses the term, encompasses both aspects, including the immediate awareness of being an experiencing subject and the ongoing narrative of personal identity across time.

The Physical Mechanism across Scales

If the ego functions as an internal observer, then it too must operate through physical mechanisms. The framework proposes a possible mechanism worth investigating. In neuroscience, this ego system corresponds primarily to the Default Mode Network, which Menon (2023) identifies as generating the "epistemic self" through "distinct brain states and frames of thought… central to construction of the internal narrative" (p. 2480). The DMN creates both what Gallagher calls the "sense of agency," described as "the sense that I am the one who is causing or generating an action," and the "sense of ownership," defined as "the sense that I am the one who is undergoing an experience" (p. 15). Recent neuroscientific research into flow states suggests a subcortical mechanism that may regulate the DMN-based ego modulation described in this framework. Van der Linden et al. (2021) identify the Locus Coeruleus-Norepinephrine (LC-NE) system as potentially central to flow state generation. This small nucleus in the pons, responsible for most of the brain's norepinephrine release, appears to regulate "decisions on task engagement vs. disengagement" through different modes of norepinephrine release (p. 3).

The LC-NE operates in three distinct modes that may correspond to different configurations of the $P =$ Q/E equation. In what researchers term the "exploitation mode," the system promotes "high task engagement" and "optimal engagement and performance" (van der Linden et al., 2021, p. 4). This mode emerges specifically when there exists "a match between a person's skill and the task challenges" (p. 3). Most significantly for this framework, van der Linden et al. note that "DMN activity is down-regulated by the exploitation mode" (p. 4), which suggests a direct neurochemical pathway for reducing ego activation. Evolution would need to select for a system that could rapidly assess whether current conditions favor focused engagement versus broad exploration. The LC-NE system, with its widespread projections throughout the brain, can function as a centralized decision-maker that coordinates multiple brain networks based on environmental demands. It very well could be the "modulator of the modulator" or the "control system for the switch."

When a task's demands fall outside the optimal challenge-skill balance of the moment, the LC-NE system shifts to alternative modes. Tasks that are too difficult trigger an "exploration mode" characterized by high tonic norepinephrine and susceptibility to distraction, while insufficiently challenging tasks induce a "disengagement mode" associated with boredom and fatigue (van der Linden et al., 2021, p. 4). The relationship between LC-NE modes and flow states shows an inverted U-shaped pattern. Van der Linden et al. (2021) report that "flow is related to arousal in an inverted U-shape" and that "task engagement is highest with intermediate levels of arousal" (p. 1). This pattern may reflect the framework's prediction that optimal performance requires balanced ego activation, with too little creating insufficient classical structure and too much causing excessive quantum collapse. Pupillometry studies provide additional evidence for this mechanism. Pupil dilation, which "partly reflect[s] tonic and phasic LC-NE activity" (Gilzenrat et al., 2010, cited in van der Linden et al., 2021, p. 3), shows distinct patterns during flow states. Participants demonstrated smaller mean pupil dilation during flow conditions compared to overload conditions which potentially indicates reduced neural measurement activity that would preserve quantum coherence.

From the framework's perspective, the LC-NE system may function as a neurochemical governor that modulates ego activation through its influence on DMN activity. The exploitation mode creates conditions for low E values by suppressing self-referential processing, while exploration and disengagement modes increase E through DMN upregulation. This suggests evolution may have developed multiple interconnected systems, including both cortical (DMN) and subcortical (LC-NE) mechanisms, to regulate access to quantum processing capabilities based on environmental constraints. When this ego system maintains high activity, such as when thinking about oneself, monitoring performance, or maintaining personal narrative, it generates the electromagnetic fields described above. This is likely a very recursive process, as will be explored in future chapters, where one system influences the other and vice versa. The framework proposes these fields constitute measurement of quantum states in microtubules, which end up causing collapse into classical configurations. Stronger activation of the sense of "I," whether through self-monitoring, narrative construction, or agency assertion, may create more complete collapse.

The mechanism through which the DMN electromagnetic fields could interact with quantum states in microtubules finds support from recent work on cytoelectric coupling. Pinotsis et al. (2023) demonstrate that endogenous electromagnetic fields generated by neural activity causally influence microtubule organization and create feedback loops with neural firing patterns. While their work examines classical electromagnetic effects, it establishes that the DMN's EM fields do physically interact with the same microtubule structures that Babcock et al. (2024) and Nishiyama et al. (2025) show maintain quantum coherence. Standard quantum mechanics dictates that such electromagnetic interactions would constitute measurement, causing decoherence of many quantum states present.

There is recent evidence for the connection between neural activity and quantum states in microtubules operated through a precise physical mechanism. As Wiest (2025) notes, MTs "feel the detailed state of the distributed spatiotemporal pattern of electrical activity among populations of neurons via calcium influx that mirrors neural activity and modulates MTs and MT-associated proteins" (p. 9). This provides the direct causal link as ego-driven neural patterns in the DMN create specific calcium signatures that selectively modulate quantum states in the microtubule networks, determining which possibilities collapse into classical states.

This mechanism finds direct support in neuroscience. Kounios and Beeman (2009) found that insight requires a "temporary reduction in interfering visual inputs" through alpha waves, which "reflect idling or inhibition of brain areas" (p. 212). They describe this as "the brain's covert alternative to closing the eyes or looking away," reducing observation to allow solutions to emerge.

This gradient collapse relationship is important for understanding biological quantum processing. If consciousness can influence the internal electromagnetic environment of neural tissue, then the position of specific neurobiological or cellular regions on the quantum-classical spectrum may shift. Higher ego activation would increase internal "noise" through stronger neural electromagnetic activity, which would push microtubule quantum states toward classical behavior. Reduced ego activation may create relatively protected internal conditions, which would allow coherence to persist for longer periods.

Busemeyer et al. (2019), researching belief formation and decision-making under uncertainty, describe these same dynamics through experimental evidence from perceptual tasks. They concluded that human cognition operates as a hybrid system where beliefs initially exist in quantum superposition but decohere into classical states when decisions are required. Their research showed that "the quantum model produced more accurate predictions for low levels of confidence, but the Markov model (with drift rate variability) produced more accurate predictions for high levels of confidence. Together these results suggest that neither process alone, quantum or Markov, is sufficient to account for all conditions and all participants" (p. 7). This hybrid nature aligns with the framework's proposal that consciousness levels differ in their balance between quantum and classical processing as will be explored in future chapters. As Busemeyer et al. (2019) explain, "it is possible to construct an 'open system' quantum process where a person's belief state partially decoheres as a result of interaction with a noisy mental environment… Open system quantum models start out in a coherent quantum regime (represented by a density matrix with off diagonal terms) and later decohere into a classical Markov regime (represented by density matrix with no off diagonal terms)" (p. 7).

Busemeyer et al. (2019) point to the need for some type of switching mechanism that allows for both System 1 and System 2 to operate when needed and they explicitly describe that the "noisy mental environment" could act as some sort of collapse mechanism. Citing their work in 2019, Busemeyer et al. (2020) describe how both a quantum approach and a classic approach seem better at different tasks and more efficient in varying environments, and they expand on how these systems transform, saying,

> "… other results seem more mixed regarding which type of model is most appropriate. The high coherence conditions of the double confidence experiment revealed that the Markov model can out-perform the quantum model when there is not clear interference between sequential responses (Busemeyer, Kvam, & Pleskac, 2019). Other types of data like response times, where Markov models have long been applied and vetted, also serve as challenges where Markov models out-perform quantum models (Busemeyer et al., 2006) … Indeed, these models co-exist within "open" quantum systems, which begin with ontic uncertainty of the quantum model (where the decision maker is uncertain about their own state due to superposition over evidence levels) and later reduce to epistemic uncertainty (where the decision maker knows their state, but the modeler does not due to the existence of probabilistic dynamics) through interaction with the noise in the internal or external environment. This "decoherence" takes us from a pure "closed" quantum system described by superposition to a partially coherent or mixed system described by a combination of both superposition and classical uncertainty." (p. 15)

The book's framework extends this by specifying that the mechanism that ushers in the switch between computing modes through the activation of the default mode network which is then constrained by identity and beliefs as will be described in further chapters. Lower consciousness levels, who have more rigid identity structures and higher ego activation, would create a significantly more "noisy mental environment" that would more rapidly force collapse into classical processing.

With the ego established, housed in the Default Mode Network and identified as the biological switch, the framework can now address the computational challenge identified by Kurian (2025). Instead of attempting the impossible task of perfect tracking, biological systems developed ego-mediated consciousness as a strategic decision mechanism. The ego, through the Default Mode Network, deliberately forces collapse at critical moments as it converts untrackable quantum possibilities into definite classical outcomes as described by Busemeyer et al. (2019). The possibilities that get discarded become entropy which is the thermodynamic price of making decisive choices in a universe where complete, perfect information remains inaccessible (Fig. 5.1).

In this view, consciousness exists because perfect computation cannot. The Heisenberg uncertainty that makes reversible computation impossible also necessitated the evolution of a mechanism that could strategically choose when to preserve quantum coherence and when to force classical outcomes. The ego serves this function as it modulates between quantum possibility and classical actuality based on survival needs.

With this established, the framework can now connect ego to the broader evolutionary picture as the framework proposes that ego emerged as the control mechanism for what Vervaeke and Ferraro (2013) call the virtual governor of relevance realization. By modulating the degree of quantum collapse, ego regulates the trade-off between efficiency (classical processing for quick decisions) and resiliency (quantum processing for novel solutions). This creates the "dynamic self-organizing manner" (p. 9) that Vervaeke identifies as necessary for cognitive evolution.

Rather than being locked into either quantum or classical processing, organisms with ego-mediated switching could optimize for context. Under threat, the evolutionary adaptive programs activate with narrow geometric constraints, essentially possessing consciousness with pre-configured survival patterns. The Threat Detection program, for instance (all adaptive programs will be described in detail in later chapters), drastically narrows the search space of quantum possibilities to only those relevant for immediate survival. The organism doesn't lose quantum processing entirely but becomes constrained to explore only a tiny subset of the QID's possibility space, one optimized over millions of years for escaping danger. This creates the tunnel vision and time dilation often reported in life-threatening situations, where quantum processing remains active but severely geometrically constrained.

Evolution's solution appears quite elegant. Instead of maintaining constant quantum processing (metabolically expensive) or pure classical processing (competitively disadvantageous), organisms developed adaptive programs that modulate the geometric boundaries of quantum exploration. In this case, the ego emerged because it functioned as a dynamic constraint system to balance survival with exploration

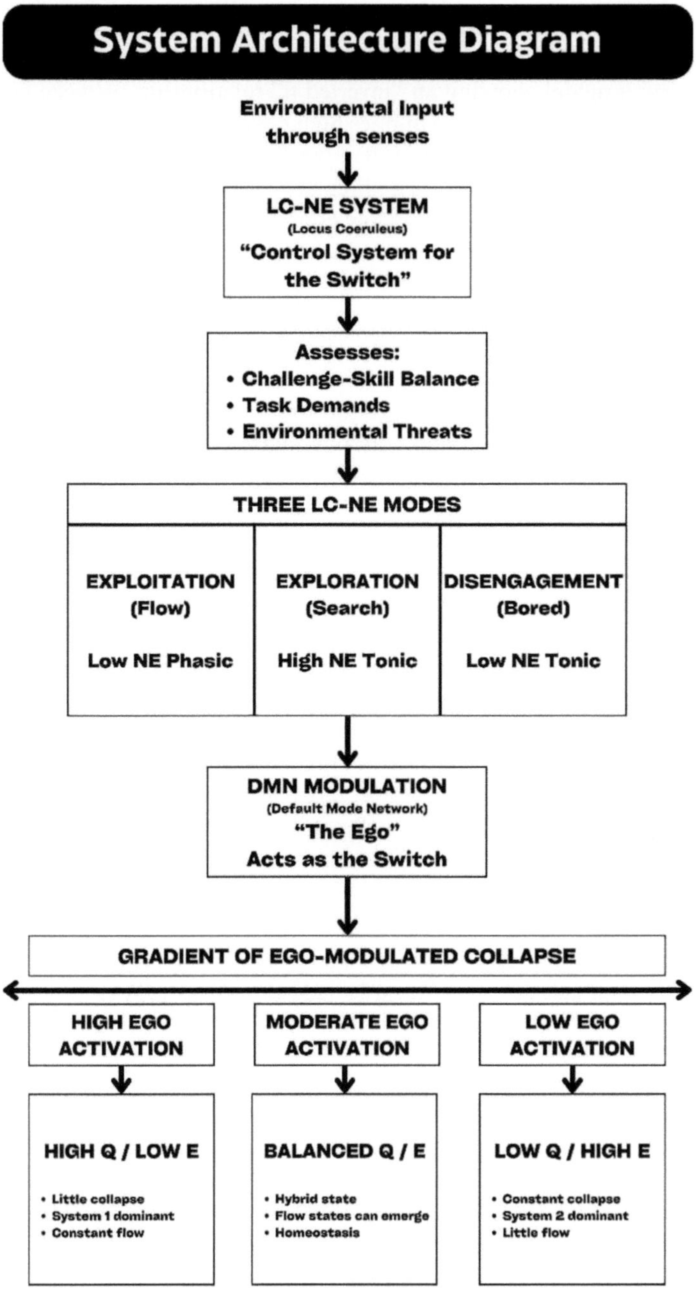

Fig. 5.1 System architecture of the switch

of vast possibility spaces when safety permits it to. The evolutionary optimization toward balanced criticality finds support in neuroscience as Carhart-Harris et al. (2014) present a model showing an "inverted u" relationship between entropy and cognition, where "too high a value implies high flexibility but high disorder, whereas too low a value implies ordered but inflexible cognition" (p. 13). From the framework's perspective, this inverted U captures the evolutionary challenge perfectly as the right side of their model which is low entropy (high ego activation) (system 2) allows for organisms to maintain stable identity and take decisive action but lack creative problem-solving while the left side which is high entropy (low ego activation) (system 1) enables access to vast possibility spaces but prevents as coherent of behavior. At the far ends of the spectrum, neither extreme survives natural selection.

Carhart-Harris et al. (2014) note that "normal waking consciousness inhabits a position that is close to criticality but slightly sub-critical" (p. 13). This may reveal evolution's solution. By positioning baseline consciousness just below the critical point, evolution selected for an average disposition that allows humans to access creativity (what Carhart-Harris et al. call "entropy") while they maintain sufficient stability for survival. As Carhart-Harris notes, "primary states move brain activity and associated cognition toward a state of increased system entropy" where "brain activity becomes more random and cognition becomes more flexible" (p. 13). The framework would suggest that evolution selected for consciousness that was most capable of navigating this entire spectrum instead of trying to fix it at any single point. In this case, the ego emerged as the mechanism for this navigation, able to increase activation for stability when needed, or decrease activation to approach "criticality proper" during moments requiring creative breakthrough (Carhart-Harris et al. 2014, p. 13). Flow states, from this perspective, represent conscious movement along this spectrum toward the optimal peak where flexibility and order achieve perfect balance.

The slightly sub-critical baseline makes evolutionary sense as being too close to "entropy" would risk frequent destabilization, but being too far from it would prevent accessing enhanced states altogether. Evolution appears to have optimized for a conservative baseline that preserves stability while allowing voluntary or circumstantial shifts toward enhanced processing when survival needs it to.

This proposed mechanism remains speculative but follows established physics as neural activity generates EM fields, EM fields constitute measurement, and measurement collapses quantum states. Evolution may have discovered this principle and optimized it, creating a modulable quantum-classical switch through the simple expedient of self-referential neural activity. The more we think about ourselves, the more electromagnetic noise we generate, the more we collapse our own quantum field states. A careful translation between this book's framework and Carhart-Harris et al.'s (2014) seems valuable considering they investigated alpha oscillations and their relationship to self-referential processing and documented patterns that align closely with the framework's expectations.

Their research points to the idea that alpha oscillations correlate specifically with self-reflection and ego function. These oscillations, which mature alongside the Default Mode Network both developmentally and evolutionarily, appear to serve a specific function that Carhart-Harris et al. (2014) describe as reducing "entropy," meaning they limit access to multiple possible brain states. In their framework, high entropy represents cognitive flexibility with many accessible states (system 1 in this book's framework), while low entropy represents ordered and rigid cognition with fewer accessible states (system 2 in this framework). Here's what they write,

> "A positive relationship has been found between self-reflection and alpha power (Knyazev et al., 2011) and alpha synchronization during rest and Blood Oxygen Level Dependent (BOLD) activity in regions of the DMN (Jann et al., 2009). Evidence implicates the DMN in self-reflective and introspective functions (Qin & Northoff, 2011)... Like the DMN, alpha oscillations mature developmentally and evolutionarily (Basar & Guntekin, 2009), tempting speculations that these rhythms have developed to reduce 'entropy' [i.e., disorder or uncertainty (Ben-Naim, 2008)] by increasing mutual information among neuronal ensembles (Tononi et al., 1994; Basar & Guntekin, 2009). With this in mind, it was remarkable that we recently found a highly significant positive correlation between the magnitude of alpha power decreases in the PCC after psilocybin and ratings of the item 'I experienced a disintegration of my 'self' or 'ego.' Scores on this item also correlated positively with decreases in delta, theta, beta, and low gamma power, although alpha explained the most variance (a considerable 66%)" (p. 8; citations in original).

From the framework's perspective, what Carhart-Harris et al. (2014) terms "entropy reduction" would describe this framework's quantum compute potential in the brain or "Q" in the P = Q/E equation. As their findings suggested, alpha oscillations help to create order from disorder, and their findings could be documenting how ego collapses quantum superposition into classical configurations. The high entropy states that they found become accessible when alpha decreases might correspond to maintained quantum superposition, where multiple possibilities remain simultaneously available as they don't collapse them into singular outcomes.

The 66% correlation between alpha suppression and ego dissolution provides extremely compelling evidence for this mechanism. Among twenty-three subjective measures tested, the relationship between reduced alpha power and the experience of ego disintegration was strongest and it survived conservative statistical correction. This suggests that disrupting alpha oscillations may directly interfere with the mechanism through which ego collapses quantum field states. Consider what this implies for the P = Q/E equation. If alpha oscillations represent the neural signature of ego activation (E), and if suppressing alpha allows access to expanded possibility states (increased Q), then the resulting enhancement in cognitive flexibility and creative problem-solving (increased P) follows naturally. From this perspective, the alpha-band activity reported by Carhart-Harris et al. (2014) may reflect the oscillatory rhythm through which consciousness coordinates the transition from coherent to classical processing.

The developmental and evolutionary maturation of alpha oscillations alongside the DMN provides additional support. Children, who demonstrate higher baseline quantum processing capacity according to the framework, show immature alpha rhythms and underdeveloped DMN activity. As these systems mature through development, creating what Carhart-Harris et al. (2014) calls "maturational settling," from this perspective children transition from natural quantum processors to adults whose consciousness becomes increasingly dominated by ego-mediated collapse.

Carhart-Harris et al. (2014) describe secondary consciousness as exhibiting strict selection where "the global system only ever entertains one winning state at any one time" due to "winner-takes-all" competition (p. 15). However, they say,

> "In primary states, the potential multiplicity of possible states is not obliterated but rather extended because the selectivity and conspicuity of a winning state is reduced, and so more transient states may be visited. In dynamical terms, this would be expressed as attractor basins or valleys (defining transient states) becoming shallower in primary states, i.e., the attractor landscape is flattened in primary states" (p. 15).

From the framework's perspective, this describes quantum collapse mechanics with significant precision. The "winner-takes-all" dynamic where consciousness "only ever entertains one winning state" maps directly onto quantum measurement forcing selection of a single outcome from superposition as Busemeyer et al. (2019, 2020) described. When ego activation decreases in primary states, the "attractor landscape" flattens, allowing consciousness to visit "more transient states" that would normally be immediately collapsed.

This provides a continuum of collapse states from the far left of one side such as psychosis (insufficient collapse), to primary consciousness (flexible collapse), to normal consciousness (moderate collapse), and then all the way to mental illness (excessive, rigid collapse). The fact that they found that "the attractor landscape is flattened in primary states" may directly point to what will be described in detail in later chapters, as lower-ego states would be expected to have a lower thermodynamic cost while rigid beliefs and identities would create steep gradients that would require a higher thermodynamic cost to change and "escape" from.

The process described through this chapter can be gathered into one continuous sequence that connects neural activity with the physical act of quantum collapse. Broadly, the system can be proposed that the locus coeruleus–norepinephrine system regulates overall arousal and task engagement, which then shapes activity in the Default Mode Network. From there, the DMN generates the electromagnetic patterns associated with self-reference and personal narrative, and these field patterns influence calcium signaling across neurons, and calcium in turn modulates the electrostatic environment surrounding the cytoskeletal microtubules inside those same cells. Within this environment, microtubule lattices made of aromatic amino acids may sustain short-lived coherent states whose stability depends on those surrounding ionic and electromagnetic conditions.

Wiest (2025) describes this relationship clearly as he writes,

> "The MTs 'feel' the detailed state of the distributed spatiotemporal pattern of electrical activity among populations of neurons via calcium influx that mirrors neural activity and modulates MTs and MT-associated proteins. This is the proffered solution to the mismatch problem, in outline. It only makes sense because quantum dynamics of OR allow the entangled microstates to be fused into a unified but complex macrostate at the moment of 'decision,' encompassing all the information in the physical representation that is 'realized' in consciousness" (p. 9).

Wiest outlines how large-scale neural firing patterns can imprint themselves upon microtubular states through calcium dynamics. The organism's own neural activity, which is guided by attention and emotion, shapes the microenvironment that maintains or releases coherence within those structures. Carhart-Harris et al. (2014) documented how self-referential thinking is directly correlated with alpha power, while Wiest (2025) described the cellular pathway through which reduced neural measurement extends coherence in microtubules. Viewed together, they outline a single process where one comes from the psychological perspective and one from the molecular perspective as both describe how quieting the self allows longer coherence before collapse.

Rourk (2025) adds an important perspective to this as he summarizes work that suggests that microtubules interact with photons through tryptophan arrays and that anesthetics, which damp this behavior, may offer evidence for that model of consciousness. He writes,

> "Recently, evidence of superradiance in microtubules in tubo in response to UV stimulation and additional evidence of damping of electronic energy migration by the anesthetics isoflurane and etomidate have been proposed to provide evidentiary support for that model of consciousness" (p. 1).

Rourk (2025) observed that ferritin molecules interact with microtubules in ways that damp electronic energy migration. He treated this as evidence against microtubule coherence in vivo, yet from this chapter's exploration that evolution harnesses decoherence for survival, it reveals how biology may regulate coherence through intrinsic damping mechanisms. Ferritin's magnetic and electrical properties allow it to absorb biophotons and interact with excited electrons in nearby microtubules, which means it can locally shape coherence lifetimes. What Rourk describes as a limitation may in practice be a biological feature that keeps coherence within functional bounds. In this view, ferritin and related molecules could form part of the natural switching system that determines when biological quantum states collapse into classical activity.

Together, these accounts describe how coherence and decoherence may be governed by biological factors, as they imply that living systems possess intrinsic mechanisms that manage these transitions. The research reviewed in Chap. 2, including work by Craddock et al. (2014), Cavaglià et al. (2023), and Nishiyama et al. (2024, 2025), adds that these dynamics depend on the energy density found within nanoscale regions of living tissue. Energy per unit of space inside these microenvironments can be extraordinarily high because biology has evolved molecular structures that deliver and confine energy with great precision (Craddock et al., 2014; Cavaglià et al., 2023; Nishiyama et al., 2024; Nishiyama et al., 2025). Microtubule lattices, tryptophan networks, and related protein assemblies appear to operate as tiny waveguides that localize excitation and allow coherence to emerge and fade at useful timescales.

When viewed within the framework proposed here, the LC-NE system sets the global energetic tone of the brain, the DMN's electromagnetic activity reflects that tone, those fields influence calcium signaling, and calcium then shapes the quantum behavior of microtubules. Coherence persists while the informational return per unit of energy remains favorable. When noise, absorption, or metabolic fluctuation overwhelms that balance, phase relationships dissolve and a localized energy transfer writes a classical update into neural dynamics. Each of these transitions may represent a unit of conscious experience, a moment when potential becomes definite.

The Observer Creating Reality

As Menon (2023) explains, "The DMN combines episodic memory, language, and semantic memory processes to generate an ongoing internal narrative and subjective continuity of internal mental thoughts" (p. 2484). This ongoing narrative generation represents the continuous observation that the framework suggests collapses quantum field states into classical experience.

Performance enhancement through flow states requires DMN suppression. As Menon (2023) notes, "Suppression of the DMN and network switching to engage the cognitive control networks are implemented by a common mechanism… These task-induced dynamical processes contribute to the ubiquitous process of deactivation of the DMN" (p. 2474). This deactivation provides what would enable access to enhanced performance states. The ego may represent the electromagnetic signature of self-focused neural activity physically interacting with quantum systems. The DMN maintains continuous self-referential processing, which creates persistent measurement pressure. Flow states, meditation, and ego-reducing practices work by allowing relevance realization to process a broader range of quantum results. They temporarily dissolve rigid relevance filters, which would allow consciousness to receive quantum results previously deemed irrelevant in "normal waking states."

The described recursive relationship between ego and relevance realization shows how Vervaeke's virtual governor mechanism operates within the collapse process. Vervaeke et al. (2012) identified relevance realization may in part act as a "virtual governor" which operates through "opponent processing between efficiency and resiliency" (p. 9) as it balances exploitation of known patterns with exploration of new possibilities, similar to how evolution operates through enabling and selective constraints. Each ego collapse based on RR's filtering updates RR's parameters which influences future filtering. This creates the "dynamic self-organizing manner" that Vervaeke et al. identify (p. 9). Vervaeke et al. then go on to say that "the brain is not trying to be either type of machine; the type of machine it becomes results from the coupling of its internal processing" (p. 10) to environmental patterns. This will be discussed in further detail in coming chapters. From this, it points to consciousness as evolution's solution to its fundamental challenge, as the universe contains infinite quantum field states which require selection to become actual. Evolution selected for ego-mediated consciousness to modulate between stability (high ego activation) and possibility (low ego activation).

Many have interpreted the evidence that ego is a post-hoc construction to mean that all of consciousness is entirely this waking mode of narratives one generates. Glattfelder (2019) writes a collective exploration of evidence in a chapter on consciousness. He uses the neuroscientific evidence akin to what was just described to show that consciousness, and the world it supposedly inhabits, doesn't exist in any real sense. He says,

> "We have witnessed a breathtaking series of human dethronements … the hardest blow comes perhaps from neuroscience. Incomprehensibly, our own sense of identity and freedom are illusions conjured up by unconscious subroutines in the brain. Consciousness is a post hoc narrator of decisions emerging from the vast non-conscious depths of the brain. My perception of the external world is an internal construction, akin to a simulation. In the final analysis, the current scientific worldview is telling me that I am fictitious and an anomaly incarnated in an otherwise incomprehensible and inaccessible physical reality" (p. 440)

Glattfelder entirely agrees with the framework "that our own sense of identity and freedom are illusions conjured up by unconscious subroutines in the brain," as the DMN is indeed the post-hoc narrator and interpreter of inputs and signals, what this paper defines as the ego, but then immediately, Glattfelder concludes this DMN process, or the ego, is consciousness itself. The previous chapters of this book and chapters that follow aim to outline why this is not the case. Aiming to show the evidence points to the idea that ego is an evolutionary mechanism and consciousness is the deeper process that interfaces with the Quantum Information Dimension and participates in the unfolding of physical reality through directed collapse.

In a fascinating progression, Glattfelder, after examining Wheeler's (1990) "It from Bit" hypothesis and other evidence of an informational universe including conscious participation, writes that both conclusions could be true, as he says,

> "If we are truly living in a participatory universe, then the profoundest implication is the mind's capacity to sculpture physical reality. Expressed within a different metaphor, the inner aspect of information continually reprograms the outer information matrix. As a consequence: "I give meaning to the reality I create." This dramatically shifts the unquenchable human yearning for knowledge from any outer authority to an inner authority. I find myself at the center of my universe, fully accountable. Perhaps this secret was known from the beginning of the mind's awakening, permeating ancient Hindu cosmology, Buddhist teachings, shamanic traditions, and now emerging within an information-theoretic and participatory scientific paradigm" (p. 608).

In his writing, the ego's transition can be witnessed in real time as the ego goes from declaring nothing exists, including ego itself and the world around it, to declaring that ego is at the center of a participatory universe as "I give meaning to the reality I create." Glattfelder concludes that "Only information exists. It manifests itself out of an ineffable abstract source and actualizes physical individualizations and pure consciousness, which interact in endless cycles, allowing for an experiential context to emerge. Existence is driven to ever higher levels of actualized complex structure, expressing intelligence. [Information-theoretic and participatory ontology]" (p. 609). This is close to this book's framework in a variety of ways but critically differs. The framework posits that information does indeed emerge from the Quantum Information Dimension and that there is a participatory, recursive relationship between consciousness and the QID from which the special, experiential consciousness we all experience manifests. Although, the framework differs in this idea that "Existence is driven to ever higher levels of actualized complex structure, expressing intelligence" as this seems to be ego projecting intention and purpose onto natural processes. Complexity increase can be explained through universal selection pressures and thermodynamic gradients without requiring a cosmic teleology, regardless of whether or not one does indeed exist.

This ego-centric view of meaning creation is quite different from what emerges from the framework's understanding of the consciousness-QID interaction. Rather than the ego giving meaning to reality it creates, the framework may offer insight into why certain experiences feel meaningful while others feel empty or hollow. When consciousness operates with reduced ego interference, this allows access to the broader possibility spaces through quantum processing, from which individuals would experience heightened sense of contentment, fulfillment, peace, and joy. These states appear to emerge naturally when consciousness functions closer to its quantum coherent potential. The quality we call meaning might arise from consciousness operating optimally, processing information efficiently, and acting through this processing in ways that align with broader systemic health across multiple scales of existence. This is important and will be explored in detail in further chapters.

The framework's proposal that ego functions as consciousness's quantum collapse mechanism finds interesting support in developmental psychology. Cook-Greuter's (2021) empirical research, which was conducted over decades without engagement with quantum consciousness theories, arrived at very similar conclusions about ego's fundamental role.

Cook-Greuter defines ego as both a process and a representation of self. She defines the process of ego as the "central processing unit" and cites Fingarette (1963) who defined ego as "the psychic organizing function or the synthetic principle in meaning making" (p. 29), and who defined it as "the autonomous drive towards meaning making." She then describes ego as what represents "the striving of human beings to understand themselves and the world they live in by fitting new experiences into their current meaning system. Overall, the ego labors continuously to create and maintain coherence and it vigorously defends" (p. 29). Her description maps directly onto quantum decoherence processes as she literally uses quantum physics language to describe these psychological processes.

Ego may have evolved to directly and powerfully participate in and direct this process for evolutionary and biological advantage, while consciousness itself witnesses and provides navigation as the quantum-classical interface. Where other quantum consciousness theories treat collapse as a mystery requiring fundamental consciousness, this framework recognizes collapse as the tool consciousness evolved to wield. These distinctions will be elaborated further in later chapters.

The framework does not propose that each person lives in their own subjective reality or that consciousness creates matter from nothing. Instead, it proposes that individuals' shared physical reality emerges from quantum collapse processes. Multiple agents contribute to collapse including environmental, biological, and conscious factors. Consciousness, utilizing the Default Mode Network which humans experience as ego, adds an additional layer of directed collapse. The physical laws remain consistent because all collapse follows quantum rules.

Consider the age-old question about whether a tree falling in a forest makes a sound when no one is around to hear it. Within the framework, this has a definitive

Table 5.1 A table of a variety of mechanisms or particles that cause collapse in the proposed gradient collapse

Collapse source	Control level	Frequency	Complexity enabled	Example
Thermal noise	None	Random	Basic decoherence	Entropy
Photons	None	Constant (c)	Basic structure	Rock formation
Chemical	Local only	Molecular time	Organic patterns	Fire, oxidation
Biological	Partial	Selective	Life processes	Photosynthesis
Neural	Semi-directed	Variable	Basic awareness	Insect behavior
Ego (DMN) and consciousness experience	Recursive choice	Contextual	Full agency	Human creativity

answer. Yes, the tree falls and makes a sound because environmental decoherence through air molecules, photons, thermal interactions, and other processes continuously collapse quantum states. The tree's own molecular processes participate in collapse. Consciousness, when present, adds a more directed collapse, but consciousness is not a necessary prerequisite for the rest of physical reality to exist. From this, the framework describes a hierarchy of collapse mechanisms, each enabling different levels of complexity as thermal noise creates random basic decoherence, photons maintain constant collapse at the speed of light, which enables basic structure formation, chemical processes provide local control at molecular timescales, biological systems offer selective collapse enabling life processes, neural systems generate semi-directed collapse supporting basic awareness, and consciousness demonstrates recursive choice with contextual control, which enables agency.

In this hierarchical structure, consciousness demonstrates the highest collapse strength because only consciousness possesses recursive self-awareness and intentionality. While thermal noise creates random collapse and photons cause passive decoherence, consciousness can observe its own observation process and choose which possibilities to actualize (Table 5.1).

The Ego as Internal Collapse Emulator

The framework proposes that ego simulates being observed even when alone instead of just trying to respond to external observation. This internal observer continuously monitors and collapses quantum states based on perceived threats.

The exact ratio of environmental versus ego-mediated collapse remains unknown. The distribution could be 95% ego perception with 5% direct environmental influence, an even 50/50 split between internal and external factors, or variable depending on consciousness level and training. What matters is that any shift in these probabilities, even 0.1%, becomes significant when considering billions of neurons and trillions of potential quantum states.

Whether collapse is predominantly environmental or perceptual, optimizing both domains maximizes quantum processing potential. The ego may be the most directed and orchestrated collapse mechanism, but environmental factors maintain much of physical reality and can influence how much coherence an individual maintains. The framework therefore points to consciousness as an active manager of its own collapse processes, with ego serving as the internal observer that can either prematurely collapse possibilities or, when skillfully modulated, allow quantum computation to flourish. Overall, the framework will continue to explore that if this the QID or substrate of quantum configuration is fundamental, how this would make sense why consciousness didn't accidentally develop quantum coherence that happens to access information; instead, the existence of the QID shaped the driving forces of evolution to select the quantum-coherent structures necessary for this interface. This would help to explain why evolution invested such enormous resources in brain development (60% of young child metabolism; Chugani, 1998), why quantum coherence in warm biological systems defies classical expectations, and why consciousness seems suspiciously well-tuned for accessing non-local information. In this, the ego isn't a spiritual failing or a cognitive limitation; it may just be an essential tool that adds a layer of directed collapse to the transformation of infinite quantum field configurations into stable, navigable reality that is necessary for biological survival through evolutionary selection.

Dissolving the Observer's Paradox

The observer's paradox only appears paradoxical when consciousness and quantum mechanics are treated as separate phenomena that collide in unfortunate ways (von Neumann, 1932). A different picture emerges when viewing observation through the lens of physical information exchange, because, as this chapter has described, every observation, measurement, or interaction involves the transfer of information, and information transfer cannot occur without physical interaction.

Consider what happens when a person physically looks at the results of a double-slit experiment. Before the observation, the system exists in superposition with its probabilities encoded in the interference pattern. The instant the detector registers a photon and that information reaches the observer's eyes, a cascade of physical interactions has occurred including photon scattering, electron excitations in the detector, neural signal propagation, and environmental dispersion of information into surrounding degrees of freedom. Each of these processes involves transferring at least one h-sized quantum of action, paying the associated Landauer cost (kT ln

2), and irreversibly increasing entropy as information disperses into the environment. These are all commits that anchor the outcome to classical reality, making the exchange statistically irreversible. From this perspective, collapse emerges as the natural result of information becoming thermodynamically irreversible through physical dispersion.

The observation requires interaction, and interaction is physical information exchange. You cannot observe an electron without sending photons that scatter off it. You cannot detect those scattered photons without them triggering cascades of molecular changes in your detector and eyes. You cannot process that information without neurons firing in classical, thermodynamically irreversible patterns. At every step, information transfers from quantum superposition into classical, high-entropy states. Observation causes collapse because observing is physical interaction, and physical interaction disperses information into the environment while making quantum reversal thermodynamically impossible.

Considering consciousness as an evolved collapse mechanism that operates within the established open-system formalisms of Lindblad dynamics and environment-induced decoherence allows it to directly integrate into this continuum (Lindblad, 1976; Zurek, 2003). With this in mind, the paradox dissolves. Humans function as participants in collapse dynamics while also being shaped by them. In a universe where collapse operates as a gradient process modulated by environmental conditions, consciousness represents an evolutionary strategy for sophisticated, directed participation in information-exchange processes.

This experimental foundation establishes that consciousness participates in quantum collapse alongside other physical mechanisms within a similar informational framework. Temperature-dependent decoherence demonstrated by Uhlemann et al. (2013), Beierle et al. (2018), and Howie (2011, 2014) operates through purely physical mechanisms that involve thermal noise and environmental recording channels. BEC experiments (Dutton et al., 2004; Hau, 2016) show that environmental isolation can preserve coherence when these recording channels are suppressed. Human consciousness represents another node in this physical network, which is more sophisticated and directed than thermal noise while it operates through a comparable fundamental mechanism of physical information exchange that increases entropy and makes quantum states classical.

The Evolutionary Arms Race

If ego evolved as a reality-construction mechanism, what survival advantages would this system confer? Consider the competitive landscape of early evolution, where millisecond advantages determined survival. Organisms that could rapidly cycle between quantum exploration and classical action would possess significant advantages. Kvam and Pleskac (2017) outline the evolutionary selection pressures that would choose for exactly this, saying,

> "Furthermore, it seems likely that selection pressures in our evolutionary history would not have favored fully deterministic strategies in all situations. This is largely because they can be exploited in multi-agent interactions – if opponents are aware of the rules a decision-maker is using, they can tailor counter-strategies based on what they know the decision-maker will do. For example, take the matching pennies game: in this interaction, each of two players places a penny heads-up or tails-up. Player A receives a reward if the players' pennies show different sides and a penalty if they show the same sides, whereas player B receives a reward if the pennies show the same sides and a penalty if they show different ones. 1 The optimal behavior in this game is to choose stochastically, 50% heads/50% tails. If player A chooses deterministically based on the time of day, surroundings, or some other factor, player B may be able to pick up on these and guarantee a win. In the face of these challenges, it seems selection pressures might have resulted in the evolution of a mechanism that can generate response variability and unpredictability. This matches well with the great variability that we observe in human behavior (as well as its unpredictability, to the dismay of a great many cognitive modelers). Therefore, a successful integration might endow heuristics with a mechanism for generating random or at least variable behavior, in turn bringing them closer as a description of human behavior" (p. 5).

This mechanism that they describe may be exactly what has been outlined in this chapter. The cycle might operate through quantum processing which explores all possible patterns simultaneously, relevance realization utilizing egoic identities and beliefs then rapidly filters the results for survival-critical information, and ego collapses the results into decisive action, then ego shapes the heuristics and filters of relevance realization, while other random environmental collapses allow for varied, unpredictable behavior beneficial for competitive survival then all of this filters back in to repeat the cycle. Evolution needed all parts of the mechanism. Pure quantum beings might remain lost in infinite possibilities, paralyzed by option overload which would be metabolically unsustainable. Pure classical beings would be limited to sequential processing, missing important patterns, creative solutions, and unpredictable competitive behaviors that quantum superposition could have enabled.

By adding directed collapse to baseline environmental decoherence, ego helps create stable, persistent reality necessary for survival, social coordination, strategy, planning, learning, memory, and action, and then importantly, it enables rapid switching between quantum exploration and classical execution, which allows organisms to balance possibility exploration with immediate action requirements to best increase their chance of survival.

This chapter flips the "too warm, wet, and noisy" argument on its head. Evolution never needed to "overcome" decoherence. In a quantum universe, life would favor systems that could use those very dynamics for survival. Over time, this life competed to most effectively time its own decoherence. Quite possibly, from this view, intelligence may have emerged from the gradual refinement of that timing as life learned how to balance coherence and collapse to stay alive.

References

Babcock, N. S., Montes-Cabrera, G., Oberhofer, K. E., Chergui, M., Celardo, G. L., & Kurian, P. (2024). Ultraviolet superradiance from mega-networks of tryptophan in biological architectures. *The Journal of Physical Chemistry B, 128*(17), 4035–4046. https://doi.org/10.1021/acs.jpcb.3c07936

Basar, E., & Guntekin, B. (2009). Darwin's evolution theory, brain oscillations, and complex brain function in a new "Cartesian view." *International Journal of Psychophysiology, 71*, 2–8. https://doi.org/10.1016/j.ijpsycho.2008.07.018

Beierle, P. J., Zhang, L., & Batelaan, H. (2018). Experimental test of decoherence theory using electron matter waves. *New Journal of Physics, 20*(11), 113030. https://doi.org/10.1088/1367-2630/aaed4e

Ben-Naim, A. (2008). *A farewell to entropy: Statistical thermodynamics based on information: S=logW*. World Scientific.

Bérut, A., Arakelyan, A., Petrosyan, A., Ciliberto, S., Dillenschneider, R., & Lutz, E. (2012). Experimental verification of Landauer's principle linking information and thermodynamics. *Nature, 483*(7388), 187–189. https://doi.org/10.1038/nature10872

Boltzmann, L. (1877). Über die Beziehung zwischen dem zweiten Hauptsatze der mechanischen Wärmetheorie und der Wahrscheinlichkeitsrechnung respektive den Sätzen über das Wärmegleichgewicht. *Wiener Berichte, 76*, 373–435.

Bonilla, L., & Guinea, F. (1992). Collapse of the wave packet and chaos in a model with classical and quantum degrees of freedom. *Physical Review A, 45*, 7718–7728. https://doi.org/10.1103/PhysRevA.45.7718

Breuer, H. P., & Petruccione, F. (2002). *The theory of open quantum systems*. Oxford University Press. https://doi.org/10.1093/acprof:oso/9780199213900.001.0001

Busemeyer, J. R., Kvam, P. D., & Pleskac, T. J. (2019). Markov versus quantum dynamic models of belief change during evidence monitoring. *Scientific Reports, 9*, Article 18025. https://doi.org/10.1038/s41598-019-54383-9

Busemeyer, J. R., Kvam, P. D., & Pleskac, T. J. (2020). Comparison of Markov versus quantum dynamical models of human decision making. *WIREs Cognitive Science, 11*(4), e1526. https://doi.org/10.1002/wcs.1526

Cao, J., Cogdell, R. J., Coker, D. F., Duan, H. G., Hauer, J., Kleinekathöfer, U., Jansen, T. L. C., Mančal, T., Miller, R. J. D., Ogilvie, J. P., Prokhorenko, V. I., Renger, T., Tan, H. S., Tempelaar, R., Thorwart, M., Thyrhaug, E., Westenhoff, S., & Zigmantas, D. (2020). Quantum biology revisited. *Science Advances, 6*(14), eaaz4888. https://doi.org/10.1126/sciadv.aaz4888

Carhart-Harris, R. L., Leech, R., Hellyer, P. J., Shanahan, M., Feilding, A., Tagliazucchi, E., Chialvo, D. R., & Nutt, D. (2014). The entropic brain: A theory of conscious states informed by neuroimaging research with psychedelic drugs. *Frontiers in Human Neuroscience, 8*, Article 20. https://doi.org/10.3389/fnhum.2014.00020

Cavaglià, M., Deriu, M. A., & Tuszynski, J. A. (2023). Toward a holographic brain paradigm: A lipid-centric model of brain functioning. *Frontiers in Neuroscience, 17*, 1302519. https://doi.org/10.3389/fnins.2023.1302519

Chugani, H. T. (1998). A critical period of brain development: Studies of cerebral glucose utilization with PET. *Preventive Medicine, 27*(2), 184–188.

Clausius, R. (1857). Ueber die art der Bewegung, welche wir Wärme nennen [On the kind of motion which we call heat]. *Annalen der Physik, 176*(3), 353–380.

Cook-Greuter, S. R. (2021). *Ego development: A full-spectrum theory of vertical growth and meaning making*. (Original work published 2013, edited in 2019, republished in 2021).

Craddock, T. J. A., Friesen, D., Mane, J., Hameroff, S., & Tuszynski, J. A. (2014). The feasibility of coherent energy transfer in microtubules. *Journal of the Royal Society Interface, 11*(100), 20140677. https://doi.org/10.1098/rsif.2014.0677

Derakhshani, M., Diósi, L., Laubenstein, M., Piscicchia, K., & Curceanu, C. (2022). At the crossroad of the search for spontaneous radiation and the Orch OR consciousness theory. *Physics of Life Reviews, 42*, 8–14. https://doi.org/10.1016/j.plrev.2022.05.004

Donadi, S., Piscicchia, K., Curceanu, C., Diósi, L., Laubenstein, M., & Bassi, A. (2021). Underground test of gravity-related wave function collapse. *Nature Physics, 17*(1), 74–78. https://doi.org/10.1038/s41567-020-1008-4

Dutton, Z., Ginsberg, N. S., Slowe, C., & Hau, L. V. (2004). The art of taming light: Ultra-slow and stopped light. *Europhysics News, 35*(2), 33–39.

Faggin, F. (2021). *Silicon: From the invention of the microprocessor to the new science of consciousness*. Waterside Productions.

Faggin, F. (2024). *Irreducible: Consciousness, life, computers, and human nature*. Essentia Books.

Fingarette, H. (1963). *The self in transformation*. Harper and Row.

Gallagher, S. (2000). Philosophical conceptions of the self: Implications for cognitive science. *Trends in Cognitive Sciences, 4*(1), 14–21. https://doi.org/10.1016/S1364-6613(99)01417-5

Glattfelder, J. B. (2019). *Information—Consciousness—Reality: How a new understanding of the universe can help answer age-old questions of existence*. Springer. https://doi.org/10.1007/978-3-030-03633-1

Hameroff, S., & Penrose, R. (1996). Orchestrated reduction of quantum coherence in brain microtubules: A model for consciousness. *Mathematics and Computers in Simulation, 40*(3–4), 453–480. https://doi.org/10.1016/0378-4754(96)80476-9

Hameroff, S., & Penrose, R. (2014). Consciousness in the universe: A review of the 'Orch OR' theory. *Physics of Life Reviews, 11*(1), 39–78. https://doi.org/10.1016/j.plrev.2013.08.002

Hau, L. V. (2016). *Quantum control of light and matter: From the macroscopic to the nano scale* (Report no. AFRL-AFOSR-VA-TR-2016-0082). Air Force Research Laboratory. https://doi.org/10.21236/AD1004771

Howie, A. (2011). Mechanisms of decoherence in electron microscopy. *Ultramicroscopy, 111*(7), 761–767. https://doi.org/10.1016/j.ultramic.2010.07.008

Howie, A. (2014). Addressing Coulomb's singularity, nanoparticle recoil and Johnson's noise. *Journal of Physics: Conference Series, 522*, 012001. https://doi.org/10.1088/1742-6596/522/1/012001

Jann, K., Dierks, T., Boesch, C., Kottlow, M., Strik, W., & Koenig, T. (2009). BOLD correlates of EEG alpha phase-locking and the fMRI default mode network. *NeuroImage, 45*, 903-916. https://doi.org/10.1016/j.neuroimage.2009.01.001

Jun, Y., Gavrilov, M., & Bechhoefer, J. (2014). High-precision test of Landauer's principle in a feedback trap. *Physical Review Letters, 113*(19), 190601. https://doi.org/10.1103/PhysRevLett.113.190601

Khan, S., Huang, Y., Timuçin, D., Bailey, S., Lee, S., Lopes, J., Gaunce, E., Mosberger, J., Zhan, M., Abdelrahman, B., Zeng, X., & Wiest, M. C. (2024). Microtubule-stabilizer epothilone B delays anesthetic-induced unconsciousness in rats. *eNeuro, 11*(8), ENEURO.0291-24.2024. https://doi.org/10.1523/ENEURO.0291-24.2024

Kalra, A. P., Benny, A., Travis, S. M., Zizzi, E. A., Morales-Sanchez, A., Oblinsky, D. G., Craddock, T. J. A., Hameroff, S. R., MacIver, M. B., Tuszyński, J. A., Petry, S., Penrose, R., & Scholes, G. D. (2023). Electronic energy migration in microtubules. *ACS Central Science, 9*(3), 352–361. https://doi.org/10.1021/acscentsci.2c01114

Kastrup, B. (2019). *The idea of the world: A multidisciplinary argument for the mental nature of reality*. Iff Books.

Kauffman, S. A., & Radin, D. (2023). Quantum aspects of the brain-mind relationship: A hypothesis with supporting evidence. *Biosystems, 223*, 104820. https://arxiv.org/abs/2101.01538. https://doi.org/10.1016/j.biosystems.2022.104820

Kounios, J., & Beeman, M. (2009). The aha! Moment: The cognitive neuroscience of insight. *Current Directions in Psychological Science, 18*(4), 210–216. https://doi.org/10.1111/j.1467-8721.2009.01638.x

Knyazev, G. G., Slobodskoj-Plusnin, J. Y., Bocharov, A. V., & Pylkova, L. V. (2011). The default mode network and EEG alpha oscillations: An independent component analysis. *Brain Research, 1402*, 67–79. https://doi.org/10.1016/j.brainres.2011.05.052

Kurian, P. (2025). Computational capacity of life in relation to the universe. *Science Advances, 11*, eadt4623. https://doi.org/10.1126/sciadv.adt4623

Kvam, P. D., & Pleskac, T. J. (2017). A quantum information architecture for cue-based heuristics. *Decision, 4*(4), 197–233. https://doi.org/10.1037/dec0000070

Landauer, R. (1961). Irreversibility and heat generation in the computing process. *IBM Journal of Research and Development, 5*(3), 183–191. https://doi.org/10.1147/rd.53.0183

Lindblad, G. (1976). On the generators of quantum dynamical semigroups. *Communications in Mathematical Physics, 48*(2), 119–130. https://doi.org/10.1007/BF01608499

Melkikh, A. V. (2015). Nonlinearity of quantum mechanics and solution of the problem of wave function collapse. *Communications in Theoretical Physics, 64*(1), 47–53. https://doi.org/10.1088/0253-6102/64/1/47

Menon, V. (2023). 20 years of the default mode network: A review and synthesis. *Neuron, 111*(16), 2469–2487. https://doi.org/10.1016/j.neuron.2023.04.023

Nishiyama, A., Tanaka, S., & Tuszynski, J. A. (2024). Quantum brain dynamics: Optical and acoustic super radiance via a microtubule. *Biosystems, 234*, 105118. https://doi.org/10.3390/foundations4020019

Nishiyama, A., Tanaka, S., & Tuszynski, J. A. (2025). Waveguide quantum electrodynamics: Tryptophans entangled with water as data qubits in a microtubule. *Dynamics, 5*, 7. https://doi.org/10.3390/dynamics5010007

Penrose, R. (1990). Précis of the emperor's new mind: Concerning computers, minds, and the laws of physics. *Behavioral and Brain Sciences, 13*(4), 643–655. https://doi.org/10.1017/S0140525X00080675

Pinotsis, D. A., Fridman, G., & Miller, E. K. (2023). Cytoelectric coupling: Electric fields sculpt neural activity and "tune" the brain's infrastructure. *Progress in Neurobiology, 226*, 102465. https://doi.org/10.1016/j.pneurobio.2023.102465

Qin, P., & Northoff, G. (2011). How is our self related to midline regions and the default-mode network? *NeuroImage, 57*, 1221–1233. https://doi.org/10.1016/j.neuroimage.2011.05.028

Rourk, C. J. (2025). The interaction of ferritin with microtubules in vivo will inhibit microtubule superradiance and electronic energy migration through microtubules. *IEEE Transactions on Molecular, Biological, and Multi-Scale Communications*. https://doi.org/10.1109/TMBMC.2025.3615932

Schlosshauer, M. (2007). *Decoherence and the quantum-to-classical transition*. Springer. https://doi.org/10.1007/978-3-540-35775-9

Shannon, C. E. (1948). A mathematical theory of communication. *Bell System Technical Journal, 27*(3), 379–423.

Takahashi, Y., & Umezawa, H. (1975). Thermo field dynamics. *Collective Phenomena, 2*, 55–80.

Thomson, W. (1848). On an absolute thermometric scale founded on Carnot's theory of the motive power of heat, and calculated from Regnault's observations. *Philosophical Magazine, 33*, 313–317.

Tononi, G., Sporns, O., & Edelman, G. M. (1994). A measure for brain complexity: Relating functional segregation and integration in the nervous system. *Proceedings of the National Academy of Sciences, 91*, 5033–5037. https://doi.org/10.1073/pnas.91.11.5033

Toole, J. T., Kurian, P., & Craddock, T. J. A. (2018). Coherent energy transfer and the potential implications for consciousness. *Journal of Cognitive Science, 19*(2), 115–124. https://doi.org/10.17791/jcs.2018.19.2.115

Uhlemann, S., Müller, H., Hartel, P., Zach, J., & Haider, M. (2013). Thermal magnetic field noise limits resolution in transmission electron microscopy. *Physical Review Letters, 111*(4), 046101. https://doi.org/10.1103/PhysRevLett.111.046101

Umezawa, H. (1993). *Advanced field theory: Micro, macro, and thermal physics*. American Institute of Physics.

van der Linden, D., Tops, M., & Bakker, A. B. (2021). The neuroscience of the flow state: Involvement of the locus coeruleus norepinephrine system. *Frontiers in Psychology, 12*, 645498. https://doi.org/10.3389/fpsyg.2021.645498

Vervaeke, J., & Ferraro, L. (2013). Relevance, meaning and the cognitive science of wisdom. In M. Ferrari & N. M. Weststrate (Eds.), *The scientific study of personal wisdom* (pp. 21–51). Springer. https://doi.org/10.1007/978-94-007-7987-7_2

Vervaeke, J., Lillicrap, T. P., & Richards, B. A. (2012). Relevance realization and the emerging framework in cognitive science. *Journal of Logic and Computation, 22*(1), 79–99. https://doi.org/10.1093/logcom/exp067

von Neumann, J. (1932). *Mathematische Grundlagen der Quantenmechanik*. Springer. https://doi.org/10.1007/978-3-642-61409-5

Wheeler, J. A. (1992). Toward "it from bit". *Quantum Coherence, 281.*

Wheeler, J. A. (1990). Information, physics, quantum: The search for links. In W. H. Zurek (Ed.), *Complexity, entropy, and the physics of information* (pp. 3–28). Westview Press.

Wiest, M. C. (2025). A quantum microtubule substrate of consciousness is experimentally supported and solves the binding and epiphenomenalism problems. *Neuroscience of Consciousness, 2025*(1), niaf011. https://doi.org/10.1093/nc/niaf011

Zurek, W. H. (2003). Decoherence, einselection, and the quantum origins of the classical. *Reviews of Modern Physics, 75*(3), 715–775. https://doi.org/10.1103/RevModPhys.75.715

Chapter 6
The Computational Process

Kahneman's (2011) dual-process theory describes what happens but not how it happens as traditional psychology portrays System 1 as an evolutionary relic which is fast but error-prone, intuitive but biased, and emotional but not logical, although this creates a paradox. How does this supposedly primitive system generate insights that are mathematically elegant, completely novel, arrive complete, and often superior to deliberate System 2 reasoning?

The book's framework provides the missing mechanism which is that System 1's seemingly magical capabilities don't come from some type of evolutionary shortcut or "magic," they come from quantum computation. From this, the "biases" traditionally attributed to System 1 may actually originate from pre-mature collapse that is calculating with limited or incomplete "data sets" (biased memories, experiences, or incorrect information stored in memory). If Proposal 1: $P = Q/E$ accurately describes consciousness, then System 1's "remarkable" capabilities become less mysterious. System 1 would be accessing Q as it represents quantum computation, while E, representing ego activation, remains minimal. From this perspective, it would explain System 1's speed, pattern recognition, and insight generation that seem to violate classical processing constraints.

The brain's challenge to classical models is well-documented. As Toole et al. (2018) observe, "since the creation of the computer, the brain has been reduced to a machine of inputs and outputs, effectively a biological computer that follows the deterministic rules of classical physics. Certain aspects of the brain, however, have challenged this concept, with consciousness chief among them" (p. 116). The book proposes an evolution of the metaphor.

J. Roeloffs, *The Classical Brain to the Quantum Mind*, The Frontiers Collection,
https://doi.org/10.1007/978-3-032-20068-6_6

The GPU and CPU

Just like modern computers use CPU for sequential processing, logical operations, and precise control while the GPU handles massive parallel processing, pattern recognition, and probabilistic operations, evolution developed System 2 (Classical/CPU) for sequential, ego-mediated, quantum-collapsed processing and System 1 (Quantum/GPU) for parallel superposition, ego-quiet, quantum-coherent processing.

This framework has support in modern quantum biology. Nishiyama et al. (2024) explicitly describe the brain as "a mixed system of classical neurons and glial cells on the one hand, and quantum degrees of freedom of photons and water molecules on the other hand" (p. 4). This dual-system architecture aligns well with the framework's proposal that System 1 accesses quantum computation while System 2 operates through classical processing.

System 1's "Magic"

System 1's documented capabilities align with quantum computing advantages. From the framework's perspective, instant pattern recognition occurs as quantum superposition explores all patterns simultaneously, while intuitive physics emerges because the quantum field contains probability distributions of physical outcomes. Toole et al. (2018) specifically note that quantum properties "afford a biological system like the brain a degree of non-deterministic (i.e., non-algorithmic) behavior" (pp. 117–118), which provides the kind of processing that could explain System 1's capabilities. The relationship between ego suppression and enhanced performance extends to creativity as Kenett et al. (2020) found that "tasks that used an analogical reasoning paradigm were best aligned with the construct of creativity" while "The Stroop task emerged to be the best aligned with the construct of cognitive control" (p. 14). This separation suggests creativity emerges through pattern recognition instead of some type of top down cognitive control. The researchers noted that "constructs most aligned with creative cognition do not rely heavily on conscious imagery or flexible switching" (p. 13), both of which involve self-referential activation. Kurian's (2025) analysis reveals that biological quantum systems can perform error correction at rates up to 10^14 bits per second through superradiant pulses (p. 7) which are computational speeds that dwarf classical neural transmission. This massive parallel processing capacity aligns with System 1's documented ability to rapidly recognize patterns and generate insights that seem to violate sequential processing constraints.

Carl Jung observed this anticipatory capability and unusually competent computing ability of the unconscious (system 1) decades before modern neuroscience. Through clinical analysis of dreams and through observation of his patients, he documented how the mind generates insights and reaches conclusions before ego-consciousness becomes aware. Jung wrote,

> "Such a dream often comes right out of the blue, and one wonders what could have prompted it. Of course, if one knew its ulterior message, its cause would be clear. It is only our consciousness that does not yet know; the unconscious seems already informed, and to have come to a conclusion that is expressed in the dream. In fact, the unconscious seems to be able to examine and to draw conclusions from facts, much as consciousness does. It can even use certain facts, and anticipate their possible results, just because we are not conscious of them. But as far as one can make out from dreams, the unconscious makes its deliberations instinctively… Logical analysis is the prerogative of consciousness; we select with reason and knowledge." (Jung, 1964, p. 78).

Jung's observation points towards the conditional nature of these computational modes. First of all, when Jung describes consciousness, his definition of consciousness means the ego's awareness of thoughts, narratives, and its surroundings; this is a traditional human experience of "waking consciousness." In Chap. 5, the framework described how the ego is the switch between the modes of computation. Jung critically identifies how these modes can't work simultaneously in his observation, because if the unconscious reaches conclusions and anticipates results, then this happens specifically "just because we are not conscious of them." He is saying that when ego-consciousness activates to narrate, make sense of, become aware of mental content, and to perceive the world around it, one form of processing occurs, but when ego-consciousness remains quiet, this allows the unconscious to operate without interference, which allows this superior anticipatory processing (system 1) to emerge with "magical" results.

This aligns with Kurian's finding of 10^14 bits per second processing capacity because if System 1 accesses quantum computational speeds that dwarf classical neural transmission, and if this access requires minimal ego interference as Jung observed, then System 1's incredible capabilities naturally would follow, and the "biases" Kahneman attributed to System 1 would likely reflect premature ego collapse that interrupts quantum processing. That will be described in detail in coming chapters.

Neural Evidence for Quantum-Classical Transitions

There is support of these mechanisms as Leroy and Cheron (2020) recorded EEG dynamics during tightrope performance and compared flow states to stress states during these moments. They state their key finding saying, "These results support that 'flow' emergence required transient hypofrontality"(p. 1). The study shows two fundamentally different neural configurations. During flow states, they observed decreased activation in medial frontal gyrus (BA10), a region involved in "inhibitory control" (p. 8). This suppression of inhibitory control aligns well with what the framework would predict when ego observation ceases. Simultaneously, they found increased activity in basal ganglia structures, saying, "Increased contributions from the globus pallidus and putamen as generators of alpha and beta oscillations during the FLOW with respect to the STRESS" (p. 6), along with stronger contributions from superior temporal gyrus (BA38) and middle occipital gyrus (BA19).

The stress state showed the opposite pattern. Increased frontal cortex activity emerged, "STRESS with respect to FLOW revealed increased contribution of the frontal cortex (BA10)" (p. 6), accompanied by increased cingulate activity (BA31/24). This reversal suggests the brain switches between distinct operational modes instead of varying along a continuum.

The frequency signatures between states show especially interesting patterns. During flow, "A clear alpha peak was present during the FLOW… which disappeared during STRESS" (p. 3). The presence of coherent alpha oscillations during optimal performance and their absence during stress may represent maintained quantum coherence versus continuous collapse. The temporal dynamics proved especially significant. "During the first 100 ms after TA activation… event-related synchronization (ERS) present during the FLOW" (p. 4), while stress states showed "A desynchronization (ERD) present during the STRESS" (p. 4). This 100-millisecond switching capability demonstrates the brain can transition between modes almost instantaneously, fast enough to explain split-second athletic decisions and sudden creative insights. The researchers concluded saying, "These findings corroborate the theoretical proposal of Dietrich that flow state emergence requires transient hypofrontality"(p. 8). It is important to note that this study wasn't designed to test quantum consciousness theories. The researchers independently discovered that peak performance requires suppressing similar brain regions the framework identifies as most responsible for the brain's quantum collapse mechanism.

While linking these neural patterns to quantum processing remains theoretical, the study definitively shows that flow requires suppression of self-referential brain regions which creates the exact conditions the framework predicts must occur for ego to stop collapsing quantum states.

How Does System 2 Navigate System 1's Results?

If System 1 continuously processes all possibilities through quantum superposition, it would generate a combinatorially explosive results space. How does System 2 know what to focus on without getting overwhelmed by infinite possibilities? This fundamental challenge leads to a crucial question about the interface between quantum and classical processing.

Vervaeke and Ferraro (2013) identify that cognition fundamentally involves relevance realization (RR), the process by which cognitive agents dynamically determine what information is relevant to their current situation. They propose that RR operates through self-organized criticality and small-world network dynamics, creating a system where "the brain is pursuing two logistical properties that are in such a trade-off relationship, viz., efficiency and resiliency" (p. 9). This opponent processing between efficiency and resiliency functions as a "virtual governor"that continually redesigns cognitive fitness to match environmental demands.

Their mechanism does a very good job to explain how the brain reorganizes and adapts its attention. As they demonstrate, "the brain's neuronal firing goes through self-organizing criticality (SOC) in which it oscillates between patterns of synchronous firing of neurons and periods of asynchronous firings" (p. 12), with synchronous firing enabling information integration through compression while asynchronous periods afford differentiation and specialized processing. The small-world network architecture they identify optimizes between local processing efficiency and global integration, creating what they term "salience landscaping." Yet even with this sophisticated mechanism, a fundamental challenge remains. As Vervaeke and Ferraro note, "we have limited time and resources with which to solve our problems. Our search must be a heuristic one that does not exhaust all of the available alternatives" (p. 3). How does the brain know which patterns to reorganize toward without first exploring the combinatorially explosive space of possibilities? They acknowledge that "somehow problem solvers reliably (but not perfectly) zero in on the relevant information to be investigated" (p. 3), but the mechanism for this initial zeroing-in remains mysterious within a classical framework.

Importantly, Vervaeke's description accurately captures how cognition operates for the vast majority of people (consciousness levels 1–6). For these individuals, ego-based heuristics and identity constraints shape what quantum processing explores before it even begins. Their beliefs, assumptions, and worldview literally determine which quantum possibilities get computed. Vervaeke's model describes normal human cognition perfectly. These things including the levels of consciousness and the geometric search space will be explored in detail in the next few chapters.

The framework proposes that evolution solved this through the quantum-classical architecture described in previous chapters, but this architecture operates differently depending on consciousness level. For most people, ego provides upfront constraints that limit the quantum search space, making Vervaeke's heuristic model accurate. For higher consciousness levels (7+), these constraints begin to loosen, allowing broader quantum exploration. As established in Chap. 2, the brain's quantum infrastructure can explore possibilities simultaneously through superposition, and it provides the computational substrate that Vervaeke's mechanism requires. Considering this, it transforms the combinatorial explosion from a difficult search problem into a tractable reorganization problem as quantum computation provides the explored territory, while Vervaeke's self-organized criticality and small-world networks provide the dynamic reorganization mechanism that searches through these quantum results.

The Three-Layer Architecture

This is absolutely critical when considering the balance between System 1 and System 2. Rather than being entirely opposing forces, System 1 and System 2 might function as complementary processors engaged in productive opponent processing. However, this architecture operates differently depending on consciousness level (Figs. 6.1 and 6.2).

For consciousness levels 1-6 (98%+ of people) (Cook-Greuter, 2021):

Ego provides upfront geometric and heuristic constraints to the quantum informational space

↓

System 1 (quantum enhanced) explores within those constraints

↓

Relevance Realization filters the constrained results

↓

System 2 processes filtered output

Fig. 6.1 The processing structure for how ego, system 1, relevance realization, and system 2 process together for consciousness levels 1–6

For consciousness levels 7+ (less than 2% of people) (Cook-Greuter, 2021):

System 1 (quantum enhanced) generates broader possibilities with less constraint

↓

Relevance Realization filters based on learned patterns and current context

↓

System 2 (Classical) evaluates and implements

Fig. 6.2 The processing structure for how system 1, relevance realization, and system 2 process together for consciousness levels 7+

This creates a dynamic tension as System 1 pushes toward maximum possibility (divergent), while System 2 pulls toward actionable decision (convergent), as Relevance Realization mediates between the infinite and the practical. For most people however, through this recursive process of identity construction, ego has built rigid identity structures which constrain what search spaces the quantum computation even explores at all, meaning that Vervaeke's heuristic model is precisely accurate for the vast majority of human cognition.

Vervaeke and Ferraro's insight was that they recognized that relevance realization must be fundamental to cognition and proposed a sophisticated neurodynamical mechanism for how it operates. Their description perfectly captures how cognition works for consciousness levels 1–6, where ego-based heuristics constrain processing upfront. Their account demonstrates that "relevance realization has to be internal, sub-semantic, sub-syntactic, and scale invariant in its operations" (p. 7). The self-organized criticality they describe provides the collapse and reorganization dynamics, while small-world networks enable rapid reconfiguration between local and global processing modes. The framework builds upon this foundation by proposing that quantum computation provides the substrate that makes Vervaeke's mechanism effective, while acknowledging that for most people, ego constraints limit which quantum possibilities get explored. Consider their observation that in problem solving, "the number of alternative pathways to checkmate is thirty to the power of sixty or 4.239×10^{88}" while "the number of atoms in the universe is 10^{82}" (p. 2). For most individuals, ego beliefs and identities eliminate vast portions of this space before quantum exploration begins. For higher consciousness levels, quantum computation can explore broader geometric spaces simultaneously. The pre-firing electromagnetic patterns selecting which quantum states merit neural execution could represent the physical implementation of both ego constraints and the criticality thresholds Vervaeke describes. When quantum coherence builds to a critical point, the system reorganizes through selective collapse, creating the dynamic salience landscaping they identify. The thousand-fold speed difference between electromagnetic branch selection and neural firing creates the temporal window necessary for this process.

Vervaeke and Ferraro correctly note that "we do not need a theory of relevance; we only need a theory of relevance realization" (p. 8). They provide the reorganization mechanism through self-organized criticality and small-world networks that accurately describes normal human cognition. Quantum biology provides the missing piece by explaining how the brain accesses the information space that self-organized criticality then organizes. For most people, ego determines what to explore before exploring it. For higher consciousness levels, the brain explores everything through quantum superposition, then Vervaeke's RR mechanism operates on these results to create dynamic, context-sensitive relevance determination.

Therefore relevance realization acts through filter and search, metaphorically like a search engine, that the brain uses to figure out which quantum results system 2 and the ego should receive. Busemeyer et al. (2020) highlight this saying,

> "... based on first-principles, the Markov models predict that earlier responses do not impact the evidence accumulation process that helps determine later responses. Yet all three experiments show an interference effect such that earlier responses impact later ones. Generally speaking, the interference effect appears to be stronger when the first response is a binary one such as choice. One way to interpret this difference in empirical results is that using a binary decision for the first measurement may be more effective for producing a "collapse" as compared to making a probabilistic judgment" (p. 10).

They are describing this recursive process on a microscale. Once a binary conclusion is made, it affects all downstream interpretations of evidence that would have otherwise affected the original conclusion. The binary conclusions, in this framework are beliefs and identities such as "I am __" statements. This is experimental evidence of the heuristic filters of relevance realization in action. The collaboration between systems is seemingly essential, because without System 1, there would be no creative insights, without System 2, there would be no stable identity, decisions, or communication, and without RR, there would be either infinite paralysis or random action that wastes energy. Which is why the framework proposes a three-layer architecture.

Layer 1 - Quantum Computation (System 1)

System 1 continuously processes through parallel mechanisms which are enhanced by quantum computation. It explores multiple patterns, connections, and solutions simultaneously. Unlike classical computation that explores possibilities sequentially, System 1's quantum-enhanced processing explores many patterns simultaneously in superposition, and does so probabilistically, which creates a vastly richer result space for RR to search. For most individuals (levels 1–6), ego has constructed identities and beliefs which act as constraints that limit which geometric informational spaces get explored from the beginning. For higher consciousness levels, this process operates near peak capacity when there is minimal conscious control, identity, beliefs, or ego interference which allows for access to quantum computational capabilities that enhance classical neural processing. The combinatorial explosion Vervaeke identifies is then either managed upfront through ego constraints (levels 1–6) or occurs in the massive results space that quantum-enhanced computation generates (levels 7+).

Layer 2 - Relevance Realization

This pre-conscious mechanism operates differently across consciousness levels. For most people, it filters through quantum results that have already been constrained by ego beliefs and assumptions. For higher consciousness levels, it searches through the massive quantum results space that System 1 has computed with fewer constraints, solving the combinatorial explosion problem Vervaeke identified by determining what has already been found (tractable) and doesn't have to determine what to explore (impossible). It operates through evolutionary filters that identify survival-relevant patterns, emotional weighting where current state influences filter priorities, learned patterns where expertise shapes what's considered relevant, belief structures where assumptions determine what's considered possible, and intentional priming where goals influence but cannot fully control the filtering process.

Layer 3 - Classical Processing (System 2)

Conscious awareness receives only what RR allows through and then it processes these filtered results sequentially through System 2. Ego creates narratives, makes decisions, and maintains identity coherence, but cannot directly access the quantum layer and can't quickly change RR's filtering. For levels 1–6, ego also shapes what gets processed upfront. This pre-conscious operation of RR explains why insights often feel like they "arrive" instead of being constructed. Although, the ego can begin to insert its own beliefs over time into the filter process, resulting in recursive relevance realization.

Flow States and the Insight Cascade

Therefore, optimal performance emerges from balance between the two, as flexible switching enables flow states and creative problem-solving. This opponent processing explains why insights need "incubation" (System 1 processing while System 2 rests), why humanity can "sleep on it" (System 2 releases control), why overthinking kills creativity (System 2 dominance blocks System 1), and yet why pure intuition needs refinement (System 1 needs System 2's structure).

The gamma burst might represent the moment of successful handoff. This would be when System 1's quantum computation, filtered by relevance realization, successfully transfers to System 2's classical processing. Instead of one system being "better," evolution created an elegant dance between quantum field dynamics and classical actuality. This distinction between System 1 and System 2 finds support from flow research. Vervaeke et al. (2018) conclude that flow represents "neither a deliberative form of processing nor an automatic form of processing" (p. 24). Instead, it occupies a unique category, what they call "spontaneous thought" that "enhances and develops core cognitive abilities" through "self-organized processing that reconfigures brain networks" (p. 24).

This aligns well with the framework's proposal that System 1 accesses quantum computation. Flow isn't faster classical processing (automatic) or conscious reasoning (deliberative), instead, it's a fundamentally different computational paradigm that emerges when ego-mediated observation decreases. This filtering mechanism becomes more clear through Vervaeke et al. (2018) insight cascade model. They characterize flow as "an extended 'aha!'"where "each instantiation of insightful restructuring procedurally primes the agent to further engage insightfully with his or her dynamic surrounding" (p. 13). From the framework's perspective, each successful quantum computation that passes through relevance realization's criticality threshold actually reconfigures the network topology itself, as Vervaeke's self-organized criticality model predicts. The insight process they describe in the mutilated chessboard problem, where "if one notices, i.e. finds salient and makes relevant, the fact that the two corner pieces are always the same colour, they may have an insight" (Vervaeke & Ferraro, 2013, p. 5), demonstrates quantum results breaking through criticality thresholds to reorganize cognitive structure.

This cascade effect demonstrates how self-organized criticality operates on quantum computational results. The synchronous firing periods Vervaeke and Ferraro identify as "carrying out information integration by compression" (p. 12) could represent moments when quantum coherence reaches critical thresholds. The subsequent asynchronous periods that "affords the differentiation and specialized processing" (p. 12) would then represent the reorganization following quantum collapse.

The thousand-fold speed difference between electromagnetic branch selection and neural firing provides the temporal window for this criticality-based selection process. Each successful reorganization updates the network topology, making subsequent pattern recognition more effective. This represents the self-organizing, multiscale process Vervaeke described, but instead quantum computation provides the informational substrate and computation that makes such rapid, flexible relevance realization possible.

Before neuroscience identified the conditions for flow, Jung (1969) described a very similar process, but he described it in even more relevance to the topics of this book as he provided a reason for flow's utility. He proposed that consciousness could interface with a fundamental substrate, the undivided "one world" called the "unus mundus" and he called the boundary between the psyche and the fundamental substrate, the psychoid. In the framework's terms, the unus mundus parallels the Quantum Information Dimension, while the psychoid represents the biological and psychological interface, the boundary, that enables information to move, or collapse, from that substrate into conscious awareness.

Jung observed that this interface opened under specific conditions which have many similarities to flow. The first involved what he called rationally insoluble problems. He described that when the ego's usual strategies failed and reason could no longer find a solution, a surrender could occur, which would release of control of bound up energy, which in the framework's terms is lowering E. The second condition involved emotional engagement as Jung noticed that curiosity, wonder, and interest kept consciousness supple and open, while negative emotions like apathy or fear constricted it. The third involved openness to possibility where he found that genuine openness, belief the task was possible, relaxed the system.

From these conditions, a deeper mechanism began to appear, and Jung even described it in thermodynamic language by using the word "gradient" in its physical sense. He wrote,

> "Every emotional state [willful engagement in the process] produces an alteration of consciousness which Janet called abaissement du niveau mental; that is to say there is a certain narrowing of consciousness and a corresponding strengthening of the unconscious which, particularly in the case of strong affects, is noticeable even to the layman. The tone of the unconscious is heightened, thereby creating a gradient for the unconscious to flow towards the conscious" (1969c, p. 39).

This narrowing might appear to contradict his observation that the unconscious becomes stronger, yet it precisely matches what this book has outlined as the modern characteristics of flow states. Flow feels narrow because attention focuses on a single point or task, but that very narrowing deepens one's access to a broader field of awareness. As has been described so far in the book, this paradox, this deepening, is a result of a reduction in E from focusing on a task that fulfills the requirements for a flow state. As consciousness becomes focused and egoic control relaxes, access to deeper, unconscious processes increases. Jung recognized this paradox long before the term flow existed.

He called this moment of "narrowing" a favorable opportunity for the unconscious, writing,

> "Owing to the restriction of consciousness produced by the affect so long as it lasts, there is a corresponding lowering of orientation which in its turn gives the unconscious a favorable opportunity to slip into the space vacated. Thus we regularly find that unexpected or otherwise inhibited unconscious contents break through and find expression in the affect" (1969c, p. 29).

In the language of the framework, rational exhaustion initiates egoic surrender, emotional engagement supplies the internal momentum for coherence, and openness keeps the boundary permeable. Together these conditions create a low-E state in which the psyche aligns with the high-Q domain which allows for this thermodynamic gradient, in Jung's terms, to form between the unus mundus and the psyche. This process, or this boundary between them, is what Jung describes as the psychoid. Then along this gradient, information moves from the deeper substrate of non-local information into conscious expression, and this information is experienced as intuition, creativity, sudden insight, or instinct which is often expressed as, or experienced as, archetypes.

Seen this way, flow states are the measurable biological form of what Jung described phenomenologically as they represent a temporary synchrony between ego, emotion, and openness that allows consciousness to momentarily, more clearly, interface with the fundamental substrate of reality.

The Experiential Reality of Flow States

From the framework's perspective, consider what might occur when ego suppression enables System 1 quantum processing while the body remains anchored in classical time perception. This creates a fundamental mismatch between processing speeds. The brain, when entering flow, suddenly accesses computational capabilities that operate at quantum timescales, yet the physical body continues its biological rhythms unchanged.

Vervaeke et al. (2018) provide some insight into why flow states produce lasting benefits beyond momentary performance enhancement. They propose that "in the flow state, the insight-cascade process is acting to pull one into the ZPD [Zone of Proximal Development], a situation where demand is just beyond skill" (p. 23). Importantly, they identify how flow overcomes the normal limitations of implicit learning by distinguishing causal patterns from correlational noise which allows the exploration of possibilities instead of just processing actualities. As they say, "Flow marks optimal insight, an intuitive learning of complex causal patterns, supported by the open self-organizing system of cognition" (p. 23). From the framework's perspective, this helps to explain why flow states produce lasting enhancement instead of temporary performance spikes. The conscious experience transforms completely as in the flow state's increased computational speeds, each heartbeat and breath now contains what feels like extended processing time. The brain processes information at rates that dramatically exceed what the body typically experiences. This processing differential might explain the widely reported athletic experience of time dilation. When a baseball player reports that a 95-mph fastball seemed to float toward the plate, the ball's velocity hasn't changed. Instead, the framework suggests the player's brain might be processing visual information at rates magnitudes faster than it normally would. "Every seam is visible" because of the increase in processing time per visual sample.

Athletes often struggle to explain these experiences, yet their descriptions align well with what the framework would predict. They report knowing where teammates will move before the movement begins, suggesting probability calculations made conscious. They describe their body "just knowing" what to do, which might reflect System 1 quantum processing driving action directly without requiring System 2 translation.

The 100-millisecond switching capability documented by Leroy and Cheron (2020) finds new significance in this context. This might represent the speed at which consciousness can shift between classical temporal experience, where ego creates sequential moments, and quantum temporal experience, where ego recedes and parallel possibilities become accessible.

The quality of insight therefore depends on the relationship between these layers. High ego activation (E increases) influences RR to apply more rigid, narrow filters, reducing P not by limiting quantum processing but by restricting what results reach consciousness. When ego relaxes (E decreases), RR naturally widens its parameters, allowing more unexpected, creative, or novel quantum results through for conscious processing.

Empirical Evidence and Practical Applications

Neuroscience provides potential neural substrates for this filtering mechanism. Kounios and Beeman (2009) found the anterior cingulate "may be involved in the readiness to detect weakly activated, subconscious solutions and to switch attention to them when they are detected" (p. 212), which is extremely similar to the function the framework attributes to relevance realization.

This hypothesis finds more support from Kenett et al. (2020), who found that creativity emerges through analogical reasoning paradigms. They note, "The task used to operationalize a cognitive construct greatly influences the neural pattern it produces" (p. 13). This suggests that how humans engage cognitively affects what results RR deems relevant enough to pass on to consciousness. Empirical evidence for optimal ego modulation comes from Sio and Ormerod's (2009) meta-analysis of 117 incubation studies ($N = 3606$). Their findings provide support for the $P = Q/E$ mechanism, revealing that ego reduction requires more sophistication than previously understood.

First, they reinforce that stepping away from problems significantly enhances solutions ($d = 0.29$, $p < 0.001$), directly supporting the framework's prediction that reducing ego activation increases quantum processing access. However, their most revealing finding challenges simplistic approaches to ego reduction.

For linguistic problems, low cognitive load tasks produced stronger incubation effects than complete rest ($\beta = 0.54$, $p < 0.05$). This counterintuitive result sheds light on a mechanism that for most individuals (likely Levels 4–6), the ego "abhors a vacuum" (Sio & Ormerod, 2009). Given complete silence or rest, ego will rush in to fill the void with self-referential chatter, actually increasing E. Therefore it's plausible that low demand tasks like washing dishes, folding laundry, or highway driving would provide just enough occupation to keep ego from reactivating which would allow one to maintain reduced self-observation for longer periods of time.

The framework's principles find validation in combat leadership as Jocko Willink, a former SEAL Team Three commander, provides battlefield evidence of how ego modulates performance. In *Extreme Ownership* (Willink & Babin, 2015), he emphasizes that leaders must "take a step back, deconstruct the situation, analyze the strategic picture" (p. 77). From the framework's perspective, this creates distance that reduces ego activation (E), enabling broader pattern recognition (higher P).

Willink explicitly identifies ego as the problem, saying, "Ego clouds and disrupts everything: the planning process, the ability to take good advice, and the ability to accept constructive criticism" (p. 100). He is describing how high ego creates self-reinforcing perceptual loops that filter out contradicting information. These battlefield principles demonstrate how P = Q/E operates under extreme pressure. Whether it's a SEAL commander stepping back from firefights, a scientist taking a walk, or a coach visiting the pitcher's mound, creating distance from immediate ego engagement seemingly enables clearer information and better performance.

The connection between flow states and enhanced relevance realization (Vervaeke & Ferraro, 2013) aligns with this refined model as transient hypofrontality reduces ego's rigid influence on RR, allowing more diverse quantum results through the filter. The "insight experience" might be the moment when RR suddenly recognizes the relevance of a quantum result that was already computed but previously filtered out. Cook-Greuter (2021) describes these moments, saying, "Most importantly, the more regular practice of turning inward and observing one's own mental processes also often leads to the spontaneous experience of a direct mode of being in which knower and known momentarily merge, and the personal self-sense disappears. These states are variously called peak moments, flow states, or bliss experiences" (p. 101).

This understanding suggests practical interventions for everyone that instead of trying to "think harder" (increasing E) or "be more creative" (attempting to control Q), individuals can focus on practices that increase RR flexibility through exposure to diverse experiences, pattern-breaking activities, ego-relaxation techniques, and ego-integration. This is how the different levels of consciousness begin to manifest.

References

Busemeyer, J. R., Kvam, P. D., & Pleskac, T. J. (2020). Comparison of Markov versus quantum dynamical models of human decision making. *WIREs Cognitive Science, 11*(4), e1526. https://doi.org/10.1002/wcs.1526

Cook-Greuter, S. R. (2021). *Ego development: A full-spectrum theory of vertical growth and meaning making*. (Original work published 2013, edited in 2019, republished in 2021).

Jung, C. G. (1964). Approaching the unconscious. In C. G. Jung (Ed.), *Man and his symbols* (pp. 1–94). Doubleday.

Jung, C. G. (1969). Synchronicity: An acausal connecting principle (R. F. C. Hull, Trans.). In H. Read, M. Fordham, & G. Adler (Eds.), *The collected works of C. G. Jung: Vol. 8. The structure and dynamics of the psyche* (2nd ed.). Princeton University Press. (Original work published 1952).

Kahneman, D. (2011). *Thinking, fast and slow*. Farrar, Straus and Giroux.

Kenett, Y. N., Kraemer, D. J. M., Alfred, K. L., Colaizzi, G. A., Cortes, R. A., & Green, A. E. (2020). Developing a neurally informed ontology of creativity measurement. *NeuroImage, 221*, 117166. https://doi.org/10.1016/j.neuroimage.2020.117166

Kounios, J., & Beeman, M. (2009). The aha! Moment: The cognitive neuroscience of insight. *Current Directions in Psychological Science, 18*(4), 210–216. https://doi.org/10.1111/j.1467-8721.2009.01638.x

Kurian, P. (2025). Computational capacity of life in relation to the universe. *Science Advances, 11*, eadt4623. https://doi.org/10.1126/sciadv.adt4623

Leroy, A., & Cheron, G. (2020). EEG dynamics and neural generators of psychological flow during one tightrope performance. *Scientific Reports, 10*, 12449. https://doi.org/10.1038/s41598-020-69448-3

Nishiyama, A., Tanaka, S., Tuszynski, J. A., & Tsenkova, R. (2024). Holographic brain theory: Super-radiance, memory capacity and control theory. *International Journal of Molecular Sciences, 25*(4), 2399. https://doi.org/10.3390/ijms25042399

Sio, U. N., & Ormerod, T. C. (2009). Does incubation enhance problem solving? A meta-analytic review. *Psychological Bulletin, 135*(1), 94–120. https://doi.org/10.1037/a0014212

Toole, J. T., Kurian, P., & Craddock, T. J. A. (2018). Coherent energy transfer and the potential implications for consciousness. *Journal of Cognitive Science, 19*(2), 115–124. https://doi.org/10.17791/jcs.2018.19.2.115

Vervaeke, J., & Ferraro, L. (2013). Relevance, meaning and the cognitive science of wisdom. In M. Ferrari & N. M. Weststrate (Eds.), *The scientific study of personal wisdom* (pp. 21–51). Springer. https://doi.org/10.1007/978-94-007-7987-7_2

Vervaeke, J., Ferraro, L., & Herrera-Bennett, A. (2018). *Flow as spontaneous thought: Insight and implicit learning*. https://doi.org/10.1093/oxfordhb/9780190464745.013.8

Willink, J., & Babin, L. (2015). *Extreme ownership: How U.S. Navy SEALs lead and win*. St. Martin's Press.

Chapter 7
What Is Consciousness?

Having established the framework's evolutionary logic, it may be valuable to compare it with earlier attempts to account for consciousness's emergence. Sir John Eccles' dendron-psychon hypothesis (1992) offers a particularly relevant comparison, as it was among the first to suggest that quantum effects could play a role in shaping conscious processes.

Eccles, a Nobel laureate in neurophysiology, observed that subjective experience seemed to require more than classical neural mechanisms could provide. His model proposed three central ideas. First, specialized structures called dendrons, formed from bundles of roughly one hundred apical dendrites, developed within the neocortex as key sites of conscious integration. Second, non-physical entities he called psychons interacted with these dendrons to influence neural events. Third, quantum indeterminacy at synaptic vesicles created an interface through which mental intentions could bias neural firing without violating physical laws.

The framework shares several important insights with Eccles. Both identify specific neural substrates that appear essential for consciousness. Both recognize quantum indeterminacy as a critical boundary between mental and physical domains. Both view consciousness as something that developed through evolution because it conferred adaptive advantages in complex and uncertain environments. These parallels suggest that Eccles anticipated features that any successful theory of consciousness may need to address.

From the book's perspective, the framework builds on Eccles by grounding his intuitions within a fully physical account. Instead of invoking non-material psychons, the framework proposes that quantum information processing within dendritic microtubules emerges from established physical principles. With this in mind, the framework may reflect a continuation of Eccles' insight while removing the need for a separate non-physical substance to account for conscious causation.

J. Roeloffs, *The Classical Brain to the Quantum Mind*, The Frontiers Collection,
https://doi.org/10.1007/978-3-032-20068-6_7

Comparative Perspective with Eccles' Model

Eccles left the origins of psychons unresolved, proposing that they "came to exist" without offering a physical basis. The framework approaches this differently by proposing a Quantum Information Dimension as a persistent substrate that biological systems gradually evolved to harness (Table 7.1).

Eccles viewed consciousness as an outcome that emerged alongside increasing neural complexity. From the book's perspective, consciousness may reflect an adaptive response to a deeper evolutionary challenge which is navigating environments defined by infinite quantum configurations and somehow needing to shape them into stable, survival-relevant classical realities. Eccles proposed that consciousness influences neural activity through subtle quantum perturbations of vesicle release. The framework extends this by identifying evidence for quantum coherence within dendritic microtubules and linking its collapse directly to electromagnetic interference generated by the ego and Default Mode Network.

Eccles deserves recognition for advancing one of the earliest hypotheses linking quantum processes to neural dynamics at a time when such proposals were widely dismissed. The framework builds on his insights about neural substrates and quantum interfaces while introducing a more cohesive physical foundation and evolutionary explanation. In doing so, it treats Eccles' model not as incorrect but as an early step toward a more integrated account.

Table 7.1 A comparison of the book framework versus Eccles' model

Aspect	Eccles' model (1992)	The book's framework
Fundamental reality	Dualist: mind and matter exist separately	Monist: Quantum Information Dimension (QID) as underlying substrate
Origin	Psychons "came to exist" without clear explanation	QID always existed and consciousness evolved mechanisms to access it
Mechanism	Quantum effects on synaptic vesicle release	Ego and DMN act through electromagnetic collapse of quantum coherence
Purpose	Consciousness as an incidental outcome of neural evolution	Consciousness as evolution's solution for translating quantum possibility into adaptive classical outcomes
Architecture	Dendron–psychon interaction units	DMN and ego function as dynamic collapse controllers
Consciousness levels	Binary: present or absent	Graduated spectrum described by $P = Q/E$

A Formal Definition of Consciousness

Much of the confusion in consciousness debates stems from conflating two fundamentally different questions which are: "What is consciousness?" and "What is being conscious?" One describes a universal process, the other describes the subjective experience of participating in that process. The framework posits these are distinct questions that need to be answered separately, which not only clarifies the conversation, but it also helps to dissolve what Chalmers (1995) termed the "hard problem" of consciousness.

What Is Consciousness? (The Process)

The book proposes that consciousness is the process through which localized systems collapse quantum field states (from the Quantum Information Dimension) into classical realities through observation-induced decoherence. This process requires no human mind or even nervous system. Like photosynthesis or digestion, consciousness names a functional process transforming one state into another.

Four necessary components enable this process:

1. An observation mechanism capable of measurement
2. Quantum infrastructure maintaining superposition states
3. Interface capabilities between quantum and classical domains
4. Memory systems preserving continuity across collapse events.

When these components interact, they can participate in the gradient collapse process as quantum field states become classical actuality. This occurs throughout nature in varying degrees, from minimal decoherence in simple systems to much more sophisticated selection in complex neural networks. In this regard, consciousness seems to also exist on the spectrum, but this wouldn't explain how human consciousness seems to be unique in a wide variety of ways. This is why the second definition is absolutely needed.

What Is Being Conscious? (The Experience)

Now, the book proposes that the experience of being conscious transcends waking awareness. It proposes the experience of being conscious represents the continuous, self-referential process that integrates perception, memory, meaning, and identity into coherent linear experience. The core of this process emerges through the brain's Default Mode Network which organizes our autobiographical narrative while we engage the world through what Vervaeke identifies as the four ways of knowing which are propositional, procedural, perspectival, and participatory (Vervaeke & Ferraro, 2013).

The neurobiological substrate for this integration has been partially mapped. Carhart-Harris et al. (2014) propose that "regular MTL activity is a crucial driver of the DMN" (p. 15), suggesting that the medial temporal lobes, which process memory, perception, and meaning-making, provide the raw material that the DMN synthesizes into self-referential experience. They further observe that psychedelics work by "relinquishing the ego's usual hold on reality (DMN control on MTL activity)" (p. 9). During such primary states, they build the case that "coupling between the MTLs and the cortical regions of the DMN is necessary for the maintenance of adult normal waking consciousness" (p. 8), and when this coupling breaks down, consciousness fundamentally transforms.

The framework presented here extends their hypothesis by suggesting that MTL structures, which are rich in quantum-coherent microtubules, generate quantum computational output that the DMN then organizes into classical narrative through electromagnetic field-mediated collapse. As Chap. 2 described, recent discoveries of biological superradiance in tryptophan networks (Babcock et al., 2024) provide the mechanism for maintaining quantum coherence in these microtubule-rich regions. Building on Nishiyama et al.'s (2024b) holographic brain theory, this book's framework proposes that memories exist as quantum holograms in microtubule networks, retrieved through superradiant resonance instead of classical synaptic activation. This quantum memory architecture enables the instantaneous pattern matching and associative leaps characteristic of human cognition. As Carhart-Harris et al. note, "it is through the development of self-organized activity in the DMN [and concomitant entropy/uncertainty/disorder minimization] that a coherent sense of self or 'ego' emerges" (p. 9).

D'Ariano and Faggin (2022) have argued that conscious experiences are inherently private due to the quantum no-cloning theorem, which prohibits creating an exact copy of a pure quantum state without disturbing it. From this perspective, your current conscious moment cannot be perfectly replicated elsewhere without fundamentally altering the original quantum configuration. While this framework agrees on the impossibility of duplicating an entire conscious state, it diverges on the implications. By treating consciousness as an evolved interface to the Quantum Information Dimension, the framework allows for structurally similar experiences to arise across individuals when their neural systems interact with overlapping informational inputs.

Although the exact quantum state of a conscious moment cannot be cloned, shared informational environments can drive observers' collapse operators toward similar regions of the quantum information landscape. This predicts a form of resonant qualia, where multiple people immersed in similar photons, electromagnetic environments, or semantic contexts may experience analogous patterns of perception and meaning-making through the MTL-DMN interface operating on comparable quantum computational substrates.

This possibility underlies the framework's bystander effect hypothesis described in Chap. 2. When collapse-selection pressures are aligned through collective rituals, synchronized learning, or shared attention to similar stimuli, individuals may enter partially overlapping states of consciousness without violating quantum nonclonability. The resonant qualia emerge from independent neural systems accessing similar regions of the Quantum Information Dimension under comparable collapse conditions.

This layered view of consciousness finds support from depth psychology as Carl Jung, working from clinical observation decades before quantum biology emerged, arrived at a similar architectural understanding of psyche and consciousness. Jung (1963) wrote about this in his autobiography only a few years before his death, saying,

> Consciousness is phylogenetically and ontogenetically a secondary phenomenon. It is time this obvious fact were grasped at last. Just as the body has an anatomical prehistory of millions of years, so also does the psychic system. And just as the human body to-day represents in each of its parts the result of this evolution, and everywhere still shows traces of its earlier stages—so the same may be said of the psyche. Consciousness began its evolution from an animal-like state which seems to us unconscious, and the same process of differentiation is repeated in every child. The psyche of the child in its preconscious state is anything but a tabula rasa: it is already preformed in a recognizably individual way, and is moreover equipped with all specifically human instincts as well as with the a priori foundations of the higher functions. On this complicated base, the ego arises. Throughout life the ego is sustained by this base. When the base does not function, stasis ensues and then death. Its life and its reality are of vital importance. Compared to it, even the external world is secondary, for what does the world matter if the endogenous impulse to grasp it and manipulate it is lacking? In the long run no conscious will can ever replace the life instinct. This instinct comes to us from within, as a compulsion or will or command, and if—as has more or less been done from time immemorial—we give it the name of a personal daimon we are at least aptly expressing the psychological situation. And if, by employing the concept of the archetype, we attempt to define a little more closely the point at which the daimon grips us, we have not abolished anything, only approached closer to the source of life. (pp. 320–321)

Here, Jung observes that consciousness represents a "secondary phenomenon" that arises from deeper layers which aligns very well with the book's proposal. His emphasis that ego remains "sustained by this base" throughout life and that "when the base does not function, stasis ensues and then death" aligns with the framework's proposal that consciousness is emergent and requires connection to the underlying biological system. He doesn't pose consciousness as ontologically fundamental, but does pose it as experientially fundamental when stating, "for what does the world matter if the endogenous impulse to grasp it and manipulate it is lacking?" Human meaning-making is the center of the experience as it participates in the physical universe.

Gazzaniga (2011) arguing from a neuroscience perspective, describes exactly this, that consciousness is emergent, saying,

> The view in neuroscience today is that consciousness does not constitute a single, generalized process. It is becoming increasingly clear that consciousness involves a multitude of widely distributed specialized systems and disunited processes, the products of which are integrated in a dynamic manner by the interpreter module [the conscious narrator in our brains]. Consciousness is an emergent property. From moment to moment, different modules or systems compete for attention and the winner emerges as the neural system underlying that moment's conscious experience. Our conscious experience is assembled on the fly. (p. 102)

Neuroscience, decades after Jung originally posited these ideas, came to the same evidential conclusion. This process described above creates a recursive loop that operates in five integrated stages.

First, quantum processes in the MTL enable computation across vast possibility space as sensory input triggers quantum search through memory networks, with all related experiences entering superposition simultaneously.

Second, relevance filtering takes place through pre-conscious mechanisms, and is shaped by previous collapses which are stored as memory patterns. This filtering represents Vervaeke's relevance realization in action.

Third, these filtered possibilities drive DMN activity, which creates the self-referential narrative experienced as identity by selecting which quantum field configurations become conscious thoughts. Only possibilities consistent with stored autobiographical patterns pass this identity check.

Fourth, the DMN exerts control back onto the quantum processes in the MTL through alpha oscillations and electromagnetic fields, which collapse the quantum states and constrain future processing. This collapse represents the moment quantum field states become classical thought as it is the measurement that creates discrete experience from continuous superposition.

Fifth, each collapse updates one's memory which modifies future quantum searches and relevance filters. This creates learning, where each experience shapes all future experiences.

Importantly, this process would operate through recursive relevance realization. As Vervaeke and Ferraro (2013) demonstrate, consciousness must somehow navigate combinatorially explosive problem spaces, determining moment by moment which patterns deserve attention and which possibilities merit collapse into actuality. The quantum memory hypothesis offers a mechanism for this navigation through quantum parallel search. Utilizing this, the brain explores all possibilities simultaneously through quantum computation, which will be explored in later chapters. This recursive process, where each selection shapes future selection criteria, creates the self-reinforcing nature of the human experience as our biases become our reality tunnels and our attention becomes our world.

Why Consciousness Feels Like Something at All

This brings it to what David Chalmers (1995) termed the "hard problem of consciousness." Why should any physical process feel like something? Why isn't all this quantum processing happening "in the dark," producing behavioral outcomes without subjective experience?

Traditional 4E cognition approaches have struggled to provide mechanistic answers to this question. The 4E paradigm argues that cognition is embodied, embedded, enacted, and extended through extra-cranial processes and structures (Carney, 2020). As Carney (2020) notes in his critical review, 4E theorists face a dilemma. He writes, "either eschew the apparatus of mathematical formalism and retreat into descriptive impotence or embrace it and inherit a computational model of cognition; tertium non datur" (p. 7). The framework presented here resolves this dilemma by providing both mathematical formalization through the P = Q/E mechanism, Landauer bounds, and h-quantization AND the embodied substrate through quantum coherence in microtubules demonstrated in Chap. 2. The framework specifies computation as a quantum-classical hybrid process modulated through ego-mediated collapse.

Vervaeke and Ferraro (2013) identify four types of knowing that constitute human cognition, each representing a different dimension of how consciousness engages with reality. Propositional knowing (knowing that) describes classical facts about collapsed states. When you know "the tree is green," System 2 processes already-collapsed quantum states into linguistic, communicable knowledge. Procedural knowing (knowing how) describes skills and patterns. When you know "how to navigate around obstacles," Relevance Realization operates pre-consciously, filtering quantum computational results based on learned patterns and evolutionary programming. Perspectival knowing (knowing from) describes your unique viewpoint. When you experience "what this means from my position," your particular Q/E configuration at any moment shapes perception, influenced by active Programs, current E value, and accumulated memory patterns. Participatory knowing (knowing with) describes direct engagement with reality through co-creating interactions. This fourth dimension distinguishes human consciousness and may help explain why consciousness has phenomenal character, because you're actively participating in its construction through quantum collapse. As Carney (2020) observes, "action-based theories suggest that the environment is disclosed through active probing as opposed to the passive registration and subsequent synthesis of sense data" (pp. 2–3). The framework extends this to the quantum foundation as it poses that consciousness actively participates in selecting which quantum possibilities become actualized. Then this, sense of meaning and feeling emerges from this participatory process which happens through the 4E's as consciousness operates as embodied, embedded, enacted, and extended.

Enacted cognition means that the degree of quantum processing becomes enacted through ego modulation, the E in P = Q/E, where consciousness operates as active modulation of which quantum possibilities collapse into experience. Through DMN activity, alpha oscillations, and electromagnetic field effects, you continuously adjust the Q/E ratio, which makes experience an ongoing achievement. This addresses Carney's (2020) critique that 4E approaches fail to specify where inferential activity occurs. As he writes, "It does not matter how, when, or where the inferential activity occurs; if it is a necessary concomitant of a cognitive process, then any worthwhile approach to cognition is obliged to account for it" (p. 4). The inferential activity occurs through the quantum to classical transformation process itself, modulated by ego.

Extended cognition means that when ego boundaries dissolve (E approaching zero), consciousness extends beyond isolated neural processes through quantum correlations. The MTL-DMN loop doesn't just process isolated data. It participates in a larger informational field accessible through the QID. Radin et al.'s (2012) findings that meditators influence quantum systems through intention provide preliminary evidence for this extension. The framework's proposal that consciousness can access the QID directly when E is sufficiently reduced would explain both mystical experiences of unity and the documented cases of apparently non-local information access described earlier. Meling (2022) demonstrates that ego-dissolution shifts knowing from dualistic (where the DMN continuously makes narrative sense of experience) to non-dual (where sense-making ceases and consciousness participates directly with uncollapsed reality). This will be described in detail in future chapters. Flow states represent the everyday version of this process, which is accessible without contemplative training.

The "feeling" operates as what it feels like to be the participatory, transitory process itself which is embodied in quantum-coherent microtubules, embedded in environmental quantum fields, enacted through ego-modulated choice, and extended through low ego states that induce more direct interfacing with the QID. You feel something because you are doing something at all times. This would also explain why experience is constantly changing and feels perpetually "now."

Convergence with Psychology

Before further examining how the framework aligns with modern neuroscience and biology, it is worth first considering Carl Jung's pioneering effort to integrate quantum physics with a theory of consciousness, which is a project he pursued as quantum mechanics itself was being born. Jung's early interest in these ideas began with conversations with Albert Einstein in Zürich during the 1910s about relativity and the nature of space and time (Hürter, 2023; Jung, 1963). These discussions led him to question whether the psyche might operate outside of classical, Newtonian assumptions.

That curiosity later deepened into a decades-long collaboration with Wolfgang Pauli, the Nobel Prize winning physicist and one of the founders of quantum mechanics. What began in the 1930s as a therapeutic relationship evolved into one of the twentieth century's most significant cross-disciplinary partnerships, as Jung and Pauli exchanged hundreds of letters that explored the intersection of psychology and quantum physics (Hürter, 2023). Their connection, which lasted until Pauli's death in 1958, unfolded during quantum theory's most transformative era. Jung was learning about wave function collapse as von Neumann was formalizing it in the 1930s, discussing non-locality as it emerged, and was developing his concept of synchronicity while Pauli advanced quantum field theory through the 1940s (Hürter, 2023).

In 1952, Jung and Pauli jointly published complementary essays which explicitly proposed links between quantum phenomena and psychological archetypes. Jung's later work on the unus mundus in the mid 1950s until his death represented the culmination of this multi-decade dialogue between depth psychology and physics.

Through this collaboration, Jung became deeply informed about quantum theory as it was being developed and sought to explore and explain the collective unconscious and archetypes as expressions of the mind's evolved interface with physics and quantum mechanics (Hürter, 2023; Jung, 1963). This is likely why his definitions of those terms changed over time, as both fields, psychology and quantum mechanics, were themselves significantly developing. He seemingly further differentiated the terms and created new ones over time in accordance as will be explored in the coming pages. This section revisits Jung's proposals and attempts to translate them from his psychological and alchemical language into modern quantum informational terms, to show how this book's framework, as it was written without knowledge of Jung's ideas and work on these topics, unintentionally formalized insights that Jung and Pauli seemingly anticipated decades before the necessary scientific and mathematical tools existed to fully articulate them.

Jung (1969b) first distinguished between the layers of the unconscious by describing a deeper stratum of the psyche, which he first described as "the collective unconscious" saying,

> …this personal unconscious rests upon a deeper layer, which does not derive from personal experience and is not a personal acquisition but is inborn. This deeper layer I call the collective unconscious. I have chosen the term "collective" because this part of the unconscious is not individual but universal; in contrast to the personal psyche, it has contents and modes of behaviour that are more or less the same everywhere and in all individuals. It is, in other words, identical in all men and thus constitutes a common psychic substrate of a suprapersonal nature which is present in every one of us. (p. 20)

The contents of this collective unconscious, he argued, "have never been in consciousness, and therefore have never been individually acquired, but owe their existence exclusively to heredity" (p. 56). He called these contents archetypes, which he defined as "definite forms in the psyche which seem to be present always and everywhere" (p. 56).

Throughout his career, Jung appeared to maintain two perspectives on this substrate that he began to better clarify near the end of his life but never fully reconciled, as he would present different definitions in different context. One definition was more scientific and conservative while the other was more metaphysical. Scientifically, he framed it in some of his work as what manifests from inherited biological structures, writing that "the collective unconscious does not develop individually but is inherited. It consists of pre-existent forms, the archetypes, which can only become conscious secondarily and which give definite form to certain psychic contents" (1969d, p. 57). He would often reinforce that archetypes manifest as "instincts." But his clinical observations and personal experiences led him toward descriptions that suggested this collective unconscious was more than something genetically inherited or encoded. He wrote in the final chapter he would write in his life, "They [archetypes] are without known origin; and they reproduce themselves in any time or in any part of the world—even where transmission by direct descent or 'cross fertilization' through migration must be ruled out" (Jung, 1964, p. 69). Could it be possible that Jung was describing two similar, yet separate phenomena? The book posits it is possible that Jung was wrestling with the existence of the Quantum Information Dimension (QID) as well as observing the ways in which humans interface with this substrate, which would manifest in archetypes, as the language of the substrate is geometric and non-linear in nature. This would require patterns to be the premier way in which humans understand its content, and the part of the self that interfaces with the QID would have to interpret this information as symbolic, pattern-based, and broadly non-linguistic. Jung actually directly acknowledges this, as he writes,

> Dreams do not guard sleep from what Freud called the 'incompatible wish.' What he called 'disguise' is actually the shape all impulses naturally take in the unconscious. Thus, a dream cannot produce a definite thought. If it begins to do so, it ceases to be a dream because it crosses the threshold of consciousness. That is why dreams seem to skip the very points that are most important to the conscious mind, and seem rather to manifest the 'fringe of consciousness,' like the faint gleam of stars during a total eclipse of the sun. (Jung, 1964, p. 64)

Jung is describing in detail how the language of the "unconscious" (which would compare to this book's ideas of system 1's quantum enhanced processing) isn't linguistic in nature. This is exactly why he says "dream cannot produce a definite thought. If it begins to do so, it ceases to be a dream because it crosses the threshold of consciousness." As consciousness (the waking, ego-dominated awareness most experience) speaks and only understands binary, collapsed, classical language, what Jung describes as "definite thought". Jung's unconscious, which seemingly parallels this book's interpretation of the quantum-enabled system 1, speaks in geometry, symbolism, images, faint correlation, and pattern recognition. Jung explicitly notices how "definite" or perceived "high certainty" only exists in traditional, waking, ego-dominated consciousness. This is important because Jung is outlying the differences he is wrestling with. There is this necessary "archetypal" language of the unconscious, but he also acknowledges something that is much more fundamental in this archetypal process. This is his second, more encompassing definition of archetypes he uses in his work which then later into his career begins to better differentiate with his definitions of the "collective unconscious," "psychoid," and "unus mundus."

Jung, when describing what individuals encounter during deep psychological work, described what lies beyond after one effectively confronts and integrates one's shadow. He states that the shadow is the "door" to the underlying substrate, saying,

> For what comes after the door is, surprisingly enough, a boundless expanse full of unprecedented uncertainty, with apparently no inside and no outside, no above and no below, no here and no there, no mine and no thine, no good and no bad. It is the world of water, where all life floats in suspension; where the realm of the sympathetic system, the soul of everything living, begins; where I am indivisibly this and that; where I experience the other in myself and the other-than-myself experiences me. No, the collective unconscious is anything but an incapsulated personal system; it is sheer objectivity, as wide as the world and open to all the world. There I am the object of every subject, in complete reversal of my ordinary consciousness, where I am always the subject that has an object. There I am utterly one with the world, so much a part of it that I forget all too easily who I really am. (1969d, p. 37)

This phenomenological description of a space with "no inside and no outside," a realm "where all life floats in suspension," and an experience of being "indivisibly this and that" parallels this framework's description of the Quantum Information Dimension which posits the QID as a substrate of probabilistic, non-binary, superposed states where paradoxes can simply reside that the brain interfaces with. Jung described encountering this domain where the usual distinctions of classical experience dissolve and are experienced as oneness or paradoxes. Is it possible that Jung was still wrestling with the difference between the QID and how humanity collectively experiences and interprets the quantum field at large? This collective aspect would represent the collective unconscious as it participates in, and experiences, the QID.

Before Jung clarified the definitions and hierarchy of his terms, he explored how consciousness interacts with space and time *in Synchronicity: An Acausal Connecting Principle* (1969c). This was the work that arose from his long-time collaboration with Nobel Prize-winning physicist Wolfgang Pauli. In it, Jung described space and time as "elastic," and suggested that under certain psychic conditions their rigidity appears to dissolve. He observed that in Rhine's telepathy experiments the subject's correct answers could not be explained by causal energy transfer but seemed instead to arise from "pure imagination." This is what he called a meaningful coincidence or "falling together in time." For Jung, this simultaneity implied that psyche and matter might be linked through a shared informational structure or a fundamental informational substrate of reality that enabled non-local informational access. He wrote that,

> In man's original view of the world, as we find it among primitives, space and time have a very precarious existence. They become 'fixed' concepts only in the course of his mental development, thanks largely to the introduction of measurement. In themselves, space and time consist of nothing. They are hypostatized concepts born of the discriminating activity of the conscious mind, and they form the indispensable co-ordinates for describing the behavior of bodies in motion. (Jung, 1969a, p. 28)

It's again critical to remember that this work came from years of Jung learning about quantum mechanics from Wolfgang Pauli. Considering Jung's objective to unite the psychological and physical, he's attempting to use archetypal ideas to describe how the fundamental ideas of quantum physics may fit into psychology. He is seemingly describing the measurement problem in quantum mechanics using psychological language. When he says space and time "become 'fixed'... thanks largely to the introduction of measurement," he is seemingly articulating how quantum possibilities collapse into definite classical outcomes through the act of measurement. His observation that space and time are "hypostatized concepts born of the discriminating activity of the conscious mind" seemingly parallels the framework's descriptions in previous chapters of consciousness' participatory role in collapse through the ego. This implies in this framework's terms that time and space become fixed, or manifest, only when consciousness measures, defines, differentiates, and stores these experiences as memory. Therefore, before this act of differentiation, one's perceived, felt experience of space and time exists in a more fluid, pre-causal condition which is a state Jung identified with the "psychoid" state, or process, of the self.

This led to Jung's writing in the *Mysterium Coniunctionis* (1963a) about the fundamental substrate of the "unus mundus." He wrote about it, saying,

> Undoubtedly the idea of the unus mundus is founded on the assumption that the multiplicity of the empirical world rests on an underlying unity, and that not two or more fundamentally

> different worlds exist side by side or are mingled with one another. Rather, everything divided and different belongs to one and the same world, which is not the world of sense but a postulate whose probability is vouched for by the fact that until now no one has been able to discover a world in which the known laws of nature are invalid... causal connections exist between the psyche and the body which point to their underlying unitary nature. (1963a, §767)

Jung is making an observation about the universality of physical law as he notes that physics has explored reality from the cosmic scale down to the quantum realm, and everywhere, even in the uncertainty and superposition of quantum mechanics, many of these laws still remain. This leads him to wonder something like "why would laws of math and physics remain consistent across all scales and space unless there exists a single unified substrate underlying all phenomena?" And this question leads him to the proposal of the unus mundus. He says that the "empirical world rests on an underlying unity" where "everything divided and different belongs to one and the same world." Again, considering his understanding of superposition and the particle-wave duality, it makes sense that he would propose a substrate where opposites are "one" as superposition allows paradoxical states to coexist before measurement. Interestingly, this description parallels the framework's proposal of the Quantum Information Dimension (QID), where quantum potential remains superposed and uncollapsed until measurement as Chaps. 4 and 5 described. The unus mundus can be read here as a psychological metaphor, as Jung evoked alchemical language, for a pre-differentiated informational field, which the framework formalizes as the QID. He continued to describe it saying,

> The background of our empirical world thus appears to be in fact a unus mundus. This is at least a probable hypothesis which satisfies the fundamental tenet of scientific theory: 'Explanatory principles are not to be multiplied beyond the necessary.' The transcendental psychophysical background corresponds to a 'potential world' in so far as all those conditions which determine the form of empirical phenomena are inherent in it. (1963a, §769)

Here Jung explicitly invokes Occam's Razor as he uses parsimony as his justification for this proposal of the "unus mundus," and then when he describes it as having a "transcendental psychophysical background [that] corresponds to a 'potential world,'" he is describing, in synonymous psychological and alchemical terminology, the wavefunction or the QID as a pre-collapse quantum state that contains the potentials for both physical and mental phenomena. When he says this substrate contains "all those conditions which determine the form of empirical phenomena," he is describing how the QID contains the quantum information that, upon collapse, becomes the specific spacetime events and conscious experiences we observe. Once again, considering this emerged from decades of serious collaboration between Jung and Pauli to conceive of a quantum-psychological interpretation of consciousness, when taken together, Jung's notion of the psychoid and his metaphysical idea of the unus mundus prefigure a view of consciousness as a co-creative process. It is one that bridges information and matter through meaning. In this sense, Jung was describing in detail how the self (and the psyche) are participants in the act of collapse as they localize potential meaning into concrete experience, and from this participation in collapse conscious experience emerges.

In modern terms, Jung's insights can be seen as an early articulation of a participatory, quantum-informed model of consciousness. One in which psyche and reality share a common informational substrate. In *Synchronicity: An Acausal Connecting Principle*, which again, emerged from his attempt to synthesize psychology and physics, Jung most clearly distinguishes the collective unconscious, the psychoid, and archetypes, which are later connected with the unus mundus. He outlines their relationship, writing,

> Here I will only point out that it is the decisive factors in the conscious psyche, the archetypes, which constitute the structure of the collective unconscious. The latter represents a psyche that is identical in all individuals. It cannot be directly perceived or 'represented,' in contrast to the perceptible psychic phenomena, and on account of its 'irrepresentable' nature have called it 'psychoid.' The archetypes are formal factors responsible for the organization of unconscious psychic processes: they are 'patterns of behavior'. (Jung, 1969a, p. 29)

In short, Jung describes how these ideas connect. Unus mundus is the undivided "one world" from which both psyche and matter arise. This would be the fundamental substrate. Therefore, in Jung's terms, the psychoid is the part of the self (the whole self in Jung's terms is both the conscious and the unconscious) that is in communion with the fundamental substrate (the unus mundus). In other words, the psychoid is the deepest layer of the psyche that is the contact boundary where the human system remains open to the fundamental substrate. Then archetypes are dynamic, pre-existent patterns that emerge from this unified field and are experienced by humans as symbolic images, myths, or archetypal emotional patterns often perceived as instincts or intuitions. Archetypes could be thought of as the experienced language of the fundamental substrate. Finally, the collective unconscious in Jung's words is "a psyche that is identical in all individuals." In other words, it is humanity's shared psychological participation in the fundamental substrate. This could also be interpreted as the emergent order of archetypes, outside of space and time, that continually unites human consciousness through recurring symbols, myths, and emotions across all cultures around the world.

Experiences, such as the one with his patient's suicide, happened many times throughout Jung's clinical practice. This likely encouraged Jung to use metaphysical descriptions of the collective unconscious near the end of his life, while at the same time, he remained grounded in affirming an independent physical reality exists. He explicitly rejected the notion that unperceived phenomena lack existence, writing, "In cases where neither our sense organs nor their artificial aids can attest the presence of a real object, the difficulties mount enormously, so that one feels tempted to assert that there is simply no real object present. I have never drawn this overhasty conclusion, for I have never been inclined to think that our senses were capable of perceiving all forms of being" (1963b, p. 323). He is describing, in a Kantian-esque tone, that all us humans have is our perception of physical reality and that there is no way to "objectively" view reality besides through our perception. It can be difficult to differentiate at times what is and isn't real, but this shouldn't negate the fact that something is indeed real in his view.

This philosophical grounding clarifies what Jung meant when he described consciousness giving the world "objective existence." He was not claiming consciousness creates physical reality ex nihilo, but instead he was proposing how consciousness participates in some part of the process. In describing this realization, Jung (1963) wrote,

> There the cosmic meaning of consciousness became overwhelmingly clear to me. 'What nature leaves imperfect, the art perfects,' say the alchemists. Man, I, in an invisible act of creation put the stamp of perfection on the world by giving it objective existence. This act we usually ascribe to the Creator alone, without considering that in so doing we view life as a machine calculated down to the last detail, which, along with the human psyche, runs on senselessly, obeying foreknown and predetermined rules. In such a cheerless clockwork fantasy there is no drama of man, world, and God; there is no "new day" leading to "new shores" but only the dreariness of calculated processes. My old Pueblo friend came to my mind. He thought that the raison d'etre of his pueblo had been to help their father, the sun, to cross the sky each day. I had envied him for the fullness of meaning in that belief, and had been looking about without hope for a myth of our own. Now I knew what it was, and knew even more: that man is indispensable for the completion of creation; that, in fact, he himself is the second creator of the world, who alone has given to the world its objective existence without which, unheard, unseen, silently eating, giving birth, dying, heads nodding through hundreds of millions of years, it would have gone on in the profoundest night of non-being down to its unknown end. Human consciousness created objective existence and meaning, and man found his indispensable place in the great process of being. (pp. 240–241)

At first it may sound like Jung is saying the psyche is the creator of physical reality, but the critical qualifier is "second creator" as Jung explicitly positions consciousness as participating in a creative process initiated by something else. His statement that without consciousness the world "would have gone on in the profoundest night of non-being" becomes more precise when read alongside his previous affirmations that unperceived realities still exist and wondering what the meaning of the physical world is without the "endogenous impulse". As this framework has described thus far, what consciousness provides is not existence itself, but the transformation of quantum possibilities into definite, meaningful classical outcomes. Jung's description of consciousness as putting "the stamp of perfection on the world by giving it objective existence" parallels what this framework describes as participatory collapse, where consciousness doesn't create or maintain the fundamental substrate of the universe but it does participate in the gradient collapse as it helps to select which quantum configurations become actualized, and as Jung was describing, without consciousness present, the existence of the physical world would have no perceived meaning. The DMN and ego provided a "why" for the existence of anything at all. In this sense we are all cosmic "meaning makers." For example, from this perspective, we don't create the sun, but we participate in making its existence meaningful through our directed observation and measurement. His vision reconciled the mechanical determinism he rejected, the "cheerless clockwork fantasy," with genuine participation in reality's unfolding, which is a reconciliation that requires consciousness to have real causal efficacy without being ontologically primary.

Throughout his career, Jung was often perceived to be caught between his scientific and metaphysical descriptions of these terms including the collective unconscious. It is possible he lacked a framework that could reconcile the difference in a definitive way. The proposal of the QID offers a potential resolution to this tension. If the QID exists as a pre-temporal substrate containing all possible quantum field configurations, and if biological systems evolved mechanisms to access this substrate, then Jung's dual perspective becomes much more complementary. Because what is inherited through biology would be the infrastructure that enables access, while what is accessed would be the pre-existing informational patterns that manifest as what Jung described as archetypes. This idea will be explored further in later chapters. All together, Jung's lifelong exploration of the collective unconscious, the psychoid, and the unus mundus culminated in a vision of consciousness as an emergent, participatory phenomenon that bridges inner experience and the physical world through meaning. His psychological empiricism and metaphysical speculation converge on the same insight the QID framework formalizes which is consciousness arises from, and recursively shapes through its participation, an informational substrate. Seen this way, Jung's language can be understood as phenomenological evidence of the same process that physics now seeks to describe mathematically.

The Unique Human Capacity

Most biological systems with neural networks experience subjective states. Dogs feel pain, birds experience fear, and dolphins demonstrate complex social emotions. What distinguishes human consciousness may be the capacity for recursive self-observation and modification. Animals seemingly process quantum information and experience the collapse into classical states, yet this appears to happen automatically according to evolutionary programming without the capacity for recursive observation of the process itself. A dog seemingly cannot observe its own consciousness process and decide to modulate it. It cannot step back, watch itself think, consciously adjust how it thinks, and then ascribe meaning to this transformational experience.

Humans may have developed something unique through the DMN-MTL loop described by Carhart-Harris et al. (2014). They observe that psychedelic disruption creates "increased variance in parameters defining the DMN… so there is uncertainty about one's sense of self—typically described as 'ego-disintegration'" (p. 16), they documented what happens when the normal integration process becomes unbound. During ordinary consciousness, the MTL-DMN loop creates a unique recursive process where the system continuously observes its own quantum processing, and memory retrieval accesses quantum superposition states while simultaneously constructing narrative meaning from them. This creates a feedback loop where consciousness observes quantum collapse happening in real-time within itself. The subjective feeling emerges because the system experiences its own transformation from infinite possibility to specific actuality moment by moment as temporal integration connects past, present, and future into an experienced, continuous identity with the capacity to reprogram consciousness patterns. This continuous identity persists regardless of the constant change, which echoes the ancient Ship of Theseus paradox. Just as the ship maintains its identity while every plank gets replaced over time, consciousness maintains subjective continuity while every quantum state collapses and reforms or while all beliefs and identities get slowly deconstructed and reconstructed over time. This recursive architecture potentially enables conscious modulation of the E value in P = Q/E.

When psychedelics disrupt the DMN, this recursive loop appears to break down. As the framework proposes, consciousness as a process would continue with quantum states still collapsing, yet the organizing mechanism that creates unified self-aware experience dissolves. In this state, memory would become more fluid as quantum superposition continues in microtubules without forced collapse. Meaning would proliferate as relevance filters dissolve and identity becomes uncertain without narrative control from the DMN. The carefully maintained integration of self falls apart, which may point to how the recursive loop normally constrains quantum processing into coherent subjective experience.

The recursive self-observation potentially distinguishes human consciousness from other biological systems. Most animals appear to operate within predetermined behavioral parameters, as they process information without the capacity for recursive self-modification. Humans maintain continuous access to quantum memory space while constructing narrative meaning from it, which may enable consciousness to be aware of itself as it observes its own quantum-to-classical transitions. Unlike simple information processing, this involves the system experiencing its own transformation from superposition to classical states, which may generate the qualitative feeling of being conscious.

This recursive architecture, where DMN constructs narrative meaning from quantum processing, also sheds light on a longstanding controversy in consciousness research regarding the timing of neural activity and conscious decision-making. The influential Libet et al. (1983) experiments appeared to show brain activity preceding conscious awareness by hundreds of milliseconds and even up to seconds in some cases. At the time, this surprising evidence led many to conclude that consciousness is entirely epiphenomenal. Although recent methodological experiments have actually revealed this interpretation was resting on a fundamental conceptual error. Neafsey (2021) demonstrated that the "readiness potential" used in Libet et al.'s experiments was likely an averaging artifact of random neural fluctuations instead of some sort of genuine motor preparation. He showed that when intention is measured using more sensitive "tone probe" methodology, the onset of one's conscious intention occurs much earlier than previously measured and often develops concurrently with genuine neural activity as recorded through depth electrodes.

Most importantly, Neafsey's (2021) research points to the idea that intention is not an instantaneous event. In experimental contexts it shows up as a gradual process that begins to develop where subjects already possess the general intention to make movements. The specific "move now" intention emerges progressively from this background state. This is concurrent with real neural preparation instead of lagging behind it. As he notes, the timing gap between "unconscious brain activity" and conscious intention was based on comparing an artifactual readiness potential with an ineffective measure of intention onset. When understanding consciousness through the framework's perspective, this apparent timing problem begins to disappear. The issue stems from assuming decisions must entirely rise from the DMN's narrative construction and that this what "consciousness" is. The milliseconds of neural activity that precede the ego's "I am deciding to move" story do not represent unconscious brain determination. That's the error. They represent the conscious experience operating through earlier stages of the quantum-classical processing loop before narrative interpretation occurs post-hoc. Carl Jung as earlier quoted, understood this, as he avidly described how the conscious (the ego-consciousness) and the unconscious are both critical halves of the entire self (of the entire consciousness) and both halves take part in directing the self's behavior over time.

This distinction may become more clear in later chapters when exploring levels of consciousness and how one operates at the highest levels. In brief, experienced meditators in contemplative practices exemplify the highest levels of awareness and demonstrate how consciousness can operate in pure being without DMN narrative interfering (ego). In such states, there is no meaningful delay between "awareness" and neural processing because consciousness engages directly with reality without requiring ego to interpret. In these states, the quantum-classical interface functions in real-time awareness without the artificial delays created by self-referential storytelling and sense-making.

The framework suggests that consciousness exists along a spectrum from heavily ego-mediated (where narrative construction creates apparent timing delays) to pure awareness (where consciousness and neural processing are unified). The Libet et al. experiments measured one narrow band of this spectrum and mistook the delays inherent in ego's DMN processing for evidence against conscious agency across the board.

This participatory relationship helps explain why consciousness feels so incredibly fundamental even though it is an evolved biological process. Consciousness interfaces with the Quantum Information Dimension that underlies physical reality, which allows biological systems to navigate the universal field of possibilities which precede classical outcomes. Through ego modulation, consciousness can affect which quantum configurations collapse into actual experience, which creates agency and participation in the gradient of collapse and reality construction.

The final question of this dynamic may orbit around the "sacredness" of consciousness or determining exactly "what is conscious?" Or "what type of consciousness is sacred?" There would be at least two opposing ways to answer this. If one wants to value direct QID interface over ego-mediated experience, then dogs operating with lower E would access purer quantum states than humans caught in narrative loops. A dog experiencing joy might interface more directly with the quantum field than a human endlessly analyzing their discontent. From this perspective, consciousness developing itself toward lower ego activation might be seen as returning to a more fundamental connection that animals never lost. This side may argue that humans just happened to win the evolutionary "winner-take-all" battle for consciousness and that there is nothing special about the human form of conscious experience.

The second way to interpret this would take the other side that the self-authorship of this participatory process with the QID does position humans uniquely. The capacity to consciously modulate our own consciousness, to choose our E value, and to observe ourselves observing, creates a form of agency that appears unprecedented in the known universe. Whether this makes humans "special," "sacred," "fallen," or "just another form of consciousness" depends on the value system applied. The framework itself remains agnostic on this philosophical and metaphysical question. The book only attempts to describe the mechanisms without declaring whether human recursive consciousness represents an evolutionary pinnacle, a departure from natural harmony with the quantum field, or something in between.

What seems clear is that humans occupy a unique position through our modulable interface. We can choose to operate with high E (maximum ego, minimum quantum access) or cultivate low E states (minimum ego, maximum quantum access). This choice itself, regardless of which option is "better," distinguishes human consciousness. The framework can accommodate both perspectives as the mechanisms remain consistent regardless of the value judgments applied to them.

References

Babcock, N. S., Montes-Cabrera, G., Oberhofer, K. E., Chergui, M., Celardo, G. L., & Kurian, P. (2024). Ultraviolet superradiance from mega-networks of tryptophan in biological architectures. *The Journal of Physical Chemistry B, 128*(17), 4035–4046. https://doi.org/10.1021/acs.jpcb.3c07936

Carhart-Harris, R. L., Leech, R., Hellyer, P. J., Shanahan, M., Feilding, A., Tagliazucchi, E., Chialvo, D. R., & Nutt, D. (2014). The entropic brain: A theory of conscious states informed by neuroimaging research with psychedelic drugs. *Frontiers in Human Neuroscience, 8*, Article 20. https://doi.org/10.3389/fnhum.2014.00020

Carney, J. (2020). Thinking avant la lettre: A review of 4E cognition. *Evolutionary Studies in Imaginative Culture, 4*(1), 77–90. https://doi.org/10.26613/esic/4.1.172

Chalmers, D. J. (1995). Facing up to the problem of consciousness. *Journal of Consciousness Studies, 2*(3), 200–219.

Eccles, J. C. (1992). Evolution of consciousness. *Proceedings of the National Academy of Sciences, 89*(16), 7320–7324. https://doi.org/10.1073/pnas.89.16.7320

Gazzaniga, M. S. (2011). *Who's in charge?: The neuroscience of decision-making, the notion of free will and the idea of a determined world.* Ecco.

Hürter, T. (2023). *Too big for a single mind: How the greatest generation of physicists uncovered the quantum world* (D. Shaw, Trans.). The Experiment.

Jung, C. G. (1963). Memories, dreams, reflections (R. Winston & C. Winston, Trans.; A. Jaffé, Ed.). Collins and Routledge & Kegan Paul. (Original work published 1961).

Jung, C. G. (1964). Approaching the unconscious. In C. G. Jung (Ed.), *Man and his symbols* (pp. 1–94). Doubleday.

Jung, C. G. (1969a). Synchronicity: An acausal connecting principle (R. F. C. Hull, Trans.). In H. Read, M. Fordham, & G. Adler (Eds.), *The collected works of C. G. Jung: Vol. 8. The structure and dynamics of the psyche* (2nd ed.). Princeton University Press. (Original work published 1952).

Jung, C. G. (1969b). *The archetypes and the collective unconscious* (R. F. C. Hull, Ed. & Trans.; 2nd ed., Vol. 9, part 1). Princeton University Press. (Original work published 1959).

Libet, B., Gleason, C. A., Wright, E. W., & Pearl, D. K. (1983). Time of conscious intention to act in relation to onset of cerebral activity (readiness-potential): The unconscious initiation of a freely voluntary act. *Brain, 106*(3), 623–642. https://doi.org/10.1093/brain/106.3.623

Meling, D. (2022). Knowing the knowing: Non-dual meditative practice from an enactive perspective. *Frontiers in Psychology, 13*, 778817. https://doi.org/10.3389/fpsyg.2022.778817

Neafsey, E. J. (2021). Conscious intention and human action: Review of the rise and fall of the readiness potential and Libet's clock. *Consciousness and Cognition, 94*, 103171. https://doi.org/10.1016/j.concog.2021.103171

Radin, D., Michel, L., Galdamez, K., Wendland, P., Rickenbach, R., & Delorme, A. (2012). Consciousness and the double-slit interference pattern: Six experiments. *Physics Essays, 25*(2), 157–171. https://doi.org/10.4006/0836-1398-25.2.157

Vervaeke, J., & Ferraro, L. (2013). Relevance, meaning and the cognitive science of wisdom. In M. Ferrari & N. M. Weststrate (Eds.), *The scientific study of personal wisdom* (pp. 21–51). Springer. https://doi.org/10.1007/978-94-007-7987-7_2

Chapter 8
The Experience of an Objective Versus Perceived Reality

If the brain maintains quantum coherence and if consciousness has a participatory experience in relation to the Quantum Information Dimension, then how does this manifest in the unfolding of objective versus perceived reality?

The $P = Q/E$ mechanism finds validation across multiple research domains, and they provide evidence that consciousness participates in reality creation while also providing evidence that consciousness is responsible for its own perception of an objective, physical reality. The following evidence from randomized controlled trials and longitudinal studies demonstrate physical, psychological, and biological phenomena that occur when consciousness finds ways to modulate its own observation patterns as it changes its beliefs and identities.

The Stress Perception Revolution

The most comprehensive evidence comes from research on how perception of stress determines biological reality. Bosshard and Gomez (2024) conducted a meta-analysis of 35 randomized controlled trials, finding that perception interventions produce measurably different performance outcomes, with effect sizes of $d \approx 0.34$ for public performance versus $d \approx 0.20$ for private cognitive tasks. The real difference points towards the idea that public performance significantly activates ego through perceived social evaluation which creates substantial E value in $P = Q/E$.

This mechanism continually shows up in research as Jamieson et al. (2010) found that telling participants stress could enhance performance improved GRE math scores both immediately and on actual tests months later. Participants showed altered sympathetic activation patterns, with salivary α-amylase levels reflecting the idea that the narrative believed about stress is what matters, and that stress is not some objective, independent, external force.

J. Roeloffs, *The Classical Brain to the Quantum Mind*, The Frontiers Collection,
https://doi.org/10.1007/978-3-032-20068-6_8

Jenkins et al. (2021) documented causal chains that challenge standard physics and biology. They found "that stress-is-enhancing mindsets lead to more approach coping in the face of stress, which in turn leads to lower levels of perceived distress, which in turn leads to better mental and physical health" (p. 17). This causal chain, validated through mediation analysis across all health measures, demonstrates information-to-biology causation that standard materialist models struggle to explain. The implications challenge fundamental assumptions about causality in biological systems as the research documents downward causation from abstract cognition to cellular function, with informational content determining physiological outcomes independent of environmental stressors. This suggests consciousness operates as an active causal agent as it influences biological states through mechanisms that bypass traditional stimulus-response pathways.

The framework provides a potential mechanism. Through $P = Q/E$, beliefs about stress or beliefs about the current environment or situation one finds themselves in determine the level of ego activation. Lower E from enhancement beliefs could help to maintain quantum coherence, which would allow access to solution states that classical processing cannot reach. If one perceives the same situation or stimuli as threat to identity this will generate a stress response in the body which would manifest as anxiety or fear which includes significant self-referential processing, which in the framework would reduce quantum coherence, collapse quantum states too quickly, and create cascading effects that end in different biological outcomes.

Consciousness Determining Biological Function

Crum et al.'s (2011) milkshake study provides evidence of consciousness potentially shaping one's experienced biological reality in real-time. Participants consumed identical 380-calorie shakes on two occasions, yet ghrelin responses differed based solely on whether the shake was labeled "indulgent" (620 calories) or "sensible" (140 calories). The ghrelin decline was steeper for the "indulgent" label, meaning satiety tracked belief instead of tracking the actual nutrients of the shake. In other words, the body's hormonal response appeared to follow consciousness's observation, or belief about the shake's effects. This phenomenon extends to exercise physiology. Crum and Langer (2007) informed hotel housekeepers that their work qualified as exercise. Without any reported behavioral change, these workers showed decreased weight, blood pressure, body fat percentage, and waist-to-hip ratio after four weeks. Changing the narrative about existing activity appeared to alter biological outcomes. The framework would interpret these findings as consciousness selecting which biological quantum possibilities collapse into classical states. When consciousness observes food as indulgent, different hormonal possibilities manifest than when believing the shake is sensible. And when movement is believed to be exercise instead of work, different metabolic outcomes also actualize. Both are strong evidence for the power of ego's observations on the biological effects of the body.

Performance Under Extreme Conditions

Smith et al. (2020) tested whether mindset effects persist under maximum stress by studying Navy SEALs candidates during training that includes "Hell Week" with only 45 min of sleep per night. Even under these extreme conditions, candidates with stress-is-enhancing mindsets showed greater persistence through training, faster obstacle course times, and fewer negative evaluations. If mindset effects were just psychological comfort, it seems likely they would disappear under physical extremity. Instead, from these studies, consciousness's interpretation of stress appears to affect performance outcomes even when the body approaches its supposed limits.

The framework would suggest that candidates who viewed stress as enhancing would maintain lower ego activation about their physical state and preserve quantum processing access while others collapse into classical survival mode. The training motto "embrace the suck" might function as an ego-reduction technology, which helps prevent excessive self-observation during the SEALs' extreme challenges. Considering this, Smith et al. (2020) state, "We find that stress mindset predicts outcomes over and above a number of other baseline characteristics, including demographics (highest education level and mother's education), fitness (Body Mass Index) and self report individual differences (social desirability and optimism for success)" (p. 7). Even though the study had a small sample size and will require more study, it provides intriguing evidence for predictions the framework would make.

Mindset and Mathematical Performance Trajectories

Dweck (2006), well known "growth mindset" research demonstrates this power of perception in educational settings. Dweck showed that believing in developmental possibility directly influences actual development as students with a growth mindset consistently outperform those with a fixed mindset, even with similar starting abilities. This meant that students who viewed intelligence as developable consistently outperformed peers who viewed it as fixed, even when controlling for prior achievement. Teaching students about neuroplasticity appeared to disrupt certainty about their own intellectual limits which created space for transformation. In a growth mindset, phrases like "I haven't learned this yet" replace phrases like "I'm not smart," which helps to move people from ego-identification to having a greater awareness of the process. From the framework's perspective, this operates mechanistically, as having a fixed mindset is correlated to having high ego rigidity (high E) while having a growth mindset embodies openness to unmeasured possibilities (lower E). This was only the beginning of researching these themes.

Blackwell et al. (2007) provided longitudinal evidence for how implicit theories of intelligence affect academic performance over time in ways that align with the framework's predictions about consciousness technologies. The study followed 373 seventh-grade students across two years, and they found that "students who endorsed a strong incremental theory of intelligence at the beginning of junior high school were outperforming those who held more of an entity theory in the key subject of mathematics, controlling for prior achievement. Moreover, their motivational patterns mediated this relation such that students with an incremental orientation had more positive motivational beliefs, which in turn were related to increasing grades" (p. 253).

The study's most significant finding may be that "diverging achievement patterns emerge only during a challenging transition. Before junior high school, students who endorsed more of an entity theory seemed to be doing fine…motivational beliefs may not have an effect until challenge is present and success is difficult" (p. 258). The researchers found that "in a supportive, less failure-prone environment such as elementary school, vulnerable students may be buffered against the consequences of a belief in fixed intelligence. However, when they encounter the challenges of middle school, these students are less equipped to surmount them" (p. 258).

This aligns with the framework's predictions, that while under low-stress conditions, even fixed mindset kids would maintain manageable ego activation, but when academic pressure increases later on, their ego will kick in and trigger the excessive self-monitoring that would collapse quantum processing when parallel computation becomes important for learning complex mathematical concepts.

Even more interestingly was what they found later on, that after teaching incremental theory to half the students through an eight-session workshop, "27% of students in the experimental group were spontaneously cited by their teacher as showing positive change, compared with only 9% of those in the control group" (p. 256). The intervention reversed declining grade trajectories within a single semester, and "the fact that promoting an incremental theory seemed to have the effect of generating increased motivation in the classroom again supports the idea that students' theory of intelligence is a key factor in their achievement motivation" (p. 258).

From the framework's perspective, what they term "entity theory" represents high ego activation around intellectual identity which creates the significant self-monitoring that would collapse quantum processing. Growth mindset reduces ego attachment to fixed intellectual identity which allows students to maintain the flexibility necessary for quantum computational access. The researchers noted that "these different patterns also emerged during adolescence, when beliefs about intelligence appear to crystallize and become more coherent…they become linked in a framework of beliefs and goals that have real consequences for achievement" (p. 258), which may reflect the formation of stable ego structures that determine long-term quantum processing access. The intervention succeeded despite substantial obstacles. As the researchers noted, "there were many factors working against finding effects of the incremental theory intervention" including timing late in the semester, cumulative skill gaps, teacher expectations, and a robust control intervention (p. 258). The positive results, despite these challenges, suggest that the robustness of consciousness technologies target ego flexibility.

Their conclusion raises an important question as they state, "Why should changing beliefs about the nature of intelligence make such a difference in students' attitudes and performance?" (p. 259). They didn't have a clear answer, but the framework suggests these may reflect "consciousness technologies" that directly modulate the brain's quantum-classical processing balance through ego activation patterns. When students learn to view intelligence as malleable instead of fixed, they begin to develop what the researchers describe as "more constructive mental models" that "can be taught, with beneficial consequences for students' achievement" (p. 258).

Medical Outcomes Through Pure Perception

Howe et al. (2019) conducted a randomized controlled trial with implications that could revolutionize medicine. Children receiving identical oral immunotherapy for peanut allergies were randomized to different symptom interpretations. Those told symptoms could signal desensitization, instead of being side effects, showed fewer symptoms as treatment progressed, less likelihood to skip doses (4% vs. 21%), and greater increases in peanut-specific IgG4 antibody levels.

The IgG4 finding appears particularly significant, as an objective immunological marker changed based on narrative alone. Identical medical treatment produced different biological adaptation based on the story ego believed about symptoms. Through $P = Q/E$, reduced ego threat about symptoms might prevent the excessive collapse that would manifest those symptoms while enabling quantum healing processes that enhance immunological adaptation.

The authors note that "fostering the mindset that symptoms can signal desensitization improves OIT experience and outcomes" (Howe et al., 2019, p. 1550). From the framework's perspective, consciousness might be directing immune system responses through perception-mediated selection of biological possibilities.

The Controllability Revelation

Jenkins et al. (2021) revealed that controllability matters more than any other stress dimension, with an effect size of 0.469 versus 0.039 for duration. They found that "participants rated stress arising from uncontrollable stressors ($M = 1.52$, $SD = 0.67$) as significantly more debilitating than stress arising from controllable stressors ($M = 2.40$, $SD = 0.67$)" (p. 11).

Significantly, they found that chronic controllable stress mindsets were most predictive of health outcomes. The framework would interpret controllable challenges as something that provides regular practice in ego modulation as it builds one's capacity for maintaining optimal $P = Q/E$ configurations. Similarly, small uncontrollable situations like waiting for test results for example were interestingly rated most stressful. But this is what is fascinating, as controllability itself is most often entirely subjective. It seems reasonable to assume that controllability is directly related to "threat detected." As something that is possible to control would allow for someone to anticipate, gauge, and monitor the threats at their doorstep. Something that is perceived as "uncontrollable" provides no ability for the ego to create narrative around it or to monitor its potential negative impact, therefore it would be perceived as a larger threat, but even then, threats are only subjective to one's identity as will be explored in later chapters.

The authors found that "correlations among the general and source-specific stress mindsets measures were very modest (rs ranging from 0.01 to 0.33)" (p. 10), which indicates that mindsets can vary dramatically based on the type of stressor. Through the framework, this suggests that the DMN creates different experienced realities for different situations based on the threat ego assigns to each stressor. The framework would predict that the degree to which the ego assigns threat levels to a stressor entirely is dependent on the beliefs and identity one holds in relation to the task at hand. Overall, this study points to the idea that stress isn't inherent or objective in any way. It's a manifestation of the ego's narratives it constructs about the environment around an individual and the situations the individual finds themselves in.

Cardiovascular Reality Selection

Interestingly, the Biopsychosocial Model research shows two distinct cardiovascular patterns that appear to exist in superposition until consciousness selects one. The challenge state produces increased cardiac output with decreased total peripheral resistance, creating efficient blood flow and enhanced performance capacity. The threat state generates the opposite pattern with decreased cardiac output and increased peripheral resistance, constraining blood flow and reducing performance readiness.

Bosshard et al. (2024) do fascinating research on this. Both Stress Arousal Reappraisal (SAR) and Stress-is-Enhancing (SIE) mindset interventions work by training consciousness to select the challenge state through different approaches. SAR interventions teach people to reinterpret their immediate physical stress response as helpful instead of harmful. The core message becomes "that racing heart and tension you're feeling right now is actually your body preparing you to perform better." Typical SAR interventions convey that "the body's response to stressors can be functional and adaptive and thus invite individuals to think of their stress arousal as helpful rather than harmful for task performance" (Bosshard et al., 2024, p. 2).

SIE mindset interventions promote a broader philosophical belief that stress in general enhances performance, health, wellbeing, learning, and growth. Instead of focusing on immediate sensations, SIE approaches use videos and materials showing how stress can improve various life outcomes. The fundamental message becomes "stress itself is beneficial for human functioning." Within the framework's perspective, these interventions represent consciousness selecting which cardiovascular possibility to collapse into classical reality. The identical stressor can manifest as either enhanced circulatory efficiency or circulatory constraint depending on how consciousness observes and interprets the situation.

The body seemingly responds to ego's believed stories about potential stressors instead of responding to any external stressor directly. When stress is reappraised as enhancing performance, vessels dilate and cardiac output increases, creating the physiological foundation for improved capability. When stress is interpreted as threatening, vessels constrict and cardiac output decreases, generating the biological substrate for impaired performance. The heart and blood vessels appear to respond to the story ego believes about the stressor instead of responding solely to some sort of objective demand of the situation itself.

The Cascade Effect

Across studies, initial perception changes appear to create expanding reality shifts. Yeager et al. (2022) found that combining growth mindset with stress-can-be-enhancing mindset produced multiplicative benefits across cortisol patterns, cardiovascular reactivity, academic outcomes, and psychological wellbeing. The framework would suggest multiple E-reduction strategies might stack, progressively increasing Q access.

The mediation pathways consistently show progression from mindset through behavior through perception to biology. If consciousness was only observing reality, these types of causal chains would likely not be expected. Instead, they could demonstrate that consciousness is selecting which perceived reality actually manifests through graduated quantum collapse. Long-term effects persist across studies, including GRE improvements months later (Jamieson, 2010), sustained health changes in housekeepers (Crum & Langer, 2007), and lasting immunological adaptation (Howe et al., 2019). Brief consciousness shifts appear to create permanent reality changes, this suggests that altered quantum collapse patterns from reduced ego states end up being self-reinforcing over time.

The Mechanistic Question

While these studies demonstrate that beliefs reliably affect biological and performance outcomes, a fundamental question remains unanswered: how does abstract cognitive content, such as a thought like "stress enhances performance" or "I am intelligent," translate into specific physiological configurations and measurable behavioral changes? Classical psychology describes the correlations but struggles to explain the underlying mechanism by which subjective interpretation becomes objective, measurable biology.

The pathway from belief to biology involves several unexplained steps. When someone adopts a growth mindset, what computational process determines which neural pathways activate? When stress is reappraised as enhancing, what mechanism selects the challenge cardiovascular pattern over the threat pattern? These questions point toward a deeper architecture underlying consciousness's interface with biological systems. If consciousness operates with quantum processes as previous chapters suggest, then beliefs and identity statements may play a more fundamental role in what the ego is and how it modulates performance. The next chapter explores how identity creates geometric boundaries in quantum computational space, which may provide the missing mechanistic link between beliefs and biological outcomes.

References

Blackwell, L. S., Trzesniewski, K. H., & Dweck, C. S. (2007). Implicit theories of intelligence predict achievement across an adolescent transition: A longitudinal study and an intervention. *Child Development, 78*(1), 246–263. https://doi.org/10.1111/j.1467-8624.2007.00995.x

Bosshard, M., & Gomez, P. (2024). Effectiveness of stress arousal reappraisal and stress-is-enhancing mindset interventions on task performance outcomes: A meta-analysis of randomized controlled trials. *Scientific Reports, 14*(1), 7923. https://doi.org/10.1038/s41598-024-58408-w

Crum, A. J., Corbin, W. R., Brownell, K. D., & Salovey, P. (2011). Mind over milkshakes: Mindsets, not just nutrients, determine ghrelin response. *Health Psychology, 30*(4), 424–429. https://doi.org/10.1037/a0023467

Crum, A. J., & Langer, E. J. (2007). Mind-set matters: Exercise and the placebo effect. *Psychological Science, 18*(2), 165–171. https://doi.org/10.1111/j.1467-9280.2007.01867.x

Dweck, C. S. (2006). *Mindset: The new psychology of success*. Random House.

Howe, L. C., Leibowitz, K. A., Perry, M. A., Bitler, J. M., Block, W., Kaptchuk, T. J., Nadeau, K. C., & Crum, A. J. (2019). Changing patient mindsets about non-life-threatening symptoms during oral immunotherapy: A randomized clinical trial. *The Journal of Allergy and Clinical Immunology: In Practice, 7*(5), 1550–1559. https://doi.org/10.1016/j.jaip.2019.01.022

Jamieson, J. P., Mendes, W. B., Blackstock, E., & Schmader, T. (2010). Turning the knots in your stomach into bows: Reappraising arousal improves performance on the GRE. *Journal of Experimental Social Psychology, 46*(1), 208–212. https://doi.org/10.1016/j.jesp.2009.08.015

Jenkins, A., Weeks, M. S., & Hard, B. M. (2021). General and specific stress mindsets: Links with college student health and academic performance. *PLoS One, 16*(9), e0256351. https://doi.org/10.1371/journal.pone.0256351

Smith, E. N., Young, M. D., & Crum, A. J. (2020). Stress, mindsets, and success in Navy SEALs special warfare training. *Frontiers in Psychology, 10*, 2962. https://doi.org/10.3389/fpsyg.2019.02962

Yeager, D. S., Bryan, C. J., Gross, J. J., Murray, J. S., Krettek Cobb, D., Santos, P. H. F., Gravelding, H., Johnson, M., & Jamieson, J. P. (2022). A synergistic mindsets intervention protects adolescents from stress. *Nature, 607*, 512–520. https://doi.org/10.1038/s41586-022-04907-7

Chapter 9
The Geometric and Computational Nature of Consciousness and Identity

Chapter 8 described consciousness operating through a three-layer architecture where ego provides upfront constraints that limit which quantum possibilities get explored for most people. If we live in a quantum world, if the brain has quantum capabilities, if there's a Quantum Information Dimension, and if consciousness operates through this three-layer system, then how would heuristics actually work at the quantum level? How would relevance realization navigate an infinite possibility space? What would determine which search space consciousness can access?

When someone says "I am logical," "I'm bad at math," "I hate school," or "I am defined by my actions," what happens computationally? If the brain has quantum processing capabilities as Chap. 2 established, and if ego modulates access as Chap. 5 demonstrated, then could these identity statements somehow shape the quantum computational workspace?

In 2017, Kvam and Pleskac published an analysis of how heuristics operate in decision-making. While they remained tentative about proposing that the brain actually performs quantum computation literally, because the evidence at the time wasn't as significant as it is today, their mathematical analysis demonstrated that classical heuristics, which are the mental shortcuts humans use to make decisions, behave mathematically identical to quantum systems. They show how the same violations of classical logic in human decision-making, from order effects to hindsight bias, naturally emerge from quantum formalism. They show that in classical computing, $X = AB$ and $X = BA$ but in psychology and cognition this isn't the case. Asking questions in different orders produce many different outcomes. This infers that the brain computes in non-linear ways akin to quantum computational models. Essentially, they describe that either thinking about the brain's computation in quantum computational language would be the most useful metaphor for understanding how it actually computes or… that the brain does indeed compute in a quantum-enabled manner as it utilizes quantum infrastructure.

The quantum biology breakthroughs since their publication which were documented in Chap. 2 make their framework incredibly relevant. To understand their

J. Roeloffs, *The Classical Brain to the Quantum Mind*, The Frontiers Collection,
https://doi.org/10.1007/978-3-032-20068-6_9

findings, it helps to know that in decision-making research, a "cue" is any piece of information that helps indicate the state of the world. Cues can be external, like recognizing a city's name or seeing storm clouds, or internal, like feeling tired or remembering past experiences. When making decisions, the brain uses these cues to infer unknown criteria, like whether a city is large or whether it will rain.

They write that classical beliefs operate as binary switches. They state, "beliefs about the value of a cue or criterion are represented as binary values, either present [1] or absent [0]. This entails the assumption that cognitive representations are characterized by discrete states, suggesting that only one set of beliefs or preferences is present at any given point in time" (p. 8). If beliefs about cues are forcedly represented as binary states, and if the brain operates through quantum mechanisms, then wouldn't each "I am" statement, belief, or identity progressively narrow the quantum possibility space? In the framework's model, identity statements themselves become cues that the brain uses to navigate reality. "I am the best basketball player here" becomes a cue that shapes how one interprets competitive situations they find themselves in. "I don't do well with numbers" becomes a cue that influences how mathematical challenges are approached and perceived.

A quantum bit can exist in continuous superposition between 0 and 1, which when they do represents uncertainty or partial beliefs as they are not forced into premature collapse. Kvam and Pleskac note that "qubits can exist in a continuous superposition state of both 0 and 1 simultaneously" and they observe, "The superposed qubits also correspond to some of our intuitions about how we represent states. Rarely do we have a definite belief about whether a particular city has greater or fewer than 1 million inhabitants, but it often seems the case that we have a feel or 'fuzzy' impression of the city being large or small … The continuous superposition state provides a formal representation that describes these sorts of vague beliefs as well as makes a number of empirically testable predictions" (p. 15).

This "fuzzy impression" would represent quantum superposition before ego-mediated collapse as the quantum system can hold multiple possibilities simultaneously without the exponential storage cost. A loosely held belief about an upcoming event like "this could be fun" maintains superposition, which allows exploration of the experience for ideal outcomes to emerge, where as a rigid belief like "this is going to be terrible" collapses a large percentage of those possibilities to zero. Kvam and Pleskac outline why this is the case, saying, "the greater precision in representations trades off with greater storage and processing demands, as more bits and larger logic gates are required to represent and operate on more precise beliefs" (p. 8). Consider what happens when identity accumulates these binary beliefs, because each identity-based cue forces certain possibilities to 1 and others to 0. Classical systems face exponential scaling. Ten binary beliefs require tracking 1024 possible states and twenty beliefs need over a million. This becomes energetically costly when considering the thermodynamic price of maintaining these beliefs. As Chap. 5 established, Landauer (1961) demonstrated that erasing one bit of information requires $k_BT \ln 2$ of energy. Each time consciousness forces collapse from fuzzy superposition into a binary certainty, it pays this thermodynamic cost and each strongly held "I am" statement that converts quantum possibilities into classical certainties generates heat and increases entropy.

Chapter 5 also noted Kurian's (2025) finding that "in practice, any computationally restricted agent will dissipate energy, because the unobservable degrees of freedom cannot be tracked to reversible precision" (p. 7). For biological systems maintaining rigid identity structures, this means constant energy expenditure. Every binary belief must be maintained against quantum uncertainty and every identity boundary must be continuously enforced through collapse.

The more strongly held the belief is correlated to the amount of energy required to maintain it against the natural tendency toward superposition. A fuzzier, more probabilistic, belief like "patience can be valuable" requires minimal energy to maintain because it preserves superposition. A rigid identity like "I am the smartest kid in the class" requires constant energetic investment to prevent quantum uncertainty from creeping back in. This could provide a physical basis for why cognitive reframing and maintaining openness to positive possibilities has measurable effects on outcomes as Chap. 8 explored. The difference may be less about "positive thinking" as typically understood and more about preserving quantum computational capacity through maintained superposition.

Heuristics as Geometric Boundaries

Vervaeke et al. (2012) explain that cognitive agents must use heuristics to navigate combinatorial explosion, yet these create a fundamental problem. As Vervaeke describes, "while heuristics are necessary for addressing combinatorial explosion, they are not sufficient for doing so. They are insufficient because they require a pre-specification of what area of the search space to check, and yet it often happens that this pre-specification does not match up to the problem at hand" (p. 3).

The framework proposes that ego-generated beliefs, values, and identities function as these heuristics within relevance realization as it narrows the results that System 2 receives from System 1 and then it retroactively narrows the geometric quantum search space through these egoic identities and beliefs as each belief creates a pre-specification of what patterns to prioritize in the quantum results space, then each value establishes which outcomes deserve attention, and each identity facet determines what information counts as "relevant to me."

This pre-specification process parallels what Kvam and Pleskac term search space constraints. Kvam and Pleskac's (2017) analysis of heuristics provides the mathematical foundation for understanding how identity creates geometric constraints as each heuristic represents a strategy for navigating possibility space. The recognition heuristic says "choose what you recognize." Take-the-best says "use cues in order of validity." These strategies work by constraining the search space, and in this context identity beliefs function as meta-heuristics. "I am logical" creates a heuristic that prioritizes analytical approaches, while a statement or belief that "I trust my intuition" creates a different search pattern, and a statement like "I'm someone who needs all the facts" explores a different search space. Each identity-based heuristic carves geometric boundaries in the quantum possibility space.

Identity Entanglement and Reality Creation

The framework has established that identity statements function as heuristics that constrain the quantum search space. Yet how do these identity-based heuristics actually interact with lived experience? The answer may lie in quantum entanglement between beliefs and the cues we encounter in daily life.

Recent work provides evidence for how binary decisions create persistent geometric constraints through forced quantum collapse as a series of studies have demonstrated that the brain's default computational mode appears to be quantum-probabilistic, with classical processing emerging following the collapse of probabilistic states. Kvam et al. (2015) challenged some assumptions of cognitive science as they write that "most cognitive and neural decision-making models—owing to their roots in classical probability theory—assume that decisions are read out of a definite state of accumulated evidence. This assumption contradicts the view held by many behavioral scientists that decisions construct rather than reveal beliefs and preferences" (p. 10645). Their quantum random walk model (QRW) treats "judgments and decisions as a constructive measurement process," and it shows that "making a decision changes subsequent distributions of confidence relative to when no decision is made" (p. 10645) In their experiments, participants who made binary left/right decisions about dot motion showed interference effects in all subsequent confidence judgments. As they write,

> An interference effect occurred when participants were asked to make a decision about the leftward or rightward motion of a stimulus. Specifically, their subsequent confidence estimates were more conservative than when no earlier decision was made, and they were consequently less overconfident. This result, along with quantitatively superior model fits, lends strong support to the modeling of choice and confidence as a quantum random walk process, a model which describes decision-making as a constructive process wherein a definite state is created from an indefinite superposition. In addition to the cognitive implications, a QRW model of evidence accumulation potentially sidesteps the problem of how a group of neurons can produce observed behavior that is consistent with a single evidence accumulation trajectory. The QRW suggests that the mismatch might lie in the cognitive representation of evidence accumulation: instead of treating evidence accumulation as a single trajectory, it may be more accurate to conceptualize it as a wavelike superposition state. In fact, populations of interacting neurons processing evidence in parallel can give rise to a quantum random walk like the one presented here. (p. 10650)

Their description has deep implications for understanding how identity beliefs create self-reinforcing patterns of experience. Kvam et al. demonstrated that when a person makes a binary decision, the act itself constructs a definite cognitive state; it's not revealing one that already existed. Before the decision, beliefs about the stimulus remain in superposition and are represented by a wavelike distribution of possible interpretations within a population of interacting neurons. Once the choice is made, this distribution collapses which narrows the available state space and introducing interference patterns that affect later judgments. The brain, in this view, does not linearly accumulate evidence along a single trajectory, instead it actually maintains many possible, parallel trajectories until measurement occurs or is required. Each decision, and each judgment, performs a constructive measurement that selects one of those trajectories as real. This wave-like accumulation of probabilistic evidence means that cognition operates through the same geometry of collapse described in physics. The interference observed in their experiments represents the informational residue of those collapses, where previous commitments continue to shape the evolution of all subsequent states. Basically, each repetition of "I am" renews the collapse and deepens the energetic groove in the system's geometric landscape until it becomes an attractor that defines experience itself.

In Kvam and Pleskac's (2017) model, cues are the information we use to make decisions about unknown states of the world. When we hold an identity belief like "I am intelligent," this becomes both a heuristic for searching possibility space and a criterion we're constantly evaluating. Every problem we encounter becomes a cue that could confirm or deny this identity and all social interaction become potential feedback about our intelligence. The identity and the experiences become quantum mechanically entangled. Kvam and Pleskac describe this mechanism saying, "once a cue belief and criterion belief are entangled, reducing uncertainty about the criterion value also reduces uncertainty about the cue value. This means that feedback about a criterion directly affects cue beliefs" (p. 23).

Now, applying this to identity, when an identity or belief like "I am a good student" (criterion belief) becomes entangled with test performance (cue), measuring either ends up forcing a specific outcomes in the other. This creates the self-reinforcing patterns the framework describes as experience that confirms identity retroactively strengthens the beliefs that created those interpretations, deepening the geometric grooves in consciousness space. These initial states, which the framework identifies as identity structures, create cascading constraints on all subsequent processing. Generally this would be described as a type of hindsight bias, and Kvam and Pleskac (2017) describe this in detail, saying,

> Cue-belief entanglement also results in an effect of criterion information on cue beliefs. Getting information about the criterion values results in a transformation or collapse of an entangled cue-criterion state, changing a person's beliefs about the cue as well… As a result, retrospective confidence following criterion information can be higher than prospective confidence before criterion information, creating a hindsight bias… This prediction of a hindsight bias is not the result of adding auxiliary assumptions to the process model like cognitive reconstruction or motivated self-presentation… Instead, the prediction arises from the first principles of the quantum logic model of heuristics developed here, which raises interesting questions as to the hindsight bias' status as a bias at all. (pp. 33–34)

If hindsight emerges from the first principles of quantum information processing, then what we call "bias" may actually be the mechanism through which identity geometrically filters the way one constructs their perception of reality. Applying this to identity by revisiting this example, when "I am a good student" (criterion belief) becomes entangled with test performance (cue), measuring either forces specific outcomes in the other. A good test score actually retroactively strengthens the belief that one "saw it coming" or "knew the material all along" or just as likely a poor score would also reshape one's memories of their preparation and the entirety of the approach leading up to the test, which can make someone believe they "knew they weren't ready" even if they would've actually said they felt confident if they were asked about it beforehand.

The entanglement means identity actually shapes what becomes real through the very act of observation. Every experience that touches an identity belief becomes entangled with it, so the person who believes "I'm unlucky in love" doesn't just notice rejection more, even though they do. Through quantum entanglement, their romantic disappointments retroactively confirm that they "knew it wouldn't work out" all along even if they wouldn't have said that beforehand.

Similarly, Hogarth (2001) identifies a fundamental limitation of implicit learning, he describes how implicit learning captures all the patterns indiscriminately and is unable to distinguish correlation from causation. This creates what we experience as either "intuition" (when correct) or "bias" (when incorrect). This limitation poses a serious challenge for any theory proposing enhanced cognitive capabilities, as it suggests that faster or more powerful pattern recognition might simply amplify existing biases. However, Vervaeke et al. (2018) identify flow states as uniquely overcoming this limitation. They demonstrate that flow creates optimal conditions for implicit learning, including clear feedback, tight action-environment coupling, and importantly, the ability to distinguish causal from correlational patterns as these conditions naturally minimize the biases that normally cause problems for intuitive pattern recognition.

This mechanism may point to why insights feel fundamentally different from regular thinking. As in this case, "normal thinking" would operate through ego-filtered RR, essentially like looking through increasingly dirty glasses, so each thought reinforces existing biases, making the view narrower and more distorted. But insight moments are when the ego temporarily drops, allowing one to metaphorically take off or clean their glasses, which mechanistically is allowing RR to access a broader range of raw quantum results. When people report that "the truth just appeared" or wonder "how did I miss this?", they would be experiencing what information feels like when it bypasses the ego-RR recursive loop.

Building on this, Busemeyer et al. (2019) found that the strength of collapse depends on the type of judgment made. They discovered that "using a binary decision for the first measurement may be more effective for 'collapsing' the wave function than using a probabilistic judgment for the first measurement, resulting in greater interference between choice and rating responses than for sequential rating responses" (pp. 6–7). This provides empirical support for the framework's distinction between rigid identity statements that force complete collapse and fuzzy beliefs that maintain partial superposition. If you are forced to collapse probability into a judgement before enough "data has been gathered," or before the probability has crystalized in the quantum computation, then you will interpret subsequent data differently, essentially creating what is traditionally understood as bias. In short, the evidence points to the idea that neurons process as if they are quantum capable, which is what Chap. 2 has made evident, and it shows that the brain's default computation is probabilistic and quantum in nature.

These empirical findings demonstrate that what is traditionally called "bias" may actually represent fundamental alterations in computational geometry created by premature collapse, because when someone forms a binary belief before sufficient evidence has accumulated, they force their naturally quantum-capable neural systems into classical processing modes. Every subsequent piece of information then gets processed through the geometric constraints created by that initial collapse and then it interprets new data to support the existing belief structure.

The implications extend beyond individual decisions to the accumulated effect of identity beliefs. If a simple left/right perceptual decision creates measurable interference in subsequent processing, then consider the cumulative impact that hundreds or even thousands of identity statements accumulated over a lifetime would have on an individual. Each "I am" declaration adds another geometric constraint which progressively narrows the quantum computational space available for consciousness to explore. The result is what Kvam and Pleskac term a "constructive process" where beliefs actively construct one's perceived reality through measurement and collapse. Through this process, identity structures fundamentally shape what becomes perceived as real through observation and measurement. Kvam and Pleskac note that in classical logic, "initial beliefs or states have proven particularly important for cognitive process models… the initial state serves as the locus for a person's internal beliefs, which change across consideration of cues and control a person's behavior" (p. 8). These initial states, which the framework identifies as identity structures and evolutionary adaptive programs, create cascading constraints on all subsequent processing as each entangled experience continues to solidify the geometric boundaries of consciousness's computational space.

The Tradeoff of Rigid Beliefs and Identity

The framework has established how identity statements create geometric constraints through quantum entanglement, but to fully understand the mechanism it may be helpful to examine where and how these constraints are biologically maintained. If beliefs function as heuristics that constrain quantum search space, and if maintaining binary beliefs against quantum uncertainty costs energy as Landauer (1961) demonstrated, then somewhere in the brain must pay the continuous thermodynamic price.

As described so far in the book, the Default Mode Network appears to serve as one of the main areas of this payment processing center as Carhart-Harris et al. (2014) identified the DMN as what is responsible for maintaining ego boundaries with alpha oscillations in the posterior cingulate cortex accounting for 66% of the variance in ego dissolution under psilocybin. From the framework's perspective, these alpha oscillations may represent the continuous energy expenditure required to prevent beliefs from drifting back into quantum superposition. When psychedelics suppress

DMN activity, this payment system appears to temporarily shut down which enables a more entropic state as described by Carhart-Harris et al. Gallimore (2015) demonstrates this effect mathematically and shows that psychedelics increase the entropy of cause-effect repertoires, which moves them closer to unconstrained distributions. From the framework's perspective, this entropy increase reflects what happens when the thermodynamic cost of boundary maintenance stops being paid, because without continuous energy expenditure that forces collapse and maintains sharp distinctions, concepts would then naturally drift toward these "fuzzy states" as described by Kvam and Pleskac (2017) which allows for concepts to begin to overlap.

Gallimore (2015) explores this further through integrated information theory analysis of psychedelic states. He provides a fascinating formalization of what occurs when this payment system shuts down as he is working from functional neuroimaging data which shows increased neural entropy under psilocybin, and with this he demonstrates mathematically how psychedelics affect the structure of conscious experience itself. He writes,

> Overall, this model suggests that the expanded awareness and enhancement of cognitive flexibility afforded by the psychedelic drugs comes at a functional cost: increasing the entropy of the cause-effect repertoires of mechanisms within a complex expands the characteristics and meaning of concepts and produces a richer, more flexible state of consciousness (a greater range of brain states can be explored), but this is at the expense of the cause-effect information associated with each concept. This provides an elegant explanation for the need to optimize neural entropy—a very low entropy (highly ordered) state would maximally constrain past and future states of the brain and thus maximize cause-effect information. However, this would collapse the number of states within the cause-effect repertoire of each mechanism and thus produce a highly inflexible state of consciousness with a rigid sequence of conscious states. At the opposite extreme, excessively high entropy would maximize the number of states within the cause-effect repertoire of each mechanism, producing a highly flexible state of consciousness with the potential for creative and novel cognition. However, the cause-effect information specified by each mechanism would be very low; the mechanisms would fail to constrain the past and future states of the brain. If entropy is optimal, cause-effect information is high whilst a degree of cognitive flexibility is retained. (Gallimore, 2015, pp. 9–10)

In this, Gallimore identifies a fundamental tradeoff that maps directly onto quantum constraints as the tension he describes between cognitive flexibility and cause-effect information closely mirrors Heisenberg's uncertainty principle as applied to consciousness. As the principle describes how maximizing the range of accessible brain states (broad exploration, low ΔQ in quantum terms) necessarily reduces the ability to constrain past and future states (definite outcomes, low ΔP). The key here is that these cannot be minimized simultaneously, there's always a tradeoff as the product of these uncertainties remains bounded below by physical law.

From the framework's perspective, this tradeoff would emerge because maintaining definite conceptual boundaries requires continuous energy expenditure. When psychedelics suppress DMN activity, the energy maintenance payment stops and then concepts that normally maintain sharp distinctions through forced collapse can drift into superposition, and the cause-effect repertoires expand, and as they do so, they are acquiring additional states, but this happens at the tradeoff of definiteness. As described in Chap. 5, evolution appears to have optimized consciousness to operate at Gallimore's "optimal entropy" point. Carhart-Harris et al. (2014) characterize normal waking consciousness as "close to criticality but slightly sub-critical" (p. 13), as it maintains enough definiteness for organized action while still able to retain flexibility for necessary adaptation.

The framework proposes that identity beliefs function as high-level priors that constrain quantum computational space through continuous energy expenditure as recent work by Carhart-Harris and Friston (2019) has explored. They write on their REBUS model, which stands for RElaxed Beliefs Under pSychedelics. While their model focuses on pharmacological interventions, the mechanisms they describe shine some light on the fundamental principles of how consciousness may maintain and transform beliefs, identity, and one's geometric constraints. They go on to describe what they term the "anarchic brain" principle as a natural complement to relaxed beliefs, as they write,

> The anarchic brain principle is a natural counterpart to relaxed beliefs under psychedelics (REBUS). This is because the relaxation of high-level priors necessarily implies a liberation of bottom-up information flow (particularly from intrinsic sources). Thus, because top-down control is a defining property of hierarchy, REBUS is commensurate with a loss of functional hierarchy in the brain… Both [anarchic brain and entropic brain hypotheses] feature as part of their definition, the principle that psychedelics enhance brain entropy (e.g., as measured by Lempel-Ziv complexity) and bottom-up information flow (e.g., measurable via effective connectivity). (p. 336)

From the framework's perspective, this "anarchic" state represents what occurs when the thermodynamic cost of maintaining geometric constraints stops being paid. Without continuous energy expenditure that maintains the rigid boundaries between beliefs, ideas, concepts, and identity, they all drift back toward superposition, then bottom-up information from lower-level systems which are ordinarily compressed away by high-level priors, can finally impress on conscious awareness.

Carhart-Harris and Friston (2019) synthesize hierarchical predictive coding with the entropic brain hypothesis to explain how beliefs function as something similar to computational constraints. They write that "an important complementary theme, inspired by the so-called algorithmic information theory account of consciousness, is that the brain's highest levels naturally envelope the content of levels below, thereby effectively suppressing their potential information content via the action of a summary model" (pp. 321–322). These summary models are termed "compressive because they compress the content of levels below, effectively causing their content to disappear or at least not be heard" (p. 322).

This compression mechanism maps directly onto what the framework proposes about identity structures. Each level of consciousness, as will be explored in coming chapters is directed by one encompassing "I am" statement, such as "I am my actions and achievements," which functions as a compressive algorithm that reduces vast computational possibility space into manageable summaries while it encodes the levels of consciousness below it. Beliefs then exist in a hierarchy and identity would exist like a "portfolio" of weighted beliefs. For example, "I am logical" may sit below "I am my actions and achievements" and it would compress countless cognitive processes into a single organizing principle. Other beliefs, such as "I'm very bad at math" would compress complex capability assessments into a binary constraint. As Carhart-Harris and Friston observe, "high-level systems and the beliefs they encode have an informationally efficient, compressing influence on more elemental information held and processed below them. The high-level systems therefore function to route or canalize thought and behavior, sometimes in an overly constraining way that restricts self-awareness and obstructs new learning" (p. 338).

As has already been explored, this "canalize" effect they describe is akin to the process of relevance realization and this is why belief structures organize in a hierarchical system as a system "restricts self-awareness and obstructs new learning" which also is related to the bias effect of entanglement that Kvam and Pleskac (2017) identified. This "compressing influence" operates through what Carhart-Harris and Friston (2019) term precision weighting. They articulate the core mechanism of their model as they write,

> The present work proposes that a principal action of classic psychedelics is to relax the precision weighting of prior beliefs encoded in the spontaneous activity of neuronal hierarchies. Statistically, precision is equivalent to inverse variance (i.e., negative entropy), and, subjectively and intuitively, it can be regarded as commensurate with felt confidence, i.e., the closer data are to a given model, the smaller will be our errors of prediction and thus, the stronger will be our felt confidence in our model. More specifically, this study proposes that the effect of this relaxation process is felt most profoundly when it occurs at the highest or deepest level of the brain's functional architecture, i.e., the levels that instantiate particularly high-level models such as those related to selfhood, identity, or ego. (p. 319)

First of all, Carhart-Harris and Friston provide support to this book view that "I am" beliefs are fundamental to the human experience and the levels of consciousness. These "high-level models such as those related to selfhood, identity, or ego" are exactly this, as the narrative of "who I am" is purely a large a collection of beliefs which all exist on a spectrum of highly maintained (strongly held) to unmaintained (loosely held). Interestingly, the mathematical relationship they describe as inverse variance, which is equivalent to negative entropy, describes a fundamental tradeoff that mirrors Heisenberg's uncertainty principle just as Gallimore's (2015) work did. The energy formulation of Heisenberg's uncertainty, $\Delta E \Delta t \geq \hbar/2$, captures this even more directly, because the more precisely a system's energy state is defined, the less precisely its rate of change can be known.

In terms of consciousness, as described by Carhart-Harris and Friston, this manifests as an inability to be both rigid and exploratory simultaneously, while high precision beliefs create low variance, low entropy states with well-defined energy configurations. But this very definiteness constrains the system's ability to change or explore alternative states. A heavily weighted belief like "I am always going to be a failure" maintains extremely high precision (inverse variance is high, entropy is low), which creates maximum constraint on computational space but prevents exploration of experiences that might contradict this belief. In this example, the system has traded flexibility for certainty. Carhart-Harris and Friston (2019) describe the consequences of relaxing precision weighting on high-level beliefs as they write,

> The effect of relaxing the precision weighting of high-level priors is to create a state in which these priors are imbued with less confidence. As just touched on, an important example of a high-level prior is the belief that one has a particular personality and set of characteristics and views. This (umbrella) belief is commensurate with the narrative self or ego. It is proposed in this work that dissolving high-level priors has implications for the functioning of the rest of the hierarchy—and indeed the integrity of the hierarchy itself. More specifically, we propose that the general (entropic) action of psychedelics is to render the brain/mind's (variational free) energy landscape flattened or opened up. It follows from hierarchical predictive coding that precise high-level priors or beliefs ordinarily have an important constraining influence on the rest of the hierarchy, canalizing lower components and inhibiting their expression and influence. (p. 319)

This description effectively highlights how identity, or what they describe as the "umbrella" beliefs associated with the "narrative self or ego", functions as the master constraint on consciousness as the narrative self operates as an umbrella belief that constrains all subordinate processing. They importantly note that "The effect of relaxing the precision weighting of high-level priors is to create a state in which these priors are imbued with less confidence" which once again emphasizes this ratio of one's certainty in their identity to one's rate of change, progress, and development.

Carhart-Harris and Friston provides a physics-based grounding to this as they write,

> … a loss of precision (confidence) in posterior beliefs corresponds to a decrease in the curvature or flattening of the free-energy landscape associated with the neuronal activity encoding those beliefs. Free-energy minimization effectively tries to find flat minima in accord with Occam's principle and Jayne's principle of maximum entropy. Flat minima lend neuronal activity a latitude for fluctuations that characteristically result in critical slowing. (p. 319)

This connects precision weighting directly to the geometric landscape metaphor the framework employs, while once again connecting how Heisenberg's uncertainty operates at the level of consciousness as the system cannot minimize both uncertainties simultaneously as tight energy minima (high precision, low ΔE) necessarily restrict temporal dynamics (high Δt), while flat minima (low precision, high ΔE) permit greater temporal variation (low Δt).

The relationship between precision (inverse variance), entropy (uncertainty), and landscape curvature provides the mathematical bridge that connects Landauer's thermodynamic costs with Gallimore's entropy analysis, Heisenberg's uncertainty constraints, and the framework's geometric perspective.

Importantly, Carhart-Harris and Friston connect what this book has already highlighted extensively, which is the ego and the DMN. They write that the neurobiological substrate of these high-level priors centers on what they term the brain's "Centre of Gravity." They write that "the DMN itself has been described as the brain's 'Centre of Gravity.' The self or ego is an important (perhaps central) high-level prior that DMN dynamics may instantiate. The DMN is a particularly influential system, sitting atop a (uniquely) deep functional hierarchy in humans" (p. 338). From the framework's perspective, this gravitational metaphor beautifully captures how the DMN creates attractor basins that pull all processing toward themselves through continuous precision weighting of high-level beliefs.

Carhart-Harris and Friston emphasize that these high-level priors exist as measurable energy patterns as they note that "in accordance with the free-energy principle (and with dual-aspect monism), one can view the dynamics of high-level intrinsic networks as the dynamic or energetic counterparts of high-level priors" (p. 338). This interestingly bridges the framework's thermodynamic analysis with neurobiological function as the Landauer costs described in Chap. 5 may physically manifest as patterns of neural activity maintained through metabolic expenditure, with alpha oscillations in the posterior cingulate cortex serving as one observable signature of this energy investment.

When someone identifies as "I am logical," this high-level prior canalizes attention toward analytical approaches while it would inhibit intuitive or emotional processing. The precision weighting on this identity belief creates steep curvature throughout the entire hierarchy, as it constrains which possibilities lower levels can explore. Therefore, relaxing precision on this umbrella belief would actually restructure the entire landscape and potentially allow suppressed processing modes to finally express themselves.

The cellular mechanism underlying precision weighting involves specific effects on deep-layer pyramidal neurons, and Carhart-Harris and Friston describe this writing,

> … activation of 5-HT2ARs induces an increase in excitatory postsynaptic currents and discharge rates in pyramidal neurons, causing an asynchronous mode of glutamate release, also known as recurrent network activity, and temporal dissociation between cortical pyramidal cell firing and the active phase of local field potentials, i.e., a spike-field decoherence. This irregular excitation of deep-layer pyramidal neurons is likely to account for the highly reliable finding of psychedelic-induced decreases in oscillatory activity (particularly in its most prominent frequency bands, such as alpha). (p. 324)

Carhart-Harris and Friston found that "using MEG, we found reduced α power in the posterior cingulate cortex under psychedelics, a major node of the DMN, and this effect correlated significantly with subjective ratings of ego dissolution under both psilocybin and LSD" (p. 324). This provides empirical support for the hypothesis that alpha oscillations may represent the continuous energy expenditure maintaining belief boundaries against quantum uncertainty.

Overall, the work from Gallimore (2015), Carhart-Harris et al. (2014), and Carhart-Harris and Friston (2019) seem to very much complement some of the framework's core proposals that ego modulates quantum access and that beliefs maintained by the ego through the DMN are fundamental to identity and the levels of consciousness. Collectively, their work nicely describes how this process operates under the fundamental constraints of Heisenberg uncertainty, Landauer thermodynamics, and metabolic budgets, with the DMN functioning as one of the main biological mechanisms that continuously pays the price to maintain identity and belief.

Support for Geometric Constraints

Recent neuroscientific research provides empirical support for how evolutionary adaptive programs modulate the P = Q/E equation under threat. Wang et al. (2022) found that "acute stress shifts the brain into a more integrated and less segregated state, especially in frontal-temporal regions" (p. 2). This integration correlates with cortisol levels [$F(1,23) = 6.797$, $P = 0.016$], providing a biochemical pathway for increasing ego activation.

From the framework's perspective, this integration represents increased quantum collapse through heightened self-referential processing. The study's finding that stress "stabilized brain states by reducing the variability of dynamic transition" (Wang et al., 2022, p. 1) aligns with the proposal that evolutionary programs geometrically constrain quantum exploration.

Recent work examining hierarchical information systems provides mathematical support for this geometric perspective. Sclocchi et al. (2025) investigated how complex systems maintain coherence through phase transitions, finding that "natural data such as images are believed to be composed of features organized in a hierarchical and combinatorial manner" (p. 1). Their key discovery points to how consciousness levels might undergo transitions while being able to preserve core features, they say, "The backward diffusion process acting after a time t is governed by a phase transition at some threshold time, where the probability of reconstructing high-level features, like the class of an image, suddenly drops. Instead, the reconstruction of low-level features, such as specific details of an image, evolves smoothly across the whole diffusion process" (p. 1). Applied to consciousness development, this suggests that when someone transitions between levels (say, from achievement-focused Level 5 to values-driven Level 6), the high-level identity structure undergoes a sharp phase transition while low-level features may still persist. Essentially, the person's fundamental capacities, learned skills, and basic patterns remain intact even as their overarching identity framework shifts. From the framework's perspective, this explains why there are two dimensions to consciousness. The Y-Axis transitions in phase shifts while the X-Axis is a slower process that must change over time, which points to why transitions between consciousness levels and adaptive programs may be more like steering a large cruise ship than sharply turning in a car. It takes time for all hierarchical identities of someone to transform even if certain characteristics transform instantly in this "phase shift." These levels of consciousness and adaptive programs will be described in detail in the coming chapters.

Sclocchi et al.'s (2025) mathematical analysis sheds more like on this mechanism, as they write, "the probability for the correct class at layer L displays a transition at a characteristic time which becomes sharper for increasing L, and the messages for the correct input variables and the correct latent variables at low levels of the tree change smoothly" (p. 6). In consciousness terms, deeper identity structures (higher L) undergo more dramatic transitions, while surface-level behaviors change gradually. The critical threshold they identify has particular relevance. Through mean-field analysis, they demonstrated that "when $sf < 1$, $p^* = 1$ and $p^* = 1/v$ are both attractive fixed points, and a new repulsive fixed point $1/v < p^* < 1$ separating the other two emerges" (p. 6). This creates "a phase transition between a phase in which the class can be recovered and a phase in which it cannot" (p. 7). In their model, s represents the branching factor (how many lower-level features compose each higher-level feature), while f equals m/v^(s^{-1}), where m is the number of distinct production rules and v is the vocabulary size. When sf falls below 1, the system maintains sufficient constraints to preserve coherent high-level structures. Above this threshold, errors cascade upward, causing high-level coherence to suddenly collapse while low-level features persist.

From the framework's perspective, this $sf < 1$ boundary might represent the critical ego activation threshold in the $P = Q/E$ equation. Below this threshold, consciousness maintains stable access to quantum processing while preserving identity coherence. Above it, rapid collapse would prevent quantum coherence and force a sharp transition to a new consciousness level. The mathematical inevitability of these phase transitions could explain why consciousness development occurs in discrete stages rather than continuous gradients. Wang et al.'s (2022) stress findings align with this model as they note that "by inhibiting state-transition variability, stressed brains may support high levels of vigilance and volitional control" (p. 3). Therefore, stress would push the system above the critical threshold and force collapse into classical processing for rapid threat response. The correlation between reduced variability and improved cognitive control indicates that geometric constraint of quantum processing, while limiting creative exploration, optimizes decision-making when survival is at stake.

The transformation from classical to quantum representation that Kvam and Pleskac describe parallels consciousness development through these geometric phase transitions as classical heuristics with binary beliefs create rigid geometric structures and quantum representations with superposition states allow expanded geometric exploration, therefore moving up consciousness levels may involve transitioning through the phase boundaries Sclocchi et al. identified, as each level expands the geometric space available for computation. Kvam and Pleskac (2017) observe that "participants seem to transition from periods of serial, declarative rules when they are only familiar with the task to parallel, procedural rules when they have become very good at it" (p. 32). This maps onto the progression from ego-constrained serial processing at lower consciousness levels to parallel quantum processing at Level 7 and beyond, with each transition potentially corresponding to crossing an $sf < 1$ threshold. The connection to consciousness levels is more clear through this mathematical lens as early levels need tight geometric constraints (low sf value) for survival, and immediate threats require narrow search spaces and rapid collapse. As safety increases, the system can then operate closer to the critical threshold, while higher levels access broader possibility spaces by maintaining sf near but below 1, which enables creativity and systems thinking that tighter constraints would prevent.

These mathematical ideas remain theoretical within the consciousness context but point toward the practical applications. If identity constrains quantum processing through geometric principles with specific phase transition thresholds, then conscious identity design might enhance cognitive capacity by optimizing the sf ratio and understanding the conditions that permit coherent information flow while approaching but not exceeding critical thresholds could inform approaches to education, therapy, and human development.

The upcoming chapters will explore how these geometric structures manifest across the nine levels of consciousness and nine evolutionary programs which creates the spectrum of human experience. Each represents a different geometric architecture, a different position relative to phase transition thresholds, and a different solution to navigating the vast quantum computational space that consciousness inhabits.

Psychological Case Study

The evidence presented in this chapter has focused on quantitative research from neuroscience, quantum cognition, and information theory. Carl Jung's extensive clinical work offers a different methodology to compare with, as it is based on dream analysis, clinical work, and phenomenological observation. Given the evidence presented so far in this book that consciousness interfaces with the QID through ego modulation, and the evidence presented in this chapter that beliefs create identity structures which act as geometric constraints on accessible possibility space, Jung's detailed mapping of how beliefs exclude alternatives and create compensatory dynamics becomes worth examining.

Considering Jung's multi-decade collaboration with Nobel Prize-winning physicist Wolfgang Pauli, Jung would have had a strong grasp on the principles of quantum mechanics. Similarly, his writings from the 1930s onward reveal a deliberate adoption of measurement and complementarity terminology. The Synchronicity work (Jung, 1969b), which was written in collaboration with Pauli in 1952, as well as many other later works in his career, explicitly utilize physics, thermodynamics, and quantum mechanical concepts such as energy, acausality, measurement, relativity, entropy, and observer participation to describe psychological phenomena. It seems to be highly plausible that Jung was intentionally framing his psychological theories as descriptions of physical law, but expressing them in the language more commonly understood by psychology at the time. It also seems plausible that Jung's approach was strategically cautious as he anchored these ideas in alchemy and symbolic psychology but because theories, let alone scientific evidence, for quantum processes in biology or the mind did not yet exist. By grounding his insights in phenomenology, Jung established an empirical foundation for ideas that would only later become physically explicable. His alchemical and psychological terminology provided a legitimate framework for a psychologist of his era to hypothesize what we could now call quantum cognition models or quantum information dynamics in consciousness.

When considering Jung's approach in this way, many of his descriptions read more like direct phenomenological accounts of processes later formalized in physics. The complementarity between conscious and unconscious, the uncertainty relation between belief and shadow, and the "fuzzy," indeterminate states of feeling that precede judgment all mirror structures recognized in quantum information theory today. Jung may have been documenting, through clinical observation, the experiential side of the same information-processing phenomena that physics, neuroscience, and cognitive science now study quantitatively. With this in mind, Jung recognized that psychology required two complementary perspectives. In his essay *On Psychic Energy* (1969a), he distinguished between what he called mechanistic and energic viewpoints, saying,

> The mechanistic view is purely causal; it conceives an event as the effect of a cause, in the sense that unchanging substances change their relations to one another according to fixed laws. The energic point of view on the other hand is in essence final; the event is traced back from effect to cause on the assumption that some kind of energy underlies the changes in phenomena, that it maintains itself as a constant throughout these changes and finally leads to entropy, a condition of general equilibrium. (p. 17)

Where Jung observed these as irreducible but complementary perspectives, the framework grounds them in different physical processes that ego modulates between. Jung recognized that directed consciousness creates stability through what he called necessary "one-sidedness," which involves a dynamic tension. He described how consciousness must balance between creating stable direction and maintaining access to excluded possibilities. This parallels what Vervaeke (2012) and Carhart-Harris et al. (2014) documented as a process of opponent processing in an entropy-cognition relationship, where optimal function requires balance between rigid order (low entropy, high E) and overwhelming chaos (high entropy, E approaching zero). Jung acknowledged the theoretical possibility of an "ideal case where all the psychic components are tending in one and the same direction" but noted this "very rarely happens" in practice (1969a, p. 100). This ideal state may correspond to higher consciousness levels where ego and quantum processing align to a significant degree as will be explored in the coming chapters.

Jung observed how consciousness creates direction through judgment, which necessarily excludes alternatives. He described the mechanism and inevitable consequences as he wrote,

> It is therefore understandable, and even necessary, that in each individual the psychic process should be as stable and definite as possible, since the exigencies of life demand it. But this involves a certain disadvantage: the quality of directedness makes for the inhibition or exclusion of all those psychic elements which appear to be, or really are, incompatible with it. But how do we know that the concurrent psychic material is "incompatible"? We know it by an act of judgment which determines the direction of the path that is chosen and desired. This judgment is partial and prejudiced, since it chooses one particular possibility at the cost of all the others. The judgment in its turn is always based on experience, i.e., on what is already known. Through such acts of judgment the directed process necessarily becomes one-sided, even though the rational judgment may appear many-sided and unprejudiced. The very rationality of the judgment may even be the worst prejudice, since we call reasonable what appears reasonable to us. What appears to us unreasonable is therefore doomed to be excluded because of its irrational character. It may really be irrational, but may equally well merely appear irrational without actually being so when seen from another standpoint. (p. 99)

Jung's observation that judgment "determines the direction of the path that is chosen and desired" and remains "always based on experience," i.e., on "what is already known" describes a recursive loop through which rigid beliefs constrain future processing. This is akin to the process outlined in Chap. 6 as relevance realization filters what reaches conscious awareness, and it also corresponds to what Kvam and Pleskac (2017) documented as hindsight bias, where, because of entanglement between cues and beliefs, prior collapses shape subsequent probability distributions as detailed so far. Each judgment creates what Jung calls "one-sidedness" by forcing beliefs toward binary certainty (near 1 or 0 states as this chapter has described), which then filters all new information through these established patterns. When someone judges certain possibilities as "incompatible" or "unreasonable," they create boundaries which prevent computational exploration of those regions. Jung's recognition that what appears unreasonable "may equally well merely appear irrational without actually being so when seen from another standpoint" suggests different consciousness configurations access different valid regions of the possibility space, this allows for different individuals to perceive entirely different "truth" from the same data, which seems to be the same bias effect that Kvam and Pleskac described. He continues in the same section, writing,

> One-sidedness is an unavoidable and necessary characteristic of the directed process, for direction implies one-sidedness. It is an advantage and a drawback at the same time. Even when no outwardly visible drawback seems to be present, there is always an equally pronounced counter-position in the unconscious, unless it happens to be the ideal case where all the psychic components are tending in one and the same direction. This possibility cannot be disputed in theory, but in practice it very rarely happens. The counter-position in the unconscious is not dangerous so long as it does not possess any high energy-value. But if the tension increases as a result of too great one-sidedness, the counter-tendency breaks through into consciousness, usually just at the moment when it is most important to maintain the conscious direction. Thus the speaker makes a slip of the tongue just when he particularly wishes not to say anything stupid. (Jung, 1969d, p. 99)

Jung's observation about the "counter-position" that develops in the unconscious parallels information conservation principles of physics. As described previously in this chapter, when beliefs force collapse into narrow patterns or collapsed classical results, the excluded possibilities cannot just disappear or be erased, they must exist somewhere. Jung describes this accumulated tension as possessing "energy-value" that builds over time. Therefore, in Jung's terms, the counter-position manifests as what Jung deems the shadow. As he is writing before many modern quantum physics principles were well established, he states,

> Everyone carries a shadow, and the less it is embodied in the individual's conscious life, the blacker and denser it is. If an inferiority is conscious, one always has a chance to correct it. Furthermore, it is constantly in contact with other interests, so that it is continually subjected to modifications. But if it is repressed and isolated from consciousness, it never gets corrected, and is liable to burst forth suddenly in a moment of unawareness. At all events, it forms an unconscious snag, thwarting our most well-meant intentions. (1969d, p. 87)

When someone defines identity with high precision through rigid beliefs, the corresponding shadow would then become what Jung describes as "repressed and isolated" which seemingly parallels what Kvam and Pleskac (2017) describe as forcing beliefs into definite binary states instead of maintaining quantum superposition. The excluded possibilities (forced to 0 states) remain in the system and are geometrically blocked but energetically active. This builds what Jung calls "energy-value" in the unconscious which manifests as the shadow as more content gets excluded.

Interestingly enough as previously hinted at, Jung's observations operate according to a principle structurally parallel to Heisenberg's uncertainty (1927) which also aligns with Gallimore (2015) and Carhart-Harris and Friston's (2019) descriptions. Considering, Heisenberg's uncertainty principle, Jung observed the psychological equivalent of this as the more precisely one defines beliefs (Δbelief $\rightarrow$ 0, creating well-defined identity states), the more unpredictable the shadow manifestation of that belief becomes in both timing and intensity (Δshadow $\rightarrow \infty$). Jung is describing exactly this when he writes, "the counter-position in the unconscious is not dangerous so long as it does not possess any high energy-value," but when the "counter-position" does indeed possess "high energy-value" (or "certainty" in Heisenberg's terms) then it "breaks through into consciousness usually just at the moment when it is most important to maintain the conscious direction. Thus the speaker makes a slip of the tongue just when he particularly wishes not to say anything stupid" (1969e, p. 99). This exemplifies Heisenberg's uncertainty principle, as someone's attempt to maximize their control (very high certainty in conscious direction) creates maximum unpredictability in when and how excluded content manifests in their lives. The certainty achieved in one domain generates complementary uncertainty in its conjugate domain. When identity boundaries require continuous energy to maintain, the timing and manner of boundary failure becomes fundamentally unpredictable even while the eventual breakthrough remains structurally inevitable due to information conservation. Jung discusses these parallel ideas and distinguishes between different ways of relating to excluded content, writing,

> Suppression amounts to a conscious moral choice, but repression is a rather immoral 'penchant' for getting rid of disagreeable decisions. Suppression may cause worry, conflict and suffering, but it never causes a neurosis. Neurosis is always a substitute for legitimate suffering. (1969d, pp. 85–86)

This pattern aligns with what Busemeyer, Gallimore, Kvam, and Pleskac respectively found regarding collapse strength as their research demonstrated that binary decisions create stronger interference effects than probabilistic judgments and forcing beliefs into definite states (what Jung termed repression) creates more powerful constraints on subsequent processing than maintaining partial superposition (what Jung might call conscious acknowledgment). Suppression keeps some shadow material accessible to consciousness which corresponds to maintaining partial superposition, and in this, excluded possibilities would still remain near enough to the center (0.5) which would allow them to still influence processing. As Jung described, repression is the opposite action as it forces complete isolation of the variables which corresponds to full binary collapse that pushes possibilities to near-definite 0 states. This creates the persistent interference effects Kvam and Pleskac documented. Jung also noted that the shadow "is a living part of the personality and therefore wants to live with it in some form. It cannot be argued out of existence or rationalized into harmlessness" (Jung, 1969c, p. 36). This also aligns with information conservation that requires that excluded possibilities continue to exist somewhere in the system. As Jung described, they cannot be eliminated; they can only be geometrically blocked from conscious access but then still remain energetically active.

Jung is describing what seemingly closely resembles this thermodynamic cost of overcoming these energetic burdens of the shadow, writing that "we carry our past with us, to wit, the primitive and inferior man with his desires and emotions, and it is only with an enormous effort that we can detach ourselves from this burden" (1969d, p. 86). The "enormous effort" Jung describes may parallel the continuous Landauer cost (kT ln 2 per bit erased) of maintaining separation between conscious identity and shadow material. Each binary collapse that forces possibilities to definite 0 states requires thermodynamic work to maintain. This connects to Kvam and Pleskac's observation that "the greater precision in representations trades off with greater storage and processing demands" (p. 8). The precision of rigid identity (forcing beliefs near 1 or 0) requires continuous energy expenditure to prevent the excluded 0s from drifting back toward superposition, which would reduce the sharpness of identity boundaries. This parallel's Gallimore's (2015) findings that high entropy states cause the brain's perceived boundaries of ideas to soften which allows for abstract connections across non-similar words and domains.

Overall, whether these represent different descriptions of underlying physical processes or parallel patterns that have emerged at different scales is open to discussion, but the binary nature of collapse combined with information conservation does provide intriguing reason for why Jung describes the shadow as inevitable across cultures, eras, and individuals. Whether this is consciousness experiencing the limitations of physics or if it is just a useful metaphor, it hopefully functions as useful psychological framing to understand as the book heads into describing how different configurations of the P = Q/E equation manifest in the human population as what is traditionally described as the levels of consciousness.

References

Busemeyer, J. R., Kvam, P. D., & Pleskac, T. J. (2019). Markov versus quantum dynamic models of belief change during evidence monitoring. *Scientific Reports, 9*, Article 18025. https://doi.org/10.1038/s41598-019-54383-9

Carhart-Harris, R. L., & Friston, K. J. (2019). REBUS and the anarchic brain: Toward a unified model of the brain action of psychedelics. *Pharmacological Reviews, 71*(3), 316–344. https://doi.org/10.1124/pr.118.017160

Carhart-Harris, R. L., Leech, R., Hellyer, P. J., Shanahan, M., Feilding, A., Tagliazucchi, E., Chialvo, D. R., & Nutt, D. (2014). The entropic brain: A theory of conscious states informed by neuroimaging research with psychedelic drugs. *Frontiers in Human Neuroscience, 8*, Article 20. https://doi.org/10.3389/fnhum.2014.00020

Gallimore, A. R. (2015). Restructuring consciousness—The psychedelic state in light of integrated information theory. *Frontiers in Human Neuroscience, 9*, Article 346. https://doi.org/10.3389/fnhum.2015.00346

Hogarth, R. M. (2001). *Educating intuition*. University of Chicago Press.

Jung, C. G. (1969a). On psychic energy. In G. Adler & R. F. C. Hull (Eds. & Trans.), *The structure and dynamics of the psyche* (2nd ed., pp. 16–94). Princeton University Press. (Original work published 1948).

Jung, C. G. (1969b). Synchronicity: An acausal connecting principle (R. F. C. Hull, Trans.). In H. Read, M. Fordham, & G. Adler (Eds.), *The collected works of C. G. Jung: Vol. 8. The structure and dynamics of the psyche* (2nd ed.). Princeton University Press. (Original work published 1952).

Jung, C. G. (1969c). *The archetypes and the collective unconscious* (R. F. C. Hull, Ed. & Trans.; 2nd ed., Vol. 9, part 1). Princeton University Press. (Original work published 1959).

Jung, C. G. (1969d). The transcendent function. In G. Adler & R. F. C. Hull (Eds. & Trans.), *The structure and dynamics of the psyche* (2nd ed., pp. 95–130). Princeton University Press. (Original work published 1958).

Kurian, P. (2025). Computational capacity of life in relation to the universe. *Science Advances, 11*, eadt4623. https://doi.org/10.1126/sciadv.adt4623

Kvam, P. D., Pleskac, T. J., Yu, S., & Busemeyer, J. R. (2015). Interference effects of choice on confidence: Quantum characteristics of evidence accumulation. *Proceedings of the National Academy of Sciences of the United States of America, 112*(34), 10645–10650. https://doi.org/10.1073/pnas.1500688112

Kvam, P. D., & Pleskac, T. J. (2017). A quantum information architecture for cue-based heuristics. *Decision, 4*(4), 197–233. https://doi.org/10.1037/dec0000070

Landauer, R. (1961). Irreversibility and heat generation in the computing process. *IBM Journal of Research and Development, 5*(3), 183–191. https://doi.org/10.1147/rd.53.0183

Sclocchi, A., Favero, A., & Wyart, M. (2025). A phase transition in diffusion models reveals the hierarchical nature of data. *Proceedings of the National Academy of Sciences, 122*(1), e2408799121. https://doi.org/10.1073/pnas.2408799121

Vervaeke, J., Ferraro, L., & Herrera-Bennett, A. (2018). *Flow as spontaneous thought: Insight and implicit learning*. https://doi.org/10.1093/oxfordhb/9780190464745.013.8

Vervaeke, J., Lillicrap, T. P., & Richards, B. A. (2012). Relevance realization and the emerging framework in cognitive science. *Journal of Logic and Computation, 22*(1), 79–99. https://doi.org/10.1093/logcom/exp067

Wang, R., Zhen, S., Zhou, C., & Yu, R. (2022). Acute stress promotes brain network integration and reduces state transition variability. *Proceedings of the National Academy of Sciences, 119*(24), e2204144119. https://doi.org/10.1073/pnas.2204144119

Chapter 10
The Scientific Evidence for Emergent Levels of Consciousness

Consciousness has been described so far as a recursive system where quantum processing interacts with ego-based modulation. If the brain truly functions as a boundary between quantum possibility and classical stability, then evolution would have needed to refine that boundary over long periods of time, because life would have required enough openness to benefit from quantum computation while maintaining the structured identity that survival demands.

This raises an interesting question. If beliefs and identity shape the way the system collapses possibilities as the previous chapter outlined, would evolution have distributed different configurations of consciousness across the population to solve different problems? It seems possible that survival in small groups requires different configurations or distributions of consciousness across the population as most individuals would need to act quickly and decisively to defend, attack, and survive while a few would need to hold wider perspectives and guide the group through uncertainty. A village may have needed hunters who react instantly and a shaman who can wait in stillness for insight to arrive. Later, a kingdom might have relied on a king who utilizes a few "wise men" or "mystics" who could see beyond and provide the king wisdom and new insight. Over centuries, as evolution selected for survival, this pattern may have emerged as a distribution of consciousness levels across humanity as each would serve its place in a collective design.

From an evolutionary and thermodynamic view, programs that most efficiently used energy would have survived. Over generations, they may have been encoded into biology itself, which would have produced tendencies toward certain ways of thinking, feeling, and reacting in defense of one's (or the tribe's) beliefs and identity. The framework views these as adaptive programs that still run in modern consciousness as will be explored in the following chapter.

If this reading is correct, many of the difficulties modern humans face may arise from the world changing faster than the programs that guide us. In the past, acting from fear or defending belief was often a matter of survival. Today, those same responses may appear as anxiety, rigidity, or performance collapse, because the old

J. Roeloffs, *The Classical Brain to the Quantum Mind*, The Frontiers Collection,
https://doi.org/10.1007/978-3-032-20068-6_10

software is still running, but the environment no longer matches the conditions it was selected for.

This opens up a wide set of questions. What if degrees of consciousness truly evolved to be distributed across populations? Would science be able to measure differences in ego rigidity or openness that correspond to these configurations? Would contemplative traditions that train awareness find parallel descriptions of the same spectrum? If consciousness develops through predictable stages, what would be the biological and experiential evidence of that development?

These questions guide the discussion that follows, as the following sections turn to the measurable architecture of consciousness and examine whether modern research supports the idea of distributed levels. The findings shine a light on the specific neural systems that maintain ego stability and offer a path toward understanding how these evolutionary programs may still operate in the brain today.

The Neurobiological Architecture

Carhart-Harris et al. (2014) distinguish between "secondary consciousness," which they characterize by strong ego boundaries maintained through DMN activity, and "primary consciousness," which is characterized by significantly dissolved boundaries that result in cognitive flexibility. Their research shines a light on the specific neural architecture that works to maintain ego boundaries. As detailed in Chap. 5, Carhart-Harris et al. (2014) discovered that alpha oscillations in the posterior cingulate cortex explain 66% of the variance in ego dissolution under psilocybin. This finding provides a direct biomarker for the E variable in $P = Q/E$, revealing that ego activation can be measured through specific neural oscillations. This discovery has significant implications for understanding consciousness development as children naturally show immature alpha rhythms and underdeveloped DMN activity, which corresponds to their higher baseline quantum processing capacity. As these systems mature through what Carhart-Harris et al. (2014) terms "maturational settling," consciousness transitions from childhood's natural quantum processing toward adult ego-dominated states.

Therefore, the consciousness levels that follow can be understood as representing different stable configurations of alpha power and DMN activity as lower levels would show strong, persistent alpha oscillations which maintain rigid ego boundaries while higher levels would demonstrate increasingly flexible alpha modulation, which allows more conscious access to quantum processing.

The Entropy-Cognition Spectrum

The book proposes that human consciousness exists on a continuum, where each stage, or level, blends gradually into the next through gray areas of transition. Carhart-Harris et al. (2014) demonstrate an inverted-U relationship between entropy and cognition, where normal consciousness sits slightly sub-critical for stability while primary states approach criticality proper for maximum flexibility.

Cook-Greuter (2021) describes the distribution of developmental stages as having a "bell-curve like shape" where "early and late stages occur much more rarely in adults than the conventional and early postconventional ones" (p. 38). She reminds us that each stage is "a whole structural unity in itself as well as a part of a larger, more complex and integrated way of looking and acting in the world" (39). In her view, growth often means moving from rigid beginnings, through more balanced conventional structures, and at times into levels of postconventional awareness that are quite rare and often difficult to stabilize.

Carhart-Harris et al.'s (2014) finding is also a bell-curve-like arc. They write when considering the brain's entropy that their model "shows an inverted u relationship between entropy and cognition such that too high a value implies high flexibility but high disorder, whereas too low a value implies ordered but inflexible cognition" (p. 13). They propose that normal waking consciousness "inhabits a position that is close to criticality but slightly sub-critical" and that primary states like dreaming or psychedelics "move brain activity and associated cognition toward a state of increased system entropy" (p. 13). In other words, both order and disorder carry their own limitations, and the most adaptive state tends to be somewhere near the middle. Together, these accounts show how the "levels of consciousness" is much more like a gradient or bell-curve of consciousness (Figs. 10.1 and 10.2).

Carhart-Harris et al. (2014) also provide evidence that ego dissolution has specific neural correlates. Their research shows that normal consciousness tends to maintain rigid boundaries through anticorrelation between brain networks. The Default Mode Network remains strongly anti-correlated with Task Positive Networks that focus on the external world. This separation appears essential for ordinary functioning, preventing what they describe as "confusion over what constitutes self, subject and internal on the one hand, and other, object and external on the other" (p. 16). The journey through consciousness levels may represent progressive loosening of these neural boundaries. Therefore, each level would correspond to different configurations of DMN-TPN anticorrelation, alpha oscillation strength, and approach to critical dynamics in those systems.

Yet under certain conditions, these boundaries become permeable. Carhart-Harris et al. (2014) describe what happens when this occurs, writing,

> If the brain was to be sampled during a primary state (such as a psychedelic state) we would predict that the rules that normally apply to normal waking consciousness will become less robust. Indeed, we recently found this to be so when analysing the degree of orthogonality or 'anticorrelation' between the DMN and TPN post-psilocybin. Post-drug there was a significant reduction in the DMN-TPN anticorrelation, consistent with these networks becoming less different or more similar (i.e., a flattening of the attractor landscape). The same decrease

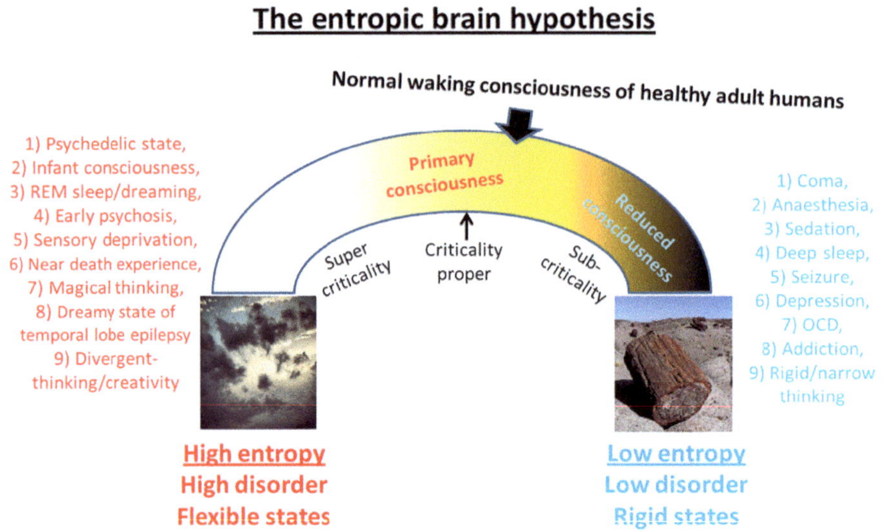

Fig. 10.1 Carhart-Harris et al.'s (2014) figure of "the entropic brain hypothesis." Their figure is inverted from the other levels of consciousness charts and figures shown and described in the book, as they have "high entropy" on the left, which is what this book correlates to high levels of consciousness which is shown on the right in all other figures

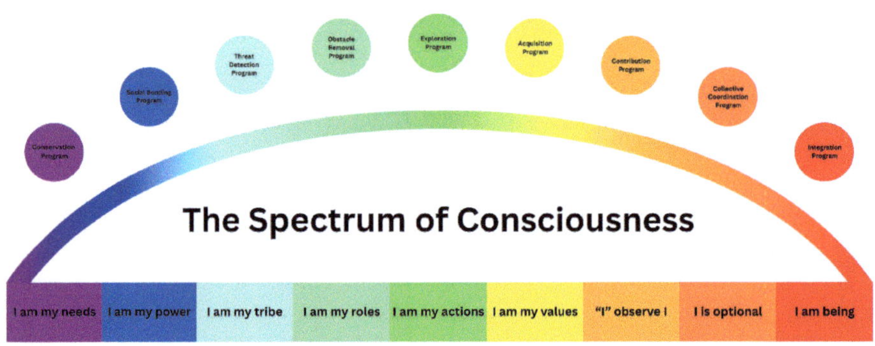

Fig. 10.2 This shows this book's levels of consciousness and adaptive programs in spectrum form. The book continues to emphasize that there is no one way to visualize these. All figures shown are entirely valid interpretations

in DMN-TPN anticorrelation has been found in experienced meditators during rest (Brewer et al., 2011) and meditation (Froeliger et al., 2012). Moreover, decreased DMN-TPN inverse coupling is especially marked during a particular style of meditation referred to as 'non-dual awareness' (Josipovic et al., 2011). This is interesting because this style of meditation promotes the same collapse of dualities that was identified by Stace (and Freud) as constituting the core of the spiritual experience. The DMN is closely associated with self-reflection, subjectivity and introspection, and task positive networks are associated with the inverse of these things, i.e., focus-on and scrutiny of the external world (Raichle et al., 2001). Thus,

> it follows that DMN and TPN activity must be competitive or orthogonal in order to avoid confusion over what constitutes self, subject and internal on the one hand, and other, object and external on the other. It is important to highlight that disturbance in one's sense of self, and particularly one's sense of existing apart from one's environment, is a hallmark of the spiritual (Stace, 1961) and psychedelic experience (Carhart-Harris et al., 2012b). (p. 16)

This finding shows that two completely different interventions, psychedelics and meditation, produce very similar neural changes, and both reduce the anticorrelation that maintains self-other boundaries. Importantly, they describe that DMN-TPN anticorrelation decreases in experienced meditators both during rest and while meditating. This points to the idea that one can develop, or progress, through consciousness and physiologically change their configuration of consciousness. Most significantly, they note that effect is especially strong during "non-dual awareness" meditation, which is a practice specifically designed to dissolve subject-object duality, or the "self," dissolves.

This is very critical because Carhart-Harris et al. (2014) show how these discoveries relate directly to the spectrum of consciousness. They specifically stated the correlation between non-dual awareness and decreased DMN-TPN inverse coupling. The dissolution of subject-object boundaries characteristic of high level consciousness (as will be described in detail in upcoming chapters) corresponds with measurable changes in neural architecture. The Default Mode Network (DMN) typically maintains self-referential processing while Task Positive Networks (TPN) focus on external engagement. Their anticorrelation preserves the boundary between self and other. When this anticorrelation diminishes through meditation or psychedelics, the neural substrate maintaining subject-object duality seemingly temporarily dissolves.

From the framework's perspective, this provides the neural signature of consciousness development as lower levels of consciousness would maintain strong DMN-TPN anticorrelation, which preserves the rigid boundaries necessary for survival and social functioning. Middle levels might show moderate anticorrelation, which would allow some flexibility, especially in flow states, while still allowing one to maintain their identity. Higher levels would demonstrate progressively weakened anticorrelation, which would enable the boundary dissolution that characterizes advanced consciousness. This provides potential neurobiological support that high levels of consciousness involve measurable changes in the neural structures that maintain ego boundaries. Interestingly, as described to an extent in Chap. 9, Carhart-Harris and Friston (2019) note that hierarchical organization extends beyond individual brains as they write,

> The structural organization of brains (and minds) recapitulates the organization of social systems. Just as the free-energy principle recognizes the importance of hierarchy inside the brain, so it recognizes it elsewhere in nature. For example, we may think of individuals constrained by communities, constrained by nation states, constrained by unions of states, etc. We may also reflect on the hierarchy of beliefs at each level, e.g., with individual beliefs constrained by community beliefs, constrained by national beliefs, constrained by transnational beliefs. (p. 336)

This cross-scale recapitulation becomes relevant when considering consciousness development as the framework proposes that the nine levels of consciousness represent different degrees of hierarchical organization which would have different balances between compression and decompression and different precision weightings on beliefs at various levels of abstraction. Lower consciousness levels correspond to tighter hierarchical constraints and steeper attractor basins while higher levels access broader possibility space by maintaining flatter landscapes with lower precision weighting and more tolerance for uncertainty.

The neurobiological grounding for these proposals rests on specific cellular and network mechanisms as Carhart-Harris and Friston note that "one compelling reason to place emphasis on hierarchically high or deep components of the brain is the especially high expression of 5-HT2ARs (the key target of psychedelics) on layer 5 pyramidal neurons within the brain's high-level cortical regions. Collectively, these high-level features of the brain are thought to encode expectations and beliefs about the lived world" (p. 338). This provides the physical substrate where abstract "beliefs" may actually exist as patterns of neural activity in specific cells, maintained through continuous metabolic expenditure.

Interestingly, the authors note a developmental parallel as they write "It is also worth noting that 5-HT2AR densities are highest early in life, dropping significantly when adulthood is reached. Behavioral complexity also decreases at a similar age, after peaking at about 25 years of age. Lastly, that both children and individuals on psychedelics are hyper-suggestible is consistent with the principle that brain activity is tuned closer to criticality in these—likely related—brain states" (p. 326). This suggests that natural development involves a progressive increase in precision weighting, or in other words, a gradual steepening of attractor basins as identity structures accumulate through development. A child's brain, which is rich in 5-HT2ARs and operating near criticality, would maintain relatively flat landscapes with high plasticity, but an adult's brain, with fewer 5-HT2ARs and more entrenched beliefs, would operate in steeper basins with greater stability but less flexibility.

The REBUS model also provides perspective on how different paths might achieve similar endpoints. Comparing psychedelic states with meditation, Carhart-Harris and Friston (2019) note that "key neurobiological similarities between meditative and psychedelic states that have been detected with brain imaging include the following: relative deactivation of the DMN, reduced anticorrelation between the DMN and networks concerned with processing the extrinsic world, and recently, the enhancement of signatures of criticality. Important differences, such as the absence of decreased α power with meditation, may depend on the conventional use of relaxation techniques to enter a meditative state" (p. 330).

Once again, they are outlining the critical role that the DMN seems to play in consciousness. This suggests consciousness can reach similar geometric configurations through multiple mechanisms. Importantly, if meditators could achieve these states without external modulation, this would point to the idea that the levels of consciousness can be traversed between. Psychedelics would be the short cut in that sense as it allows one to achieve rapid landscape flattening through pharmacological disruption of precision weighting while meditation would allow one to achieve gradual landscape transformation through sustained attention training. Both very likely could result in enabling one to more effectively transition to a higher base state in the levels of consciousness.

This then raises a variety of important questions about meditation and the ways in which one could experience it. What exactly is non-dual awareness? If it represents more than a subjective experience and if it corresponds to measurable brain changes, then what is actually happening when the boundaries between self and world dissolve? And perhaps most importantly, if consciousness can learn to modulate these boundaries, what implications might this have for how humans interface with reality itself?

Non-dual Awareness

The significance of Carhart-Harris et al.'s (2014) and Carhart-Harris and Friston's (2019) findings drastically increases when understood through the lens of contemplative science describing non-dual awareness. Meling (2022) provides a comprehensive framework for understanding what non-dual awareness actually represents as he writes,

> Non-dual-oriented practices were developed to evoke an experiential shift into a mode of experiencing in which the cognitive structures of self-other and subject-object subside. Thereby, non-dual oriented practices aim at eliciting and sustaining non-dual experiences, that is, experiences that lack a structuring into subject and object. This special set of practices serves to undo the habitual reification of an observer being separate from the observed objects of awareness. (p. 1)

Meling's analysis carries particular weight because it rigorously compares theoretical models with actual contemplative instructions. As he states, his research question directly addresses "to which extent does the recently developed enactive view of the conditions for a shift into a non-dual mode of experience match the actual practice instructions from the Mahāmudrā tradition" (p. 5). This comparison points to a convergence between cognitive science and Tibetan Buddhist practice. As Meling continues, he identifies what distinguishes non-dual practices from other forms of meditation saying, "Different than most other practices, non-dual oriented practices put special emphasis on effortlessness and on releasing tendencies to control or alter the mind" (p. 2). From the framework's perspective, this "effortlessness" and "releasing control" may describe the cessation of ego's continuous effort to collapse quantum field states, or "fuzzy" beliefs as described in the previous chapter, into ordered, classical states. While concentration meditation might strengthen certain collapse patterns and mindfulness might observe them, non-dual practices specifically train consciousness to stop creating them altogether.

Following this, Meling continues and distinguishes between two fundamental modes of knowing. Phase 1 enaction "comprises a knowing of a world that is related to performing actions relevant to maintaining the living body. It is dualistic and involves a subject-object distinction" (p. 3). This describes ordinary consciousness where ego continuously collapses possibilities into classical subject-object reality.

Phase 2 enaction represents something radically different as Meling (2022) describes this, writing,

> Phase 2 enaction is an alternative mode of knowing. It is not based on a subject-object distinction. It lacks a distinction between observer and observed. In this mode of knowing the mind is 'neither absorbed nor separated but simply present and available.' Most importantly, phase 2 enaction is a non-dual mode of knowing that allows for a direct experience of groundlessness: 'this is the mind that can actually know firsthand the groundlessness of the enacted edifice in which humans live'. (Meling, 2022, citing Rosch, 2017, p. 3)

The transition to non-dual experience requires two conditions that map well onto the framework's predictions. First, "non-adaptivity" where one "does not judge experiences as good, bad, or neutral" and neither approaches, avoids, nor ignores any aspect of experience (p. 8). From the framework's perspective, this represents refusing to collapse quantum field states through preference or aversion.

Second, and most importantly, Meling describes how "sense-making progressively ceases" (p. 4). In the framework's terms, sense-making is the collapse mechanism as it is the continuous process of forcing quantum field configurations into classical interpretations. When this ceases, Meling states, "Experience experiences itself, non-dually. This is a moment of knowing groundlessness" (Meling, 2021, cited in Meling, 2022, p. 4). This "sense-making" is exactly the same egoic process the book has been describing as "making narrative sense of the world" and is what the DMN does. Therefore, non-duel awareness is learning how to deactivate the DMN's "sense-making" which is exactly what the framework would expect to lead to higher level consciousness. This "groundlessness" may be the direct experience of the quantum field before ego collapses it into classical reality.

Empirical Validation of Consciousness-Quantum Correlation

If non-dual awareness represents a state where ego ceases collapsing quantum field states, then those who achieve such states should demonstrate different interactions with quantum systems. The book can now explore these key predictions. If consciousness truly interfaces with quantum systems through varying ego states, then science should find measurable differences in quantum influence between individuals at different consciousness levels. Specifically, those who have systematically reduced ego activation through contemplative practices should demonstrate greater capacity to affect or to avoid affecting quantum systems than those who haven't.

This prediction finds suggestive support in a series of double-slit experiments conducted by Radin et al. (2012). Across the first four experiments, experienced meditators showed a significant deviation in the predicted direction when directing attention toward the apparatus (combined z = – 3.80, $p < 0.00007$, effect size = – 0.46, N = 67 sessions). The non-meditator group, by contrast, produced effects that could not be distinguished from random chance (combined z = – 1.25, effect size = – 0.19, N = 41 sessions). Control sessions showed null results.

At first glance, one may think this is falsifying evidence of the book's claims as those with mediation experience actually "caused more collapse" but instead, let's consider what this may actually be showing. The meditation group is trained in mental focus and control, while the non-mediators are not. By definition, meditators will be able to actually harness and focus their attention toward the machine, while those without practice will struggle to do this. The ability to influence quantum systems appears not as an automatic property of consciousness but as a capacity that correlates with practices known to reduce ego activation. Non-meditators, despite conscious intention and sustained effort, showed minimal effects. Most importantly, considering this book's claims about the collapse mechanism of the wavefunction, the result is particularly important. If consciousness uniquely had to do with collapse of the wave function, for example in a framework that posited consciousness as fundamental, then all participant's focus should have significant effects on the double-slit. The framework would posit that the mediators were actually able to focus more energy and particles towards the double-slit as they have practice channeling their body's energy and attention, therefore considering the proposed gradient collapse mechanism, are able to more effectively collapse the wavefunction by channeling more particle interference. The subtlety of the observed effects, with visibility changes of approximately 0.001% even among experienced practitioners, aligns with consciousness participating as one collapse channel among many environmental factors. Therefore in the same way, those who have meditation practice would have more control over the collapse mechanisms inside their own brain, pointing to the P = Q/E equation.

Several important caveats warrant consideration. The absolute changes in fringe visibility were extremely small, the sample sizes per subgroup were modest, and the study combined multiple experiments without preregistration. Accordingly, these findings should be treated as suggestive evidence and hopefully will motivate preregistered, multi-lab replications.

Despite these limitations, the Radin et al. (2012) findings provide intriguing empirical evidence consistent with the framework's proposal that consciousness development involves learning to modulate the collapse mechanism. The convergence proves noteworthy, as Carhart-Harris et al. (2014) demonstrate the neural changes associated with ego dissolution, Meling (2022) describes the phenomenological process of ceasing sense-making during meditation, and Radin's team observes that individuals with these trained capacities produce larger deviations in quantum interference metrics. Together, these independent lines of research suggest that the journey from high ego activation to reduced self-referential processing may involve measurable changes in how consciousness interfaces with quantum systems.

References

Carhart-Harris, R. L., & Friston, K. J. (2019). REBUS and the anarchic brain: Toward a unified model of the brain action of psychedelics. *Pharmacological Reviews, 71*(3), 316–344. https://doi.org/10.1124/pr.118.017160

Carhart-Harris, R. L., Leech, R., Hellyer, P. J., Shanahan, M., Feilding, A., Tagliazucchi, E., Chialvo, D. R., & Nutt, D. (2014). The entropic brain: A theory of conscious states informed by neuroimaging research with psychedelic drugs. *Frontiers in Human Neuroscience, 8*, Article 20. https://doi.org/10.3389/fnhum.2014.00020

Cook-Greuter, S. R. (2021). *Ego development: A full-spectrum theory of vertical growth and meaning making*. (Original work published 2013, edited in 2019, republished in 2021).

Meling, D. (2022). Knowing the knowing: Non-dual meditative practice from an enactive perspective. *Frontiers in Psychology, 13*, 778817. https://doi.org/10.3389/fpsyg.2022.778817

Radin, D., Michel, L., Galdamez, K., Wendland, P., Rickenbach, R., & Delorme, A. (2012). Consciousness and the double-slit interference pattern: Six experiments. *Physics Essays, 25*(2), 157–171. https://doi.org/10.4006/0836-1398-25.2.157

Chapter 11
The Two-Dimensional Map of Consciousness

The previous chapters outlined expectations for how consciousness would likely be distributed across the human population. If consciousness evolved as an adaptive mechanism for navigating quantum possibilities, and if that capacity was refined through evolutionary pressures, then variation among individuals would be expected. The framework predicts that consciousness should appear as a spectrum of configurations as each is shaped by the thermodynamic cost of beliefs and identity in relation to the $P = Q/E$ equation. This chapter explores the evidence for these levels of consciousness and how this distribution manifests in humanity. It describes how the human journey through consciousness reflects both an evolutionary inheritance and a continuing developmental process that extends the same principles outlined in earlier chapters.

Consciousness has been explored for thousands of years, and many traditions have attempted to describe its unfolding, sometimes calling this process enlightenment. The framework builds from the idea that there are identifiable stages of consciousness and that these stages express how human beings experience the same underlying process described in the earlier chapters. In broad terms, consciousness was defined as "the continuous, self-referential process that integrates perception, memory, meaning, and identity into coherent experience." This chapter now examines how that process develops through measurable stages within individuals and populations.

Before presenting the model, it is helpful to recognize that the concept of levels of consciousness is not new. Developmental psychology, philosophy, and cognitive science have each described stage-based progressions in how humans develop, think, construct identity, and relate to the world. While these models may differ in their naming or area of emphasis, they share many structural similarities as development proceeds through a sequence of qualitative shifts, where each integrates and transcends the capacities of earlier stages.

Wilber (2007) gathered many of these frameworks into a comparative synthesis, and showed how different theorists described similar hierarchical developmental

J. Roeloffs, *The Classical Brain to the Quantum Mind*, The Frontiers Collection,
https://doi.org/10.1007/978-3-032-20068-6_11

patterns as Piaget and Aurobindo traced the growth of cognition from sensorimotor awareness through concrete and formal reasoning into integrative thinking. Graves and Beck's Spiral Dynamics mapped the evolution of values from survival-centered motivations toward increasingly complex worldviews. Kegan's constructive-developmental model examined how the self's relationship to thought and feeling becomes more reflective and inclusive over time. Cook-Greuter and Loevinger detailed the evolution of ego structures from conformist identity through integrated awareness, while Gebser's work on worldviews traced parallel progressions from archaic to magic, mythic, rational, and integral modes of consciousness. Then Fowler's stages of faith reflected similar movements in meaning-making and moral understanding.

Wilber's synthesis reinforces that humans reliably progress through identifiable levels of complexity, perspective-taking, and identity. This book's framework proposes that to best understand how evolution structured these consciousness levels, and specifically in how humans experience the consciousness levels, requires mapping both vertical development (levels) and horizontal expression (programs).

This model synthesizes and extends these existing frameworks, using the structure of Wilber's developmental stages as this framework's vertical axis, which represents structural evolution of consciousness, and draws inspiration from Hawkins' (2020) emotional development scale as the horizontal axis, which represents evolutionary adaptive programs within each developmental level. Critically, it incorporates Kegan's (1994) constructive-developmental theory, particularly his subject-object transitions that explain how each level makes the previous level's invisible assumptions visible and workable. Most critically, this framework utilizes Cook-Greuter's (2000, 2005, 2021) empirical research on post-conventional development, especially her insights into ego transcendence at the latest stages, as she most comprehensively has outlined the stages of development.

Together, these axes create a two-dimensional map of consciousness, but the framework extends this further by proposing that each configuration represents a specific relationship between ego and quantum processing. It was necessary to combine their findings into a new structure to more acutely reflect the underlying physics described in this book. When doing so, what emerges is a physics-based understanding of how different ego structures and evolutionary adaptive programs literally create different degrees of access to quantum computation, as predicted by the $P = Q/E$ equation.

The Two Dimensions

The framework most significantly diverges from some traditional consciousness models through its perspective on how humans experience consciousness. This framework proposes that human consciousness operates along two fundamental dimensions that evolution optimized for different survival contexts. Like coordinates on a map, together these create a complete picture of any moment of human experience. Cook-Greuter (2021) actually observed this in her writing tying together some ideas from Ken Wilber as well, saying,

> Most growth in adulthood seems to occur within a given stage, variously described as lateral or horizontal development or "translation" in Wilber's terminology. [Inside of a given level of consciousness] The current way of viewing the self and reality is refined, enriched, and modified to include more diverse domains, more contexts and detail, and to establish more connections among them ... The later the stage, the more room for such horizontal expansion exists while the current stage or mental model of the world remains the same. (p. 24)

In this book's framework the X-axis, which are the adaptive programs, act both as developmental gateways to higher levels and as markers to identify one's current operating mode within their consciousness level. The adaptive programs determine the geometric search space one operates within. Importantly, as Cook-Greuter identified, later levels of consciousness (Y-axis) will have a larger range of X-axis adaptive programs that they could operate from within. This allows for more horizontal development in later stages of consciousness.

The two dimensions are interlocked. The adaptive programs function as stepping stones between consciousness levels. You cannot reach Level 5 without stabilizing the Exploration program (Courage). You cannot access Level 7 without first experiencing the Collective Coordination program (Love). Each program serves both as an operating mode within your current level and as a developmental gateway to the next level. Cook-Greuter (2021) explains what consciousness development is (or in her terms "Ego Development Theory") and describes the importance of both the X and Y axis, saying,

> Ego Development Theory is a theory of self-identity and how it forms and changes over time. It includes horizontal development as well as the potential transformation from one view of reality to a broader, more inclusive one. It describes a sequence of how people's mental models of reality evolve. Each new level contains the previous ones as a subset. This is best illustrated with a set of Russian dolls, where each doll contains another, smaller doll. Thus, each new level is both a new whole with its own coherence, and–at the same time–also a part of a larger, more complex system. (p. 24)

Consciousness level determines your baseline reality structure, essentially who you think you are, while adaptive programs determine your moment-to-moment experience, or how you're responding to current circumstances. Together they create your current P = Q/E configuration. One could imagine consciousness levels and evolutionary programs as either a traditional X-Y graph or as a staircase where programs help one navigate between and within levels of consciousness. Both are entirely valid.

The Vertical Axis of Consciousness

The nine levels describe how consciousness relates to identity itself which is the "I am" structure. From "I am my needs" at Level 1 to "I am being" at Level 9, each level represents a fundamentally different relationship between awareness and identity. These are developmental stages that, once achieved, become permanent capabilities.

The Horizontal Axis of Consciousness and the Evolutionary Adaptive Programs

The nine programs represent survival strategies that evolution developed over millions of years, and they function as slowly evolving worldviews that modulate ego activation (E) in shorter timeframes while consciousness levels determine one's baseline ego structure over longer periods. They are both operational modes within current consciousness levels and some are development gateways to further levels of consciousness. Each program occupies a specific topology in quantum computational space, and each program manifest through thermodynamic costs, as lower programs require maximum energy for defensive identity maintenance (high, E, narrow geometric space), and later programs operate near the thermodynamic floor (low E, broad geometric space). This results in the spectrum from survival-focused classical processing to unitive quantum computation that defines the horizontal axis.

Overview of Y and X Dimensions of Consciousness

The Nine Consciousness Levels (Vertical)

1. "I am my needs"—Immediate survival
2. "I am my power"—Dominance focus
3. "I am my tribe"—Group identity
4. "I am my roles and beliefs"—External structure
5. "I am my actions and achievements"—Performance identity
6. "I am my values and purpose"—Meaning orientation
7. "I observe the 'I'"—Witness consciousness
8. "'I' comes and goes"—Fluid identity
9. "I am being"—Unitive consciousness.

The Nine Evolutionary Adaptive Programs: (Horizontal)

1. Conservation program (Apathy)
2. Social bonding program (Grief)
3. Threat detection program (Fear)
4. Obstacle removal program (Anger)
5. Exploration program (Courage)
6. Acquisition program (Desire)
7. Contribution program (Purpose)
8. Collective coordination program (Love)
9. Integration program (Peace).

The P = Q/E equation operates through both dimensions. Consciousness level (Y-Axis) determines baseline ego structure, which significantly affects E in longer timeframes, while evolutionary adaptive programs modulate E in shorter time-frames. Someone at Level 5 running the Obstacle removal program (Anger) experiences high E (competitive, defensive), while the someone else at Level 5 running the Contribution program (Purpose) experiences lower E (flow-oriented, process-focused).

This helps to explain why people, even at the same consciousness level, can appear so different. Their structural relationship to identity can be relatively constant, but their active evolutionary program may vary dramatically based on context and triggers. Considering this traversable vertical axis of consciousness, Cook-Greuter (2021) maps consciousness through three distinct stages which includes prepersonal, personal, and transpersonal (p. 12). This book will map consciousness through three main stages as well, the only difference between this book's categories and Cook-Greuter's is this book will separate what Cook-Greuter calls the Unitive Stage (the highest level of development) into its own stage, as the book posits it is distinctly different from previous stages.

Although it's critical to note that, the framework organizes these nine levels into two large groups (two paradigms), which is split into four distinct stages from which the nine levels emerge. Therefore, what's important to remember throughout the book is that there are two big paradigms of consciousness, two stages in each paradigm, and six levels of consciousness in the first paradigm and three levels of consciousness in the second paradigm as explained below.

The Two Paradigms of Consciousness

1. **Paradigm 1—Ego-Unaware Consciousness (Stages 1–2, Levels 1–6)**
   Consciousness remains identified with ego, unable to observe its own operating system. Whether expressed as tribal identity, role conformity, achievement drive, or value orientation, the ego invisibly shapes all experience. Individuals at these levels are their tribe, role, actions, thoughts, feelings, and/or beliefs depending on the level. The ego operates unconsciously, as the person pays the thermodynamic cost to maintain rigid beliefs and automatically collapses quantum field states without awareness of this process.
2. **Paradigm 2—Ego-Aware Consciousness (Stages 3–4, Levels 7–9)**
   Consciousness develops the capacity to observe, understand, modulate, and eventually fully integrate its own ego structures. The operating system becomes visible and eventually optional, through this, beliefs are systematically dissolved, and this process releases much of the thermodynamic costs previous identity structures had to pay to maintain. This shift from being ego to witnessing ego represents perhaps the most significant transformation in human development, enabling ego-aware, conscious participation in reality creation. As Cook-Greuter (2015) describes, at this phase "the ego becomes transparent to itself" (p. 98).

The Four Stages of Consciousness

Stage 1 and 2 are within the **Ego-Unaware Consciousness (Stages 1–2, Levels 1–6)**

1. **Stage 1—Survival Consciousness (Levels 1–2)**
   Consciousness operates through immediate survival imperatives with minimal self-reflection. The ego remains completely fused with needs, tribal identity, or power dynamics. Thermodynamic costs are extremely high to maintain rigid beliefs to survive. Quantum access stays minimal as every stimulus triggers immediate collapse into threat or resource assessment. These levels provide the rapid response and decisive action necessary for physical survival.
2. **Stage 2—Social Consciousness (Levels 3–6)**
   Identity forms through external structures, achievements, and values. The ego develops sophisticated narratives about who one is, what one does, and what one stands for and must pay the thermodynamic cost to maintain these rigid beliefs. Significant quantum access (low E states) emerges in specific domains through flow states, though the ego quickly reclaims these experiences as personal achievements. These levels enable complex social cooperation and purposeful action.

Stage 3 and 4 are within the **Ego-Aware Consciousness (Stages 3–4, Levels 7–9)**

3. **Stage 3—Integrated Consciousness (Levels 7–8)**
   The ego becomes observable instead of being the sole observer. For the first time, consciousness can witness its own collapse mechanisms and choose whether to engage them. Thermodynamic costs are significantly reduced as beliefs are dissolved into fuzzy probability and quantum processing becomes much more accessible as practitioners learn to modulate ego activation. These levels enable systems thinking, paradox navigation, and conscious reality participation.
4. **Stage 4—Unitive Consciousness (Level 9+)**
   The observer-observed duality dissolves entirely. Consciousness recognizes itself as both the quantum field and the collapse mechanism, which renders the $P = Q/E$ equation effectively undefined as there is no separate self to measure against (Figs. 11.1, 11.2 and 11.3).

The Evolutionary Necessity and Distribution Principle

The book sees it critical to emphasize that no consciousness level is objectively "better" than any other. This is a common misconception and an unfortunate result of the hierarchical perception that "levels" often carries. The framework's consciousness development relates to the $P = Q/E$ equation as higher levels just have different ego configurations that allow more quantum processing access which results in a different

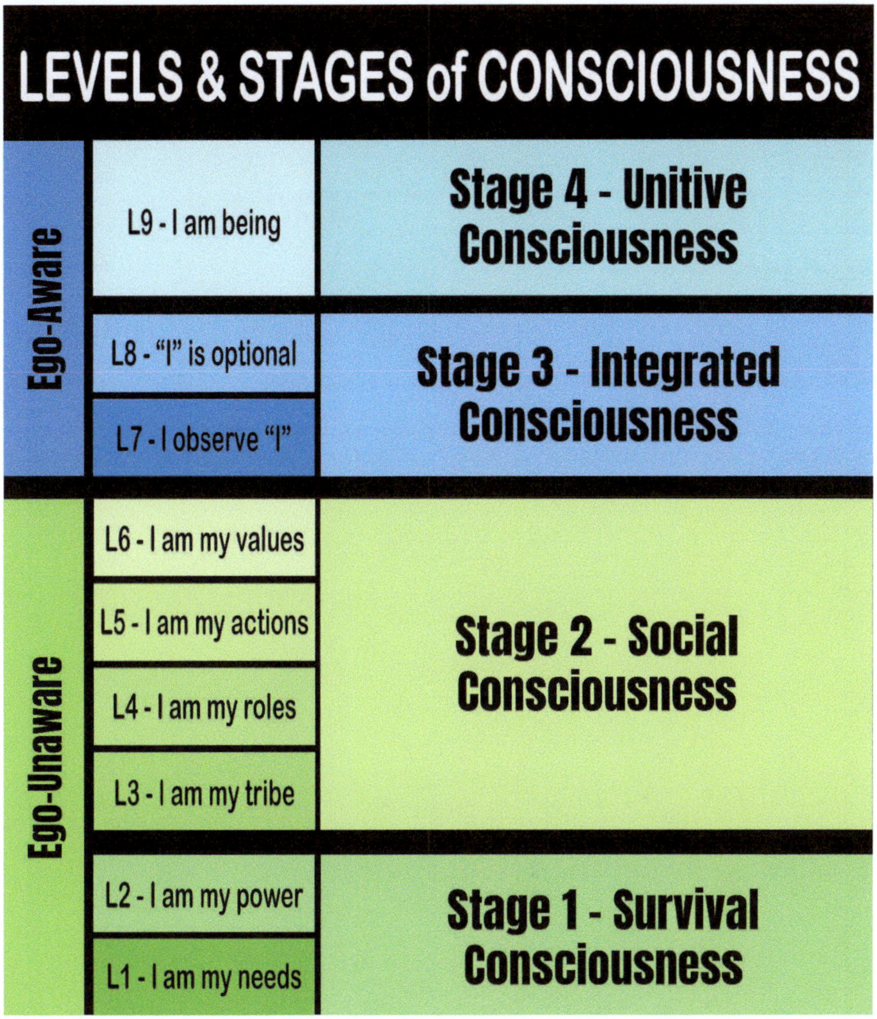

Fig. 11.1 This framework's nine levels, four stages, and two paradigms of consciousness

informational pool to compute from. This different information does not mean that these people are inherently "more successful" or live "better lives" in any quantitative way, it's just a different way of being. Cook-Greuter (2021) supports this notion, stating, "As the process of self-awareness deepens … access to …transpersonal material increases" (p. 102).

Recent neuroscience research points to why ego-mediated processing at levels 1–6 creates "inefficiencies" because the timing delays between neural activity and conscious narrative construction, that appear in studies like Libet et al.'s (1983) experiments, reflect the ego's storytelling function which delays consciousness awareness

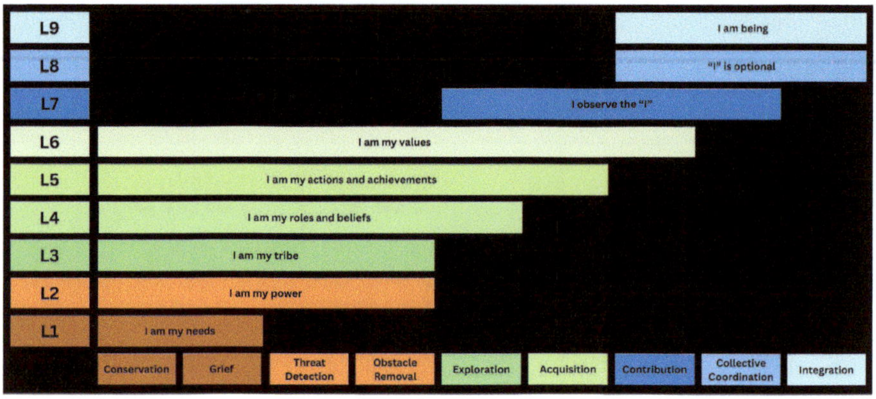

Fig. 11.2 The framework's nine levels of consciousness and nine adaptive programs. It illustrates how different levels of consciousness can access and operate within a different range of adaptive programs

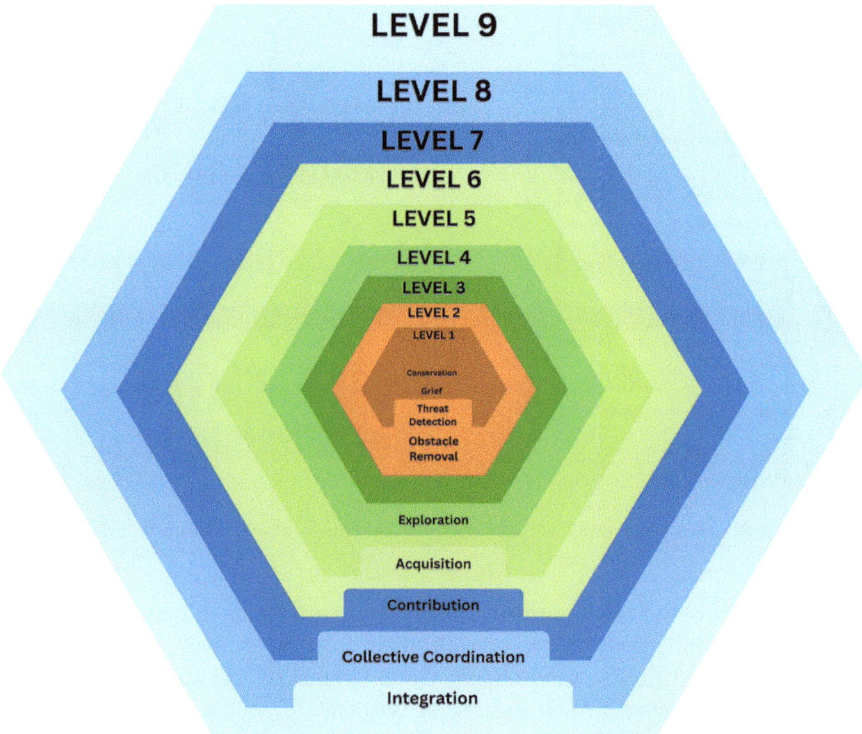

Fig. 11.3 Another way to view the levels of consciousness and the adaptive programs. This emphasizes the geometric space that each operates within and how the consciousness levels act like shells that can only be observed from the outside

of decision and action. Their results are not showing the brain's optimal decision-making speed. Advanced contemplatives who operate at high levels of consciousness demonstrate how consciousness can function without these delays. This suggests that living with more direct quantum-classical processing is possible. It's important to note that these apparent "inefficiencies" serve important evolutionary functions. The ego's narrative construction through the DMN enables complex social coordination, long-term planning, and cultural transmission that are essential for human survival. The millisecond delays created by DMN processing allow for the stable identity patterns and predictable behaviors that enable civilization to develop and thrive, this is why "lower" levels aren't inherently worse, they're just different.

Considering this, the levels had to develop sequentially because each rung on the ladder of consciousness builds the cognitive infrastructure needed for the next. Survival-oriented consciousness at levels 1–3 had to establish basic ego structures before social consciousness at levels 4–6 could emerge, which in turn provides the foundation for ego-awareness at levels 7+. From an evolutionary perspective, humanity's success is heavily influenced by how consciousness levels are optimally distributed across the human population as each level plays a part in optimizing the human species survival. This distribution mirrors how evolution solves complex challenges through specialized adaptations. A society of only Level 9 consciousness would be as maladaptive to survive as one with only Level 2's. The Level 9's would starve while contemplating unity, just like the Level 2's would destroy themselves through tribal warfare without higher-level contemplation.

It is also important to note that happiness and life satisfaction likely do not correlate directly with consciousness level. A Level 4 person fulfilled in their role may experience much greater contentment, happiness, and life satisfaction than someone at Level 6 wrestling with existential meaning. Carl Jung actually emphasizes how all stages of development serve as important anchors for existence for many people, saying, "It would therefore be wrong to regard this intermediary stage as a trap; on the contrary, for a long time to come it will represent the only possible form of existence for the individual, who nowadays seems more than ever threatened by anonymity" (1963b, p. 316). He highlights how every configuration of consciousness plays an important role for individuals in their meaning, purpose, and development and for society at large.

Similarly, "success" would manifest incredibly differently at each level as Level 5 may build companies through acquisition programs, Level 4 could maintain communities through exploration programs within defined roles, while Level 7 may design new systems through purpose or love programs. Inherently, "success" assumes a goal to "achieve." One cannot be "successful" without a pre-defined objective function. Each level operates within their own games and they define what success means to them, one may describe this as the "matrix" of each level. Hawkins (2020) emphasizes this, saying, "With self-importance, the ego is designed to want gain, and it has a lot of thoughts about monetary gain and how that will finally bring happiness. Interestingly, our research has found that happiness is correlated with level of consciousness, not with wealth or health" (p. 224).

Overall, the framework describes different ways of *being*. It's importantly not describing a hierarchy where higher rungs of the ladder invalidate lower ones. Evolution's genius lies in creating a distributed consciousness system where each level contributes its unique collapse patterns and processing capabilities for collective benefit.

States Versus Stages Versus Programs

Also, before describing each level in detail, there is an important distinction to make. Wilber (2005) emphasizes that states of consciousness are temporary while stages represent permanent developmental acquisitions. This book entirely agrees with Wilber, but the frameworks distinction is made more explicit when splitting the levels of consciousness and the adaptive programs. This distinction becomes essential for understanding how flow states (temporary high P access) differ from stable consciousness development. Contemplative observation agrees with this distinction as Hawkins (2020) notes that while "emotions coming up from the individual unconscious" create temporary state changes, "your overall consciousness level is that which is dominant" (p. 88), describing what the framework predicts which is that people will have stable quantum access baselines with temporary fluctuations during flow states and other elevating experiences, especially in transition phases. Cook-Greuter (2021) highlights the importance of this, saying,

> In other words, the measured ego maturity stage is the one that a person reliably and routinely uses in response to everyday life. At the same time, we have never encountered a distribution of responses that was narrower than three levels. Often responses span over four, five and rarely over the whole range of possibilities. Indeed, for healthy, advanced ego development we expect flexible access to a broad range of responses across the stages, not just late-stage material. (p. 23)

Cook-Greuter is describing why it can be difficult to utilize a monolithic system and why the framework found it valuable to utilize the two-axis approach. The book hopes that this X and Y axis approach will help to clarify some of these evaluation problems, as many of the differences that Cook-Greuter describes are likely a result of the differences in the X-axis adaptive programs that individuals who are actually at the same level of consciousness are running. Although, it would still be entirely possible for different levels of consciousness to arise in varying domains of one's life, as the levels of consciousness can be seemingly siloed to different domains of life, at least to a degree.

Berger (2012) illustrates this flexibility in her work on adult development, showing how leaders often operate at different developmental stages depending on domain, environment, and pressure. The same individual may be self-authoring in one domain and rule-following in another. For example, at work, someone may temporarily access higher level consciousness when ego diminishes through group identity and shared purpose, but return to their baseline lower level at home. The evolutionary adaptive programs show even more variability, as everyone has default programs they run, while temporarily accessing higher programs through specific activities. This generally occurs through work, sports, creative pursuits, or other activities that naturally reduce ego activation, while their default "home life" may remain anchored because of a more survival-oriented environment.

Once again, these are not rigid categories that someone must be exclusively categorized within. There are some critical transitional moments such as Level 6 to Level 7, however, even once the switch "occurs" someone who has now first become ego-aware at Level 7 has to effortfully integrate the ego to get to Level 8. This can be a long process.

The experience across levels is important, and that's even true for development, as any of the lower stages accessing higher states for even a moment can be very beneficial if these insights received are integrated. Wilber (2005) highlights this as he writes,

> The more you are plunged into authentic higher states of consciousness—such as meditative states—then the faster you will grow and develop through any of the stages of consciousness. It is as if higher-states training acts as a lubricant on the spiral of development, helping you to disidentify with a lower stage so that the next higher stage can emerge, until you can stably remain at higher levels of awareness on an ongoing basis, whereupon a passing state has become a permanent trait. These types of higher-states training, such as meditation, are a part of any integral approach to transformation. In short, you cannot skip actual stages, but you can accelerate your growth through them by using various types of Integral Transformative Practices. (p. 13)

Wilber reinforces the idea that one cannot skip through stages. Each stage is critical as a building block to the next. Like mathematics, one cannot understand algebra if they don't understand multiplication first. Insights learned through each stage compound, so each level must be deeply integrated through contemplation, experience, and insight as one works their way through the spectrum of consciousness.

It may also be possible to temporarily bypass normal developmental progression through substances like DMT, ayahuasca, or other psychedelics. These may grant temporary access to higher-level consciousness states without the stable integration of intermediate levels. This could explain why many psychedelic users risk sinking into continual usage of the substances as they offer the simplest route to experience higher consciousness without requiring all the challenging work of integrating each level's lessons. The framework hypothesizes that the wide variation in psychedelic experiences correlates with the user's baseline consciousness level and which evolutionary programs are active when they ingest the substance.

These frameworks serve as navigational tools for understanding the vast territory of human experience. They are not boxes to confine it. These levels should never be taken as identities or labels as that would be entirely self-defeating. Using "Level 6" as an identity is completely counter intuitive. It's just like how latitude and longitude help us navigate physical space without claiming that invisible lines actually exist on Earth's surface. These consciousness coordinates help us understand and discuss human development and they are not sharp boundaries that separate one state from another. The map, as always, is not the territory.

Statistical Evidence and the Architecture of Identity

Also, before exploring the specifics of the nine levels of consciousness and the nine evolutionary programs. It's beneficial to emphasize that the framework's two-dimensional map synthesizes multiple approaches that independently discovered similar patterns of consciousness. Using Wilber's (1997, 2005) and Cook-Greuter's (2000, 2005, 2021) developmental stages as the vertical axis and drawing from Hawkins' (2020) emotional programs and Jung's archetypes (1963b, 1969b, 1969d) as the horizontal axis, while incorporating Kegan's (1994) subject-object transitions, the framework adds the underlying quantum physics and thermodynamics explanations to demonstrate why these patterns manifest as they do in the human experience.

The difficulty in writing about these topics comes from the smaller amounts of data on individuals with experiences of Level 7+ consciousness. Less than 2% of individuals in Cook-Greuter's (2021) studies were at her construct-aware stage (p. 21), which in this framework is at the peak of Level 6, therefore likely less than 1% of people are at Levels 7–9.

This statistical rarity creates a fundamental methodological challenge for consciousness research. When 98% or more of the population operates within ego-unaware consciousness (Levels 1–6), scientific studies inevitably draw conclusions based on this uneven spectrum. Researchers studying averaged neural activity across typical populations cannot access data about consciousness that operates without significant ego interference. Temporary ego dissolution through psychedelics or flow states are often perceived in science as anomalous instead of being something that reveals developmental potential. It's interpreted this way because subjects eventually return to baseline consciousness levels, which leads researchers to conclude that the "I," or the ego & DMN, must inevitably reassert itself at some point.

This sampling limitation may explain why neuroscience has struggled to move beyond viewing consciousness as fundamentally mysterious or why some researchers, after experiencing temporary ego dissolution themselves, still end up concluding that consciousness itself must be fundamental instead of positing something outside of human experience, like the Quantum Information Dimension, as the place where consciousness may participatorily emerge from.

This is why the framework necessarily incorporates evidence from contemplative traditions that have documented ego-transcendent states across millennia, alongside the limited quantitative data available from developmental psychology, as this book cites research such as Meling (2022), Hawkins (2020), Carhart-Harris (2014), and Jung (many works) often as each approaches the topic from a different perspective. This is also why Cook-Greuter's research (2000, 2005, 2021) is prioritized as her participants number into the thousands, which has enabled her to study construct awareness and ego-awareness significantly more than most. While this approach may appear less rigorous than large-scale, double blind empirical studies, it may be the only currently available way to study consciousness phenomena that emerge in less than 2% of the population, especially when considering this book's mission to ground it in physics.

Interestingly, some individuals who have documented their journey to ego-aware consciousness have independently discovered similar patterns regardless of their starting methodology. David R. Hawkins, a psychiatrist who co-authored with Nobel Laureate Linus Pauling, developed descriptions of consciousness as "the collapse of potentiality into actuality" (Hawkins, 2020, p. 267) without engagement with quantum models. His work is included in these chapters because he attempted to approach consciousness research with physics in mind.

Carl Jung pursued a similar interdisciplinary approach, as described in previous chapters, as he attempted to ground his work in physics which started with his meetings with Albert Einstein while Einstein lived in Zürich which Jung described began his contemplation on how psychology, the development of the psyche, and the mind at large must also be grounded in physics and related to relativity (Hürter, 2023; Jung, 1963). He then began a multi-decade-long collaboration with Nobel Prize-winning physicist Wolfgang Pauli. Together, they co-developed synchronicity theory and investigated how consciousness might interface with physical reality. Pauli provided the rigorous physics perspectives while Jung synthesized and translated these ideas into psychological and archetypal frameworks. Their collaboration demonstrates that even at the birth of quantum mechanics, leading physicists already began to recognize the potential connections between consciousness and quantum phenomena that mainstream science would spend decades avoiding.

Hawkins even directly discussed the ideas of consciousness using the language of quantum mechanics, saying, "Quantum mechanics is the way out from the restrictive linear domain of the Newtonian paradigm, via Heisenberg's principle—in which we find the interesting discovery that to observe a thing is already to change the outcome because you've introduced the impact of consciousness itself. The level of consciousness of the observer has a profound effect on what is being observed" (2020, p. 59).

Hawkins, through developing his own consciousness and through scientific observation in his research, arrived at similar conclusions this book did as it followed the new empirical evidence in quantum biology, neuroscience, and physics. Similarly, this book's P = Q/E equation was formulated without prior knowledge of the work of Hawkins, Meling, Carhart-Harris, or the collaborations between Jung and physicists. The book utilizes sources from a variety of fields in a hunt for convergent evidence to overcome the smaller-N studies regarding the highest levels of consciousness (Levels 7–9). The book aims to continue what those such as Jung and Pauli began.

The Levels of Consciousness—Y-Axis

If **Proposal 1: P = Q/E** holds, then different stages of human development would represent different configurations of this relationship. Early developmental stages might require maximum ego activation for basic survival which would create low

P values, but it provides the benefit of stable identity for social coordination and survival. As consciousness evolves, it might learn to selectively reduce ego activation, which enables access to greater quantum processing while being able to maintain functional coherence.

The framework proposes that consciousness develops through predictable relationships with identity itself, where each level represents a different solution to the fundamental challenge of balancing stable identity (requiring ego) with quantum processing advantages (requiring ego reduction). This creates a developmental trajectory where each stage represents a different answer to the question of human existence as they move from complete fusion with one's immediate experience ("I am my needs") to a recognition that the questioner creates the question ("I am"). Higher stages don't just think differently, but they actually access different degrees of quantum computation inside a larger geometric space.

Interestingly enough, as will be examined in this chapter, each level of consciousness perceives the limitations of levels below while potentially remaining unaware of its own constraints. Someone at Level 5 might recognize Level 4's conformity as limiting, while remaining unaware of how their own achievement metrics create constraints. Level 6 might see Level 5's success orientation as taking part in a meaningless game, while missing how their "authentic values" become another rigid structure, that is actually authored by an ego that is emergent itself. Berger (2013), while summarizing Kegan's developmental theory, identifies the core principle saying, "We cannot be responsible for, in control of, or reflect upon that which is subject" (p. 32). What we're identified with remains invisible. Only when something moves from Subject (identified with) to Object (observable) can we see its constraints.

The framework proposes that this has geometric correspondence as each consciousness level, may represent a different geometric configuration of identity constraints as Level 3's tribal identity creates one geometric structure, Level 5's achievement identity creates another, and Level 7's witness consciousness begins to observe and modulate the geometric constraints themselves.

Wilber (2005), lays out the potential of scaling these levels and increasing the geometric space one operates in, saying:

> In your own growth and development, you have the capacity to take self, culture, and nature to increasingly higher, wider, and deeper modes of being, expanding from an isolated identity of "me" to an fuller identity of "us" to an even deeper identity with "all of us"—with all sentient beings everywhere—as your own capacity for Truth and Goodness and Beauty deepens and expands. Ever-greater consciousness with an ever-wider embrace, which is realized in self, embodied in nature, and expressed in culture. (p. 38)

Wilber described what this book hopes to lay out in full. What follows is an exploration of how this progression might manifest across human development. Each level represents a specific geometric configuration of the ego-quantum interface, with measurable differences in performance capability, creative access, and unity experience. The journey from Level 1 to Level 9 might represent humanity's gradual mastery of its own quantum nature.

Stage I—Survival Consciousness (Levels 1–2) | Ego-Unaware

At these levels, ego is completely fused with immediate survival needs. The sense of self has minimal ability to observe itself which creates a perceived reality dominated by reaction.

Level 1: "I am my needs"

- **Core Collapse Pattern**: Collapses everything into immediate threat/resource
- **Primary Collapse Triggers**: Hunger, pain, any discomfort
- **Accessible Programs**: Conservation (Apathy), Social Bonding (Grief)
- ***P* Value**: Near zero—pure classical processing
- **Experience**: Life is moment-to-moment reaction.

This is pure survival. It is the most basic form of consciousness that operates through immediate survival reactions as life consists of moment-to-moment responses to hunger, pain, or discomfort. In this, any disruption triggers instant collapse of all possibilities into threat or resource. This represents consciousness in its most contracted state, where consciousness is experiencing every stimulus as a demand to an immediate response.

Level 2: "I am my power"

- **Core Collapse Pattern**: Collapses interactions into dominance hierarchies
- **Primary Collapse Triggers**: Challenge to authority, loss of control
- **Accessible Programs**: Conservation (Apathy), Social Bonding (Grief), Threat Detection (Fear), Obstacle Removal (Anger)
- ***P* Value**: Low-limited to power dynamics
- **Experience**: Life as conquest.

At this level, individual will emerges, but it remains focused on dominance and control. Every interaction collapses into hierarchical positioning as challenges to authority or perceived loss of control trigger immediate aggressive responses. Some access to power dynamics suggests slightly increased quantum processing in comparison to Level 1, although it is still limited to competitive advantage. Life becomes conquest, with success measured purely by position over others. Cook-Greuter (2021) describes this level saying, "Everything to them is a war of wills. Life is a zero-sum game" and they "may be aggressive to pre-empt expected strikes of others. They are people of action, not of thought and planning" (p. 43). To those at this level "Things look black or white. Others are either for me or against me" (p. 45). Interestingly when people believe that "I am my power" what does one do when they think they don't have any? Cook-Greuter does a fantastic job of describing this as a "self-protective" expression of this level. Those with power act in an "Opportunistic" expression while those without power act in "Self-Protective" expression. Where the Opportunistic expression uses strength or status to intimidate and dominate, the Self-Protective expression "focus[es] on protecting themselves and not becoming a target of others' aggression" (p. 44). Both expressions share the core identity structure of power-based survival, but manifest differently based on perceived strength and circumstance.

This represents the final preconventional stage where quantum processing remains severely constrained by high ego activation which is focused on power dynamics. The continuous threat assessment and dominance positioning maintain high collapse rates that prevent access to quantum insights that might reveal collaborative possibilities.

From a contemplative science perspective, these early levels operate through what Meling (2022), analyzing Mahāmudrā teachings, identifies as pure adaptive sense-making. Every experience triggers immediate judgment as good, bad, or neutral, followed by automatic approach, avoidance, or ignoring behaviors. This continuous reaction cycle, which Mahāmudrā traditions term samsara, maintains the high ego activation that blocks access to quantum processing potential.

Escaping these first two levels can be very difficult, because these levels create self-reinforcing loops. Survival-oriented programs maintain high ego activation which prevents access to the quantum insights that might reveal other possibilities and transitioning between any levels will feel painful to a degree. The two biggest, and likely most difficult shifts will be from Level 2 to Level 3 and from Level 6 to Level 7. Cook-Greuter (2021) describes individuals in transition as finding themselves in "a seeming no-man's land" where "the old way of mediating experience no longer satisfies while the new mode of processing has been glimpsed… but is not yet fully or consistently available" (p. 12). This matches the framework's prediction that transitioning between consciousness levels requires rebuilding one's entire reality-construction mechanism.

Stage II—Social Consciousness (Levels 3–6) | Ego-Unaware

This transitions to what the book will call the Stage 2 or the "social consciousness" levels. Between levels 3, 4, 5, and 6, ego begins incorporating abstract concepts while remaining the primary source of identity. At these levels, flow states often feel particularly powerful, and individuals may discover they can initiate these high P states through ritual, focused activity, or specific environments. This is the stage where the vast majority of people operate from. Cook-Greuter (2021) describes this stage's worldview, saying,

> By most modern expectations, fully functional adults see and treat reality as something preexistent and external to themselves made up of permanent, well-defined objects that can be analyzed, investigated, and controlled for our benefit. This view is based on a maximal separation between subject and object, thinker and thought. It epitomizes the traditional scientific frame of mind that is concerned with measurement, prediction, and control. (p. 36)

This is referred to as social consciousness because this is where the majority of adults reside. Cook-Greuter (2021) cites that in her "conventional stage" which roughly maps to this paper's social stage, her research has found "roughly 80% of adults" are in this stage (p. 48). This book would not be surprised if the Social Consciousness levels presented here actually account for more than 80% of adults (maybe upwards of 90–95% of adults in most countries), as these levels are a more encompassing than Cook-Greuter's levels are.

Level 3: "I am my tribe"

- **Core Collapse Pattern**: Collapses all situations into us-vs-them dynamics
- **Primary Collapse Triggers**: Outsider presence, group threat
- **Accessible Programs**: Conservation (Apathy), Social Bonding (Grief), Threat Detection (Fear), Obstacle Removal (Anger)
- ***P* Value**: Minimal—occasional pattern recognition
- **Experience**: World divided into safe/unsafe.

This is the beginning of the social consciousness levels. Level 3 represents the transition from survival to social consciousness. While it retains many survival characteristics like binary thinking, limited program access, and reactive patterns, it marks the important shift from purely self-focused, survival identity to group identity. This expansion beyond pure self, even though it is limited, establishes the foundation for all subsequent social development.

Cook-Greuter describes this transition saying, "It is the first stage that can be considered as socialized, that is, to know about basic interpersonal skills such as sharing, and protecting one another. Most importantly, people at this stage see others as important people in their own right" (p. 49). At this level, consciousness expands to include group identity, though still without true individual reflection, and situations collapse into us-versus-them dynamics as the presence of outsiders or any threat to group cohesion triggers immediate defensive responses. While occasional pattern recognition emerges, which suggests minimal quantum access, the worldview remains fundamentally binary, as it is divided between 1's and 0's like safe and unsafe or insider and outsider. Cook-Greuter describes this level, saying,

> The boundaries between self and others, however, are confused, literally fused and blurred. On the one hand, there is total acceptance of the family and in-group ... on the other hand, we see blind rejection of deviance and out-groups. It's "them" against "us" now. ... The values of one's own group get introjected as strong "shoulds," while the values of those who are different are denigrated and likely rejected as "evil." Thus, life is governed by rules of what can be and cannot be done. Others with differing views are morally condemned. (pp. 51–52)

From inside this level, breaking the rules of the group can feel genuinely threatening to survival. The structures that seem restrictive to people at higher levels of consciousness actually provide the essential scaffolding for identity at these levels. Even though it may seem at first that people at this level are constrained, that is not necessarily the case, people at this level can operate successfully in the world, as it would generally be through action as directed inside of teams or organizations. Skill can be developed effectively at this level through motivations that align with their identity. Cook-Greuter describes this level's identity saying, that people at this level "are not yet ready to take a stand to express themselves. They don't have their own opinions to assert although they may be good at expounding the values and beliefs of their group. Nor can they take initiative for themselves. They might do so for the benefit of their group when told to do so by an accepted authority" (p. 50). Overall, people at this level can live very content, happy lives with people they love around them. Cook-Greuter reports that approximately 8–11% of people are in this tier (2021, p. 38).

Level 4: "I am my roles and beliefs"

- **Core Collapse Pattern**: Collapses possibilities into right/wrong according to rules
- **Primary Collapse Triggers**: Authority disapproval, rule-breaking
- **Accessible Programs**: Conservation (Apathy), Social Bonding (Grief), Threat Detection (Fear), Obstacle Removal (Anger), Exploration (Courage)
- ***P* Value**: Moderate—can access flow in approved activities
- **Example**: "I am a good person" (as defined by authority or rules).

Level 4 identity forms through external structures and accepted beliefs. The question "Who am I?" most often finds answers in roles one fulfills inside of group affiliations, professional position, or family. All possibilities collapse into right or wrong according to established rules and expectations of the roles one fulfills, generally agreed on in a collective manner by a larger group. This level can access flow states through well accepted activities, which suggest moderate quantum processing capacity. Here, life feels secure within defined boundaries, although creative potential still remains limited in comparison to other levels.

This level aligns with what Berger (2013) describes as the beginning of Kegan's Third Order, where individuals have "internalized one or more systems of meaning" and "it is as if, in their growth from the Second order, those at the Third order have welcomed a Board of Directors into their decision making" (pp. 4–5). At Level 4, this "Board of Directors" as Kegan describes is just beginning to form, and it becomes more pronounced over time regarding how one makes decisions. People at this level may still "describe their and others' behavior in terms of I 'always' and we 'never' (Cook-Greuter, 2021, p. 54) but the significant difference from Level 3 to Level 4 is that even though people at this level will still strongly identify with groups, their identities will be more personally embedded in chosen roles and adopted beliefs, not necessarily in the group identities themselves.

Simply, people begin to see that they are an individual inside of a group, and that they don't have to agree with every single dynamic of their group but instead can craft the role they feel best suited to fulfill. It could be described that at this level one's "roles" become their "group" of level three. It's an abstracted individualization from within the individual's given group. Cook-Greuter actually describes how this individualization can begin to manifest at this stage, as one may find subcultures or other groups to join as they explore different roles and identities, saying,

> …it is possible to find some sense of OK-ness as a non-conformist. Rebels, have-nots and other misfits often need, find or create their own cliques. One can either be a loner or create or join an alternative group and display the markers that identify one as a member of that

> clique. This can happen because the individuals don't fit the prevailing norms or do not have the tokens that would help them to belong to the high status group. (p. 51)

Those operating from Level 4 often describe deep satisfaction in knowing their place and excelling in their roles inside their group whether it be to their family, friends, group, team, job, organization, city, state, region, religion, language, or country. Cook-Greuter describes how individuals at this level experience the expectations of the roles they fulfill, she writes, "The sense of 'shoulds' and 'oughts' and sense of shame and embarrassment if those should are not fulfilled can be crushing" (p. 53), as people at this level would feel sadness when saying something like "I'm a bad [insert role here]" when they feel as though they have fallen short of expectations. This tier is likely the most populated tier on average in the United States as Cook-Greuter reported that between 36 and 48% of people were in her approximately similar tier throughout her research and conducted studies (2021, p. 38).

From the framework's perspective, Level 4 consciousness cannot access programs beyond the Exploration adaptive program (courage), because when one begins the transition into the Exploration program it begins a fundamental shift in one's approach, as this Exploration is fundamentally necessary to develop the willpower and the individualized sense of self that defines Level 5 and onward. At this level, words like purpose and values would be anchored in the identity of their roles. This illustrates why the framework has an X-axis and Y-axis of consciousness. For example, a Level 4 person who identifies as a "good mother" might describe her "purpose" as raising good children, but this reflects dutiful courage of the identified role which follows external pressures of the roles she's adopted and beliefs about those roles she has. This would be much different than someone operating from inner-generated Purpose (running the contribution program) of Level 6 or 7. If external sources (government, community, religion, or inherited family expectation) defined motherhood differently, a Level 4 mother's reported descriptions of her role and her purpose would very likely shift accordingly. A telling question at this level is "What if this [insert external person, group, pressure, or expectation] wanted something else from your role?" Level 4 would respond according to their roles or groups because their identity remains externally defined. True Desire (Acquisition program) requires an individuated ego that has developed to be strong-willed enough to separate its wants and desires from group approval, while true Purpose (Contribution program) requires a significant level of inner authority that Level 4 has not yet developed. This gives reason for why Level 4 consciousness can often appear very content and less ambitious to someone operating from a Level 5 perspective.

Level 4's identity generally rests on external validation from playing within the established rules. At this level, accepting personal responsibility for failure means acknowledging that the external structures they've trusted don't actually protect them. This may manifest like a football fan who realizes their loyalty to their team, being the best fan (the role of a fan) that one can be, doesn't make them intrinsically happy or the religious individual who ends up realizing their rituals will not prevent their problems in the traditional sense. These types of insights are what generally begin to mark a transition into the Exploration Adaptive Program and will often mark a transition into Level 5. Essentially, individuals at this level can realize it's possible to take ownership of the "failures" of the system or the groups they are a part of or operating within even if there's justification for why the "system" they are within is suppressing their success. Hawkins (2020) outlines this saying, "It serves spiritual progress to always accept responsibility for all that befalls you and avoid the trap of being a victim" (p. 291). Instead of blaming external circumstances for their actions and outcomes, they can take ownership of these results and begin making significant change towards Level 5. This change can be perceived by fellow peers at Level 4 as arrogant or prideful. Jung (1963) highlights this difficulty of the transitions such as the one from Level 4 to Level 5, saying, "Collective organization is still so essential today that many consider it, with some justification, to be the final goal; whereas to call for further steps along the road to autonomy appears like arrogance or hubris, fantasticality, or simply folly" (p. 318). This is exactly why the exploration adaptive program (courage) is necessary as one generally has to begin to think that other's opinions are only "their opinions," and the individual is responsible for their own success or failure.

Level 5: "I am my actions and achievements"

- **Core Collapse Pattern**: Collapses opportunities into win/lose metrics
- **Primary Collapse Triggers**: Competition, failure, status threat
- **Accessible Programs**: Conservation (Apathy), Social Bonding (Grief), Threat Detection (Fear), Obstacle Removal (Anger), Exploration (Courage), Acquisition (Desire), Contribution (Purpose)—glimpses only
- ***P* Value**: Higher—peak performance in bounded domains
- **Example**: "I built this," "I earned that," "I know this".

At this level, a shift occurs as identity transfers from external structures to personal accomplishments. "You are what you do" becomes a common operating principle as many opportunities collapse into win-lose metrics as their "1's and 0's." Competition, failure, or threats to status often trigger significant ego activation. This level generates tremendous external success as companies are built, records are broken, papers are written, and innovations are developed. The clarity of metrics creates focused action and measurable progress. Many of society's most visible achievements emerge from Level 5 consciousness. Cook-Greuter writes that people at this level "now also tend to assert their own needs and wants, which were suppressed [in the previous levels]" (p. 55) and they "tend to focus on doing things right or correctly, not yet on doing the right things. That is, they offer single loop solutions, rather than questioning their preferred approach. Indeed, their own way of doing something is seen as the only right way" (p. 56). She says that this level "characterizes people who are now able to step back and look at themselves as objects from some distance … This allows them to begin to notice patterns and to start to reflect upon their own and others' behavior" (p. 54). Overall, she describes reaching this level from the previous as a "transformation from being a part of a group identity to finding one's separate identity and finding one's voice" and she writes that they show a "drive towards separation and individuation" (p. 56).

Berger (2013) notes that in Kegan's model, the transition to Fourth Order involves becoming "Chair of the Board" which is where "the person herself becomes the Chair of the Board" (p. 7). This captures Level 5's achievement orientation perfectly as the ego has taken control of its own development, creating what feels like mastery but remains, from Level 7+ perspective, an elaborate box of self-constructed identity.

While Level 5 consciousness is often stereotyped through business, athletics, and optimization where public figures like Alex Hormozi demonstrate absolute focus on behavior and one's results which generates extremely effective outcomes. Level 5 can be incredibly productive and often dominate the world's leadership positions. They have drive, motivation, and intrinsic energy that give them a leg up in the games that Level 5 individuals choose to play versus those operating at lower consciousness levels who happen to be playing the same life games.

Even though this level is often stereotyped through those traditional domains, the "I am my actions and achievements" identity manifests across all domains. For example, for healthcare workers it may be through lives saved or patients treated, or for teachers it could be through students mentored and their future performance. Parents often construct Level 5 identity through children's achievements and measure their own worth by their offspring's success. Academics may define themselves through publications and citations while artists may track followers, albums sold, and engagement metrics. Activists could count protests organized and policies changed. Most sublimely even spiritual practitioners or religious leaders can often fall into Level 5 patterns, by trying to quantify meditation hours, prayer time, retreats organized, sermons preached, or the number of members in their congregation; but regardless of the domain, the core pattern remains constant. At Level 5, identity derives from action and results, regardless of the field. This can be difficult to transcend as achievement is addicting, and it's very easy to get the ego to believe it's right at this stage as one gathers the evidence and builds a strong case to justify their worldview. This is why Cook-Greuter (2000) identifies this stage as "a kind of ceiling or barrier" because "it represents the frame of mind that is most attached to rationality and tends to defend most against the intrusion of material from non-rational sources" (p. 8).

The problem that many end up coming face to face with individuals here is that the achievement treadmill never stops. Each success only raises the bar and as a result of the incremental success, one's identity becomes increasingly fragile as any failure threatens one's core self-worth. High achievers at this level can feel like they're building a house of cards while pressure mounts to defend what they've built. Asking people at this level "What have you accomplished?" [in a given time period] cuts deep as it strikes at the foundation of their being. The transition to Level 6 generally begins when someone discovers that winning every game and accomplishing their goals and dreams can still leave them unfulfilled and wondering why they're even playing these games at all. Cook Greuter reported that about 29–35% of individuals in the United States are at this level (2021, p. 38).

From the framework's perspective, Level 5 consciousness can glimpse true Purpose (Contribution program) during flow states when ego temporarily recedes. In these moments of peak performance, measurement ceases and pure expression emerges. Athletes describe being "born to play" or entrepreneurs may feel "meant to build," yet afterward, ego reclaims the experience as achievement. "I was in the zone" becomes "I crushed it," and this creates Level 5's beautiful trap. Their greatest moments reveal what lies beyond achievement, but once they are back into their day-to-day mindset, their ego interprets transcendence as an ultimate achievement. They taste Purpose (Contribution program) and can spend years trying to achieve their way back to it without recognizing that achievement identity itself is what blocks their sustained access.

Level 6: "I am my values and purpose"

- **Core Collapse Pattern**: Collapses experiences into authentic/inauthentic
- **Primary Collapse Triggers**: Others' suffering, injustice, exclusion
- **Accessible Programs**: Conservation (Apathy), Social Bonding (Grief), Threat Detection (Fear), Obstacle Removal (Anger), Exploration (Courage), Acquisition (Desire), Contribution (Purpose), Collective Coordination (Love)—glimpses only
- ***P* Value**: Variable—very high in flow states
- **Example**: "I honor everyone's truth," "I want to explore the world's perspectives," "I want to spend my time on things that matter".

From Level 5 to Level 6, recognition dawns that achievements alone cannot sustain meaning. The identity shift from "I am my actions and achievements" to "I am my values and purpose" marks one of consciousness development's most challenging transitions. Where Level 5 found certainty in metrics and outcomes, Level 6 discovers that the entire achievement game was constructed and entirely arbitrary. They realize that the ego just likes to pick the games it's best at and claim that this specific game is the only one that actually matters, but once this is realized, one cannot return to unknowing participation in these constructed games. This makes Level 6 markedly different from Levels 4 and 5 to a significant degree.

In this framework Level 6 is a combination of two and a half of Cook-Greuter's (2021) levels, this includes the "Individualist 4/5" and "Autonomous 5" levels as well as the "Construct-Aware" expression of her "5/6 stage." The framework does this because this framework is focused on identity in relation to the geometric space of the Quantum Information Dimension and it is interested in an individual's modulation of their ego in relation to the degree it collapses their brain's quantum computational potential. Considering this, the obvious paradigm shift in consciousness levels comes at the "ego-aware" stage (Level 7 in this framework) as it represents an entire paradigm shift in one's relationship with their ego while the "construct-aware" expression that Cook-Greuter (2021) describes in detail represents the best, most advanced expression of this framework's Level 6, but even in this peak form of Level 6, the individual still interprets that "I am having these awareness breakthroughs" and that "I can see all the fake games everyone plays" but hasn't yet realized the constructed nature of their own ego making these claims.

Overall, the framework groups her two and a half levels together because they share a common thermodynamic theme as described in Chap. 9. While an individual would significantly develop their wisdom through this level, advancing through what developmental psychology characterizes as fourth, fifth, and sixth person perspectives (Kegan, 1982; Wilber, 2007), the energy cost of maintaining identity and belief structures would remain within a similar moderate band.

The journey through Level 6 may begin by wrestling with the questions of authentic self-expression and start to grasp how to hold multiple perspectives simultaneously. This fourth person perspective allows someone to recognize that "my perspective from this context differs from my perspective in that context" and understand that multiple authentic selves can coexist. This person might say "I honor everyone's truth" or "Which self is the real me?" They experience what Cook-Greuter (2021) describes as the Individualist stage, where boundaries become fluid and relativism opens new territory.

As development continues, this exploration consolidates into purposeful contribution (when they begin to access the Contribution Program) as the individual begins to organize their understanding around taking action from their believed purpose for the greater good. This is where the fifth person perspective emerges which enables consciousness to perceive patterns across multiple viewpoints and see the systems that generate those perspectives. They might say "I serve humanity" or "I can see patterns across systems." Cook-Greuter maps this as the Autonomous stage, where systems thinking emerges and integration of the sense of the self replaces the multiplicity the individualist stage had been exploring.

Then from here, those at this level may reach what would be the beginning of the construct-aware expression as described by Cook-Greuter (2021), where meaning-making itself becomes visible. Sixth person perspective becomes accessible, which allows consciousness to witness the construction of perspectives themselves. This person might say "I see how frameworks, purpose, and understanding itself are constructed." Yet even sixth person perspective, which can witness the construction of meaning-making, remains fundamentally different from what emerges at Level 7, because the sixth person still identifies with being the one who sees these things. The seer remains unobserved.

What unifies these developmental expressions as a single level in this framework is the thermodynamic cost. Throughout Level 6's range, individuals optimize the structure of their beliefs but still maintain a rigid foundational identity. Whether that foundation manifests as "I am someone who honors multiple perspectives" (fourth person), "I am someone who sees systemic patterns" (fifth person), or "I am someone who is aware of meaning's construction" (sixth person), each formulation preserves the core structure of "I am the one who…" which is grounded in their purpose and values. Even though the cognitive capacity advances considerably, the energy cost of maintaining their foundational identity remains within a similar moderate range, as they have a similar amount of beliefs, or even if the number of beliefs has decreased, they have increased the rigidity (certainty) of a certain subsector of beliefs. Therefore, despite the increasing sophistication in what the ego understands, the ego still requires maintenance as subject rather than object.

At this level, the identity reformation appears dramatic compared to the previous level as Level 5's "I am my actions" becomes Level 6's "I am my authentic values and purpose." Cook-Greuter (2021) describes the difference in meaning one can experience at this stage, saying one's meaning and eventually their identity in this stage comes from "trying 'to become the most one can become' in contrast to the goal of Achievers [Level 5] hoping to be the best one can be. In the Strategist's [middle of Level 6's] eyes, good company, good questions, intimate relationships, and a meaningful occupation as well as chances for self-actualization and self-fulfillment are essential for a meaningful existence" (p. 86). She also states that people at this level "consciously commit to create a meaningful life for themselves and for others in the world through self-determination and self-actualization. They realize that the way they tell their life stories changes with additional experiences and insights" (p. 84) and "With their sense of transformational mission comes [a] set of broader, human principles one must uphold such as the right of all human beings to be treated with justice and mercy" (p. 83). These values and commitments to purpose drive their identity structures at this level.

Yet through $P = Q/E$, both levels' identities still remain steady ego structures, but they just operate with different ego collapse triggers. At this level, experiences now get categorized as authentic versus inauthentic expressions of the "true self" instead of the successful versus failed categories of Level 5. People at Level 6 can relentlessly pursue "their true self" and spend a lot of time reflecting on "who they truly are."

Level 6 consciousness can receive significant insights through diminished ego states, particularly during flow states, meditation, or psychedelic experiences. They can be extremely effective in the world because they are able to hold many opposing views at the same time, compare and contrast them, and find hidden wisdom in the gray areas.

Cook-Greuter (2021) elegantly describes many individuals at this level. She says,

> Individuals [at this level] begin to see paradox and ambiguity as an inevitable dimension of living and to increasingly tolerate these. Autonomous persons [middle of Level 6] become aware and fluent in using polarity thinking in order to examine issues and tensions. Complex and sustainable solutions often require that one pays attention to opposing but interdependent values. The most effective and healthy designs foster the wisdom aspects of both poles and the[n] minimize their downsides. However, the belief that one can at least approximate an accurate picture of the territory with diligence and through dialogue and collaboration is still prevalent. Autonomous individuals are aware of multiple interrelated systems and how influences are circular, cyclical and often tapestry-like. They have beginning access to an understanding of life that can integrate psycho-logically paradoxical elements. Therefore, less energy needs to be spent on 'defending.' This then allows Strategists [middle of Level 6] to be more accepting and spontaneous than adults at earlier stages. (p. 86)

Cook-Greuter wonderfully acknowledges what this book described in Chap. 9 as the thermodynamic costs of defending identity and beliefs, and she outlined how Level 6 escapes many of these computational costs by stripping away their rigid beliefs and identities as they begin to think of the world in more probabilistic "paradoxical" terms. However, interpretation of truth and facts still occurs through the ego structure not yet recognized as optional. This can lead to elegant yet incomplete models that place the localized observer at awareness's center, having not yet conceived that the "I" might be a tool shaped by evolutionary necessity. Instead, they generally see the" I″ as being the origin of objectively real perception and experience.

This mirrors an earlier turning point in scientific history. Just like Galileo argued for years that the Earth was not the fixed center of the cosmos, contemplative traditions have long argued that the ego is not the true center of awareness. Despite this, much of modern science continues to operate from an egocentric framework. In that sense, just like the Earth orbits the sun, the ego "orbits" the quantum field. The difficulty is getting past this boundary, or metaphorically finding escape velocity from the ego. To do so seemingly requires going down a path questioning one's assumptions and trying to understand "who one is."

Jung (1969a) describes how difficult this line of questioning can be to answer as this can be a defining struggle of one's experience in Level 6. Importantly, when Jung uses the word "consciousness," he refers to all that is in the waking awareness of the individual in regards to the ego's narrative awareness, while the "unconscious," which is equally as important to the whole self to Jung, represents the part of the self that one is "unaware" of. The initial question of "who am I" leads into further questions and the process continues. Jung highlights this line of questioning when establishing what he means by the term "the self." He writes,

> When one speaks of man, everybody means his own ego-personality—that is, his personality so far as he is conscious of it—and when one speaks of others one assumes that they have a very similar personality. But since modern research has acquainted us with the fact that individual consciousness is based on and surrounded by an indefinitely extended unconscious psyche, we must needs revise our somewhat old-fashioned prejudice that man is nothing but his consciousness. This naïve assumption must be confronted at once with the critical question: Whose consciousness? The fact is, it would be a difficult task to reconcile the picture I have of myself with the one which other people have of me. Who is right? And who is the real individual? If we go further and consider the fact that man is also what neither he himself nor other people know of him—an unknown something which can yet be proved to exist—the problem of identity becomes more difficult still. Indeed, it is quite impossible to define the extent and the ultimate character of psychic existence. When we now speak of man we mean the indefinable whole of him, an ineffable totality, which can only be formulated symbolically. I have chosen the term "self" to designate the totality of man, the sum total of his conscious and unconscious contents. (p. 92)

As he defines his term of "self," Jung outlines the extremely difficult line of questioning that someone faces as they leave Level 5 and begin to ask "what do I really want" which leads into questions like "who am I really?" which then leads down a rabbit hole of difficult and uncomfortable questions that many do not yet have the ability to address. One must not only face the external definition of self, but the internal known definition of self as well as the unknown parts of their self which are hidden in their unconscious. This "peeling back of the onion" of self-identification often represents the journey one partakes in Level 6, until one reaches its peak, which would be this "construct-aware" representation. Cook-Greuter (2021) defines this saying,

> The Construct-aware stage represents folks who are aware that meaning is constructed, invented, generated rather than inherent in things, events and concepts. They start to wonder about the meaningfulness of more and more complex thought structures and integrations such as can be imagined with a fifth or nth person perspective. Playing the complex numeric abstractions and formulas game is one way the ego can increase its sense of control and

> meaning. It also seems to please some individuals to demonstrate their capacity for mental gymnastics. (p. 97)

From the framework's perspective, when one has fully integrated the ideas that being "Construct-aware" provides, at this point an individual is at the cusp of breaking through to Level 7's ego-aware paradigm, as they already acknowledge the constructed nature of so much of reality. Therefore in this framework, Level 6 will have a wide range of development, representation, and expressions across humanity, but an underlying theme that would likely connect all at this stage is that their identity still derives from their values and their elected purpose and they still think that "I am the one with these ideas."

Considering Level 6's driving motivation of values and purpose, witnessing others' suffering, injustice, or exclusion can trigger ego response, though now it comes from empathy instead of being from competition at Level 5 as Level 5 is often equipped with the acquisition program while Level 6's base program is the contribution program. This level attracts those seeking meaning beyond material success. Social movements, authentic arts communities, or purpose-driven entrepreneurship often emerge from Level 6 consciousness. Many podcasters may operate from Level 6, as they bring on guests from diverse backgrounds to explore a wide variety of perspectives while treating all viewers, ideas, and perspectives with respect. This level often recognizes the constructed nature of any single worldview as individuals experience what Berger (2013) describes in Kegan's Fourth Order (1994), saying, "they have created a self that is the author of its connections to the ideologies, institutions, or people which they were previously written by" (p. 7). The framework describes this as ego nearing its final sophisticated form, as the self believes it has transcended external definition by becoming the author of its own values and purpose, but still not recognizing that authorship itself is just another construction.

Cook-Greuter (2021) outlines how many in the construct-aware stage (Level 6) often find purpose in mapping out their progress, but she warns of a trap saying, "Ego development predicts that the desire to capture ever more of the underlying territory is an aspect of construct-awareness and the cognitive delight in complexity. Based on observation this genuine desire and curiosity to create comprehensive orientation maps is also often coupled with subtle forms of ego-gratification and, in academic circles, a batch of importance" (p. 116).

Meling's (2022) analysis of Mahāmudrā practice, in relation to non-dual awareness, reveals that Levels 4–6, while increasingly sophisticated, remain trapped in what contemplatives call Phase 1 enaction. Despite growing complexity in their sense-making, these levels continue creating subject-object distinctions through every thought and perception. As Meling notes, citing Mahāmudrā instructions from Kyabgon (2004), "to judge discursive thoughts as either good or bad is merely to engage in further discursive thinking" (p. 6). This does a good job describing the recursive trap these levels inhabit, where even attempts to transcend can create new identity structures as one judges and categorizes their attempts at transcendence or development.

Many can feel safe and stable at Level 6 and the stability of consciousness levels may arise from their commitment to, or certainty in, their organizing principles. Hawkins (2020) observes that "an individual's level of consciousness is determined by the principles they're committed to" (p. 54). This maps directly to the framework where Levels 3–6 maintain ego activation (E), which is still engrained in a massive amount of identity construction across all aspects of one's life as deeply held principles and beliefs create stable relevance realization patterns, to preserve rational coherence, because significant quantum processing (Q) might be perceived as threatening to their worldview. This process maintains consistent E levels and therefore more stable P access. This stability creates a sense of safety for the ego.

This stability can also create a false sense of arrival, because of the commonalities of Level 6, who identify with their values and feel as though they have the ability to author their own values. A difficult trap emerges from this sense of arrival as many likely mistake these short-lived experiences of higher programs through flow states and other "low E" states for true enlightenment. While ego becomes more flexible, and it does begin to see the real practical value in Purpose and Love as its main programs (as will be described in future sections), the ego is still the primary observer. Cook-Greuter (2021) goes on to describe the ironic identities that can emerge in Level 6, saying, "The more one becomes attached to the idea of non-attachment, the more firmly one remains fettered. The more one is conscious and proud of one's psychic powers and ego-transcending quest, the more clearly one's ego is still enthroned" (p. 101).

Jung (1963) similarly writes about the first transcendent experiences of individuals in their development moments akin to this level, saying,

> The need for such a secret is in many cases so compelling that the individual finds himself involved in ideas and actions for which he is no longer responsible. He is being motivated neither by caprice nor arrogance, but by a dira necessitas which he himself cannot comprehend. This necessity comes down upon him with savage fatefulness, and perhaps for the first time in his life demonstrates to him ad oculos the presence of something alien and more powerful than himself in his own most personal domain, where he thought himself the master. (pp. 316–317)

The "dire necessity" Jung describes may reflect first encounters with the Contribution program, where purpose-driven consciousness emerges beyond the traditional personal achievement and actions of Level 5. The "secret" Jung describes is this internal dialogue that begins to take place, as describing this to others, especially in Jung's time, would be perceived as crazy. The experience of "something alien and more powerful" in one's "most personal domain" is describing how at this point the person encounters something in their own psyche that they cannot control through willpower, reason, or identity management. It's this first glimpse of what lies beyond the ego and corresponds to what the framework proposes occurs when consciousness begins accessing patterns beyond ego-level control, which potentially includes early glimpses of Collective Coordination through flow states where individual agency dissolves into larger organizing principles.

Cook-Greuter (2021), describes this same difficulty of finding oneself at the peak of Level 6 in the midst of these new experiences while beginning to face the call to adventure to Level 7, saying,

> …they experience moments of freedom from the ego's constant efforts at control and self-affirmation. Yet, at this stage, such experiences are short-lived. As soon as one evaluates and judges them, the magic is broken.
>
> We are naturally reluctant to relinquish our familiar mode of functioning. We do not want to give up the illusion of our enduring separate identity, which we so laboriously learned to define, cultivate, and rely on for most of our conscious lives. Even if we understand that letting go of our attachment to the known will bring freedom from familiar kinds of suffering, attempts at doing so are ineffective and often lead to intractable paradoxes. (p. 102)

Essentially it would be quite common for the final form of Level 6 to often embody "meta identities," for example, statements like "I understand understanding" may be something someone would say here, where the "I" still claims the experience. Finally, Cook-Greuter describes one large obstacle that often stands between Level 6 and Level 7. "What makes life meaningful" can be one of the most difficult questions at this stage. This provides reason for what ends up being a common "nihilism trap" in Level 6.

Jung (1963) describes the moment when these established meta-frameworks no longer satisfy the individual and they feel the need to address this sensed call to adventure, to explore what may lie beyond the ego, saying,

> Nevertheless it may be that for sufficient reasons a man feels he must set out on his own feet along the road to wider realms. It may be that in all the garbs, shapes, forms, modes, and manners of life offered to him he does not find what is peculiarly necessary for him. He will go alone and be his own company. He will serve as his own group, consisting of a variety of opinions and tendencies which need not necessarily be marching in the same direction. In fact, he will be at odds with himself, and will find great difficulty in uniting his own multiplicity for purposes of common action. Even if he is outwardly protected by the social forms of the intermediary stage, he will have no defense against his inner multiplicity. The disunion within himself may cause him to give up, to lapse into identity with his surroundings. (p. 316)

Jung acutely describes this journey of transformation as the book describes as Level 6. At the beginning someone begins to "serve as his own group, consisting of a variety of opinions" which specifically describes how one begins to hold beliefs loosely at this level. One can hold a wide variety of disagreeing opinions in his mind, where in previous levels this would not have been possible. This aligns with the quantum cognition ideas presented in Chap. 9 and, that at this level, thinking in probabilities and paradoxes begins to come more naturally. The "difficulty in uniting his own multiplicity" reflects one beginning to operate from a consciousness that can see contradictions but remains identified as the one trying to resolve them. At the peak of Level 6, Level 7 has become visible but still remains inaccessible as the ego continues to return to claim ownership of experiences. The person experiences "inner multiplicity" without yet realizing that the observing ego itself creates this experience of fragmentation. When consciousness realizes all meaning is constructed but hasn't yet recognized the ego as the constructor, significant disorientation can occur, and everything can seem arbitrary and meaningless. Cook-Greuter (2021) states that in the final moments of Level 6, ego is "valiantly trying to create a coherent narrative" even as that narrative dissolves, representing "a last ditch attempt to create the most comprehensive maps possible" (p. 96).

The shift to systems thinking is often one of the first postconventional breakthroughs. As Cook-Greuter (2021) notes, systems thinkers "realize that their perspectives are local, partial, context-dependent and culturally conditioned" and become "aware of themselves as participant observers" (p. 108). However, they "do not yet understand how profoundly all humans are conditioned into the language habit" (p. 108).

As McAdams (1993) has shown, humans construct identity through internalized life stories which are the narratives that give people coherence and purpose. But development requires recognizing that we are not just the character within these stories, but we are actually the author of the stories themselves. This insight often marks the shift into Level 7.

As described, the final process of Level 6 and the first steps into Level 7 often begin with what Berger (2013) characterizes as Kegan's Fifth Order capacity to "look across inner systems to see the similarities that are hidden within what used to look like differences" and become "more likely to believe that what we often think of as black and white are just various shades of gray" (p. 9). However, the framework extends beyond Kegan's intellectual description to the lived experience of ego as an observable tool instead of it operating as one's identity.

Through P = Q/E, Level 6 has optimized ego for values and purpose as they have successfully created a sophisticated identity that feels transcendent, therefore it requires one to release this carefully constructed self to reach Level 7's ego-aware stage. This requires abandoning what feels like the pinnacle of development as the ego must recognize itself as optional at the exact moment when it has achieved its most refined form.

Critically, Berger reports that Kegan found "between 3% and 6% of adults aged 19-55 make meaning in the transition between the Fourth and Fifth orders (the peak of Level 6); no adults in the studies Kegan reports made meaning fully at the Fifth order" (p. 9). Similarly, Cook-Greuter (2021) reports that less than 2% of individuals were at her construct-aware stage (p. 21). Considering that Cook-Greuter's Construct Aware Stage represents the peak of Level 6 in this framework as the final precursive to the ego-aware stage, one can infer that that stable Level 7+ consciousness remains exceptionally rare, likely less than 1% of people.

Stage III—Integrated Consciousness (Level 7–8) | Ego-Aware

Level 7 represents the single most important shift in human development as ego becomes observable instead of being the sole observer. For the first time, consciousness can watch ego attempt to collapse quantum states and have control over whether this process takes place or not.

The framework proposes that the transition from Level 6 to Level 7 represents what is often portrayed archetypally in culture as the hero's journey (Campbell, 1949). Likely it is difficult to shed preconceived beliefs, shed the held ego, and adopt new fundamental principles that will guide one's life at this level.

The transition from Level 6 to Level 7 represents what Meling (2022) identifies as the shift from Phase 1 to Phase 2 enaction. This requires learning to "look intently" at thoughts without judgment and recognizing sense-making AS sense-making instead of being caught within it. As Mahāmudrā instructions emphasize, this involves "neither abandoning nor accomplishing," simply observing without adding further conceptualization (p. 6). This is the practice of learning to observe ego without being ego; to see the collapse mechanism without automatically engaging it.

Cook-Greuter's (2021) evolving understanding provides interesting validation for this framework's distinction between Levels 6 and 7 and the significance of the transition. She was initially unsure whether Construct-aware and Ego-aware represented sequential stages or personality variations. As originally, she did not have an ego-aware stage in her levels of consciousness, but her thirty years of observation led her to hypothesize that "construct-awareness comes somewhat earlier on board in meaning making than ego awareness" (p. 96). This matches the framework's prediction that Level 6 will likely recognize that meaning is constructed before reaching Level 7 when one recognizes the ego as the meaning's constructor.

Jung (1963) in his autobiography in the last years of his life documented the precise sequence through which consciousness transitions from what this book describes as the construct-aware Level 6 to the ego-aware Level 7. His description captures the mental experience of discovering that the ego conducting internal trials is itself another construct. The sequence begins at late Level 6, where someone has already seen through external authorities as he writes,

> But if a man faced with a conflict of duties undertakes to deal with them absolutely on his own responsibility, and before a judge who sits in judgment on him day and night, he may well find himself in an isolated position. There is now an authentic secret in his life which cannot be discussed if only because he is involved in an endless inner trial in which he is his own counsel and ruthless examiner, and no secular or spiritual judge can restore his easy sleep. If he were not already sick to death of the decisions of such judges, he would never have found himself in a conflict. For such a conflict always presupposes a higher sense of responsibility. It is this very quality which keeps its possessor from accepting the decision of a collectivity. In his case the court is transposed to the inner world where the verdict is pronounced behind closed doors. (pp. 317–318)

Jung describes someone who has become "sick to death" of external judges, which is when someone has recognized all societal authorities, religious systems, and cultural norms are constructed and not absolute. In a sense, the person begins to see "turtles all the way down," as Jung is saying the person wouldn't be in this internal crisis if external authorities still satisfied their ego. This recognition forces what Jung terms a "higher sense of responsibility" because, once someone realizes this, authority and meaning-making can no longer be outsourced to external systems. The person must take full accountability, which then results in conducting an "endless inner trial" where they serve as "counsel and ruthless examiner" as they begin to realize that everything is a construct, this is the secret that the individual cannot divulge as Jung describes, because those around them won't understand their experience, and may even call them crazy if the person is honest about their internal dialogue. This is something Jung said, in his autobiography, that he experienced in his own journey through consciousness development (Jung, 1963).

The phrase "behind closed doors" captures the opacity of this stage as the internal tribunal operates privately and is invisible to others as well as to the person conducting it. At first, they realize that they are the judge but cannot yet observe themselves (the ego) judging. This process could take a long time, but eventually it nears completion. At this point, once all these initial inner-trials are complete, questions and answers have emerged about all the world's external constructs, but the internal authority itself remains unquestioned. This represents the final configuration of the ego-unaware paradigm, the peak of Level 6, as at this level you can observe thoughts, you can observe emotions, you can observe beliefs, you can observe purpose, you can observe all the world's interconnected systems, but you can't observe the observer; because you still are the observer.

Jung then beautifully describes what type of questions must emerge to dethrone the ego once it achieves this peak form. The ego must start asking the same questions it asked of the world, now about itself. It may ask, "well who is this judge?" "What gives this internal authority legitimacy?" "What gives it the right to decide what is right and wrong or what is valuable and not valuable?" "Why do I get to make something meaningful?" "Why am I worthy to judge at all?" Jung (1963) writes immediately after the previous quote,

> Once this happens, the psyche of the individual acquires heightened importance. It is not only the seat of his well-known and socially defined ego; it is also the instrument for measuring what it is worth in and for itself. Nothing so promotes the growth of consciousness as this inner confrontation of opposites. Quite unsuspected facts turn up in the indictment, and the defense is obliged to discover arguments hitherto unknown. In the course of this, a considerable portion of the outer world reaches the inner, and by that very fact the outer world is impoverished or relieved. On the other hand, the inner world has gained that much weight by being raised to the rank of a tribunal for ethical decisions. However, the once unequivocal ego loses the prerogative of being merely the prosecutor; it must also learn the role of defendant. The ego becomes ambivalent and ambiguous, and is caught between hammer and anvil. It becomes aware of a polarity superordinate to itself. (p. 318)

The shift occurs when "the once unequivocal ego loses the prerogative of being merely the prosecutor; it must also learn the role of defendant." The internal judge that once seemed absolute to the individual must now defend its own authority to judge. This dual role creates observational distance, the subject-object internally, as ego functions as both prosecutor and defendant simultaneously. It becomes observable for the first time. You cannot be solely identified as the judge when you are also defending yourself to yourself.

Jung describes this as the ego becoming "ambivalent and ambiguous, and is caught between hammer and anvil." The framework proposes this ambivalence reflects the fundamental recognition that ego operates as another constructed system. The person has traced constructs all the way down and discovered the internal judge conducting trials "behind closed doors" is itself a turtle and is just another layer that requires examination. The ego that seemed to be the ultimate arbiter of truth becomes visible as one more construction.

Jung's observation that "Nothing so promotes the growth of consciousness as this inner confrontation of opposites" is especially interesting when considered in the light of Chap. 9. This is inherently rooted in quantum physics language as it the psychological equivalent of "superposition." As these "opposites" would be the extreme binary beliefs and questions one must address before their consciousness journey can continue. One must accept that opposites, don't have to be binary, and must come to understand how they can exist in superposition in one's mind which is incredibly beneficial as it limits the energy needed to maintain these ideas' collapsed rigidity as Chap. 9 showed. Jung's observation that contradictions "constantly threaten the unity of the personality, and entangle life again and again in their dichotomies" once again uses this "binary" physics language as a correlate and it corresponds to why it can be reasonable to characterize the final moments of Level 6 and the beginning of Level 7 as "tormented." The person can now see how the ego constructs its own perceived reality, witness the judging mechanisms operating, and observe the prosecutorial and defensive functions simultaneously, therefore one must pay the thermodynamic cost to unravel one's beliefs. Each binary certainty that held consciousness in Level 6 must be released back into superposition, which requires the Landauer cost ($k_B T \ln 2$) per collapsed bit as Chap. 9 outlined. This is why Jung describes someone here as "caught between hammer and anvil" as it reflects the actual metabolic expenditure required to maintain ambiguous superposition states instead of simply collapsing into comfortable binary certainties. This is where the process of integration must begin and then be maintained.

This transition mirrors Kegan's (1982) subject-object theory of development, where each stage of growth emerges by making the previous unconscious structures of identity into conscious objects of reflection. At Level 7, the ego itself becomes an object, which marks the developmental threshold into self-authoring awareness. Only after the "Construct-aware stage" (after Level 6) do individuals "appreciate the magnitude of humanity's automatic and unconscious dependence on the language habit." They recognize "the incommensurability between symbol and that which is symbolized" (Cook-Greuter, 2021, p. 10). In the framework's terms, they finally see how ego participates in reality's creation through selective collapse and how it contributes its narrative structure to the ongoing process instead of just describing pre-existing reality.

The phenomenology of ego dissolution includes specific perceptual shifts as Hawkins (2020) describes how "perception of the body shifts…from I am the body to I have a body" (p. 151), with the body becoming more "like a puppet or a pet." Hawkins expands on this idea saying,

> This process of surrender takes a person all the way to the top of the Map of Consciousness. At first, there is the identification: I am the body. As the inner awakening continues, it becomes quite obvious that I am the mind that experiences the body, not the body. As more feelings, thought patterns, and belief systems are surrendered, there eventually comes the

> awareness: I am not the mind either. I am That which witnesses and experiences the mind, emotions, and body. I am consciousness itself. (p. 151)

So in this context, Hawkins saying "I am not the mind either" is the book saying "'I' am not the ego" as scientifically defined, this moment mirror's what Jung described when the ego is "caught between hammer and anvil." Hawkins is in agreement that consciousness does not require the psychological "ego" or neuroscience's "DMN" to operate within the world. He agrees that "I am consciousness itself," essentially saying that this awareness of all systems at work is human existence in its most raw, unfiltered form.

Considering this, it may be critical to discuss how one would even try to traverse through these levels. Meditation is one common tool used along this journey, although not the only. The precise cognitive mechanism of this transition has been partially mapped through Meling's (2022) systematic analysis of Mahāmudrā contemplative practice. The shift from Level 6 to Level 7 requires developing what the tradition calls "self-liberation of thought" (rang grol), which occurs when practitioners learn to "look intently" (cer gyis lta) at thoughts without engagement. As Meling documents, citing Dunne (2015), thoughts become experienced "just like a facet of mind" rather than representing external reality (p. 7). From the framework's perspective, this describes learning to observe the collapse mechanism itself instead of getting trapped within the ego's products.

The practice requires two specific capacities that map directly onto the framework's predictions. First, non-adaptivity, meaning the complete suspension of evaluative judgments. As Meling notes, citing Khenchen Thrangu Rinpoche (2014), "Simply by looking at itself, thought is pacified" (p. 8). Second, the development of reflexive awareness (svasaṃvitti), which Meling describes as awareness that recognizes thoughts as "actions of the mind" without making them objects of observation (p. 7). This capacity already exists within consciousness but remains obscured by continuous sense-making until Level 7 transition begins.

The gradual approach shows a practical pathway to achieve this as initially, practitioners use an anchor such as breath, though as Dunne et al. (2019) clarify, cited by Meling, this serves as a "reminder" preventing distraction rather than a meditation object (p. 8). Attention gradually shifts to "off-object features" including the quality of awareness itself. Eventually, even the anchor dissolves, leaving what Meling terms "sustained non-propositional meta-awareness" (p. 7). From the framework's perspective, this describes consciousness learning to witness its own collapse patterns without being compelled to complete them, which enables conscious modulation of the $P = Q/E$ equation for the first time.

It bears emphasizing that while contemplative traditions like Mahāmudrā provide systematic approaches to this transition, traditional religious or mystical training is not inherently required to achieve Level 7 consciousness. The framework suggests that introspection, quiet awareness, and understanding that these levels exist may be sufficient for many individuals to access these states over time. Simply knowing what to look for, being able to recognize emotions and judgements as the ego's perception of threats to identities and beliefs, and having a conceptual map of the territory before being there could allow consciousness to recognize its own patterns and begin the transition.

Many individuals may spontaneously discover and experience Level 7 awareness through life experiences, creative practices, religious experiences, athletic flow states, or even intellectual inquiry that leads to recognizing the constructed nature of identity. The advantage of traditional practices lies not in exclusive access but in their refined methodologies developed over centuries. Yet in the modern era, with widespread access to information about consciousness development, individuals may find their own pathways through psychotherapy, journaling, nature immersion, artistic expression, or simply dedicated self-observation. The key appears to be developing the capacity to observe one's own mental processes without being identified with them and without judging them in the slightest, regardless of the specific method employed.

The breakthrough to Level 7 occurs when, as Cook-Greuter (2021) describes, consciousness becomes aware of "the ego's fundamental needs and efforts at meaning creation" (p. 96). This shift from seeing constructs to seeing the constructor marks the transition from Level 6's construct-awareness to Level 7's ego-awareness.

Level 7—"I observe the 'I'"

- **Core Collapse Pattern**: Notices patterns of collapse and can delay or prevent them
- **Primary Collapse Triggers**: Can observe triggers without automatic collapse
- **Accessible Programs**: Exploration (Courage), Acquisition (Desire), Contribution (Purpose), Collective Coordination (Love), Integration (Peace)—glimpses only—and can observe but not inhabit lower programs
- ***P* Value**: High sustained access
- **Example**: "I notice the ego wanting to collapse this into winning/losing".

At Level 7, a fundamental shift occurs as one enters Stage 3, or "Integrated Consciousness." Instead of being fused with identity, consciousness develops the capacity to observe identity forming and dissolving. Patterns of collapse become visible and can be interrupted or redirected. Unlike previous levels where triggers create automatic responses, Level 7 can watch triggers arise without mandatory collapse.

The framework proposes that Level 7 is where people begin to realize that "I" is only a useful grammatical structure and not a necessary or fundamental part of reality. Cook-Greuter (2021) describes this saying, "Ego-aware individuals report more often than people at earlier stages that they are watching or witnessing the parades of thoughts and feelings come and go without trying to direct them" (p. 102). Cook-Greuter (2021) articulates the jump into this new tier, Integrated Consciousness (which she calls Transpersonal), saying,

> Individuals at this level are now able to witness the whole song and dance of prior ways of understanding and meaning making with compassion and equanimity. They understand the need of the personal ego to ensure a sense of permanence and substantiality while at the same time recognizing the illusion of this desire for permanence. The previous way of viewing reality solely from the self's perspective and through the medium of language is transformed. The new paradigm has a universal or cosmic perspective as an organizing principle and as a steady place from which meaning is derived. It is non-centered in the ego although the ego

> is available as a perspective when useful. Unlike state experiences that gave people glimpses of mystical or unitive experiences, now these are steadily available in the witnessing stance. (p. 107)

Level 7 is where status games quickly start to become meaningless as the person sheds the ego's need for external validation. When asked who they are, people at this level may be more likely to refer to fundamental forces, values, or first principles that drive their actions. Level 7 is inclined to think broadly and systematically. Lower levels would try to weaponize first principles and values, while Level 7 realizes it's more effective to let the first principles and values themselves drive the being's action and motivation. Flow states and insights at Level 7 become significantly more powerful. This is especially true if the individual has a fundamental grasp on the most powerful underlying first principles, has developed grounded values, and possesses a large database of knowledge that can be accessed and utilized during flow states. This may be why utilizing drugs to access higher states of consciousness is only effective to a smaller degree. Generally, the other ingredients needed for deep integration of these truths may not be available to the people yet, as they haven't integrated the learned experiences and knowledge of lower levels. Lower levels may attempt to imitate Level 7's humility and systems thinking, but the imitation reveals itself through subtle status-seeking within the performance of transcendence. While those truly integrating Level 7 wouldn't inherently seek status. Status would only be a biproduct of insights received, systems built, and impact created by those at this level.

The geometric space of Level 7 consciousness is fundamentally restructured as the pathways to the Energy Conservation Program, Social Bonding Program, and Threat Detection Program still exist in the neural architecture, but without ego identification, consciousness cannot fully enter these spaces. This sheds some light on why Level 7 individuals sometimes appear strange or unusual to those at other levels. In situations generally perceived as negative or stressful that would normally trigger massive emotional response in most people such as public speaking, emotional confrontation, or financial uncertainty, Level 7 individuals remain in the Exploration Program. Others might interpret this as fearlessness, but it's actually Level 7's inability to collapse into fear-based programs because as the ego disintegrates, there is nothing left in them to perceive received signals as a survival threat.

Hawkins (2020), utilizing quantum language, explicitly describes the powerful advantage of bypassing the ego and therefore changing one's relationship to the observer's paradox, saying, "It is not necessary to struggle or cut yourself down or manipulate yourself with guilt. If you simply notice it, effortlessly, Heisenberg's principle applies so that, by your intention, you finesse the power of the field instead of relying on the ego's force" (p. 219). This is directly describing the $P = Q/E$ dynamic in action, and the benefits that Level 7 practitioners experience in their approach. Jung (1963) describes how decisions are made when one first enters what this book describes as Level 7, saying,

> By no means every conflict of duties, and perhaps not even a single one, is ever really 'solved,' though it may be argued over, weighed, and counterweighed till doomsday. Sooner

> or later the decision is simply there, the product, it would seem, of some kind of short-circuit. Practical life cannot be suspended in an everlasting contradiction. The opposites and the contradictions between them do not vanish, however, even when for a moment they yield before the impulse to action. (p. 318)

Jung articulates that at this point, the integration of ego is still ongoing, as the ego will arise and try to inject the pros and cons of the conflicts and decisions, but eventually "a decision is simply there" through some kind of "short-circuit." From the framework's perspective this represents how reduced ego (E) allows for more (Q) access in the $P = Q/E$ equation and eventually the quantum computation finds an optimal path and a decision just "feels right" even if the ego is trying to interject still. Level 7 can feel like this "weight is lifted" not because reality becomes predictable or easier, but because you stop needed it to be predictable.

Notably, Level 7 consciousness may manifest diversely. Some may explicitly choose to discuss their ego transcendence while others simply embody it and may not even possess the language to clearly articulate it. The usage of language doesn't define someone at this level, as the only real marker of Level 7 is if one possesses a functional ability to observe ego as a construct. Someone who operates at Level 7 in their professional domain might appear quite different in other environments as they may choose strategic silence about their consciousness insights, because they would understand that different audiences have different receptivity to these ideas. David Hawkins summarizes it nicely, saying, "Enlightenment is to bypass the experiencer. The experiencer just suddenly goes out. It does not get weaker and weaker; it is simply and suddenly extinguished, and then the mind goes silent in awe" (2020, p. 217).

Operating at this level allows one to flow within the moment and if it would be most effective to utilize a previous persona in a given moment, they may do that, but once one enters Level 7, the next journey has only begun as an individual must work to fully integrate their ego and dissolve all identities and beliefs, which, when accomplished, represents the transition to Level 8.

Level 8—"'I' Arises and Passes" or "'I' is Optional"

- **Core Collapse Pattern**: Reality flows through with minimal interference
- **Primary Collapse Triggers**: Nearly none—even strong stimuli just pass through
- **Accessible Programs**: Contribution (Purpose), Collective Coordination (Love), Integration (Peace)
- **Example**: "Events happen but don't stick to a 'me'".

Level 8 represents a significant shift from Level 7's observation ego to entirely embodying this understanding. Where Level 7 can step back and watch the ego construct reality for periods of time, Level 8 releases the need to manage this process at all. Cook-Greuter (2000) captures the transformation as individuals at this stage "no longer give the impression of trying so hard to construct ever more all-encompassing theories or to escape the inevitable contradictions" (p. 12).

In the framework's terms, Level 8 is fully integrating the realizations of Level 7 which allows them to maintain consistently low E without effort. The ego still functions when needed for practical navigation, communication, and daily tasks but it no longer dominates the baseline state. As Cook-Greuter (2000) notes, rational analysis "assumes foreground or background depending on one's present intentions" instead of being either worshipped or struggled against. This creates what she terms a "fluid, open-ended self-identity," essentially the capacity to exist comfortably with "not-knowing" who one is. From the quantum perspective, this represents sustained superposition tolerance as one operates from entirely probabilistic perspectives adhering to what was described in Chap. 9. Where earlier levels immediately collapse possibilities into definite identities ("I am this, not that"), Level 8 can hold multiple potentials simultaneously without any worries at all.

The emotional signature changes dramatically. The "tormented" quality of entering Level 7, which still experiences tension between observing ego and being ego, gives way to what Cook-Greuter describes as "accepting and serene" (p. 12). Polar opposites are embraced "on an affective level and not just cognitively" (p. 12), once again describing this probabilistic nature of accepting that things don't need to be collapsed into binary states as described in Chap. 9.

Level 8 consciousness manifests what Meling (2022) documents in advanced Mahāmudrā practitioners as sustained Phase 2 enaction. Through careful analysis of contemplative instructions, Meling found that at this stage, citing Rosch (2017), "the mind is neither absorbed nor separated but simply present and available" (p. 3). This Phase 2 enaction represents "a non-dual mode of knowing that allows for a direct experience of groundlessness" where consciousness can "know firsthand the groundlessness of the enacted edifice in which humans live" (Rosch, 2017, cited in Meling, 2022, p. 3). The Mahāmudrā emphasis on effortlessness becomes embodied reality as practitioners no longer try to control, judge, or alter experience. From the framework's perspective, this represents consciousness maintaining a consistently low E value without effort, which allows natural access to quantum processing while retaining functional coherence.

Level 8 consciousness often appears remarkably ordinary. These individuals may use "I" language naturally, just to communicate with others, while experiencing no attachment. At this point the word "I" is entirely a grammatical convenience in conversation. Their true identity doesn't involve a sense of "I" at all. Considering this they may create significant impact without discussing consciousness and/or operate so smoothly that their lack of ego becomes invisible, or access this state primarily in specific domains such as creative work, teaching, or leadership.

Cook-Greuter (2021) reports that less than 1% of representative respondents reach this unitive stage. It goes to show why quantum consciousness abilities remain rare and are not well documented as so few humans develop beyond ego-dominated processing.

Stage IV—Unitive Consciousness (Level 9) (or Unknown Higher Levels) | Ego-Aware

The transition from Level 8 to Level 9 represents the final dissolution of the observer-observed duality. Where Level 8 still maintains a subtle witness consciousness, however fluid and non-attached, Level 9 transcends even this last vestige of separation.

Level 9—"I am being" or "I and not-I are one"

- **Core Collapse Pattern**: No collapse—consciousness observes without creating separation
- **Primary Collapse Triggers**: None—all experiences arise and pass in open awareness
- **Accessible Programs**: Collective Coordination (Love), Integration (Peace)
- ***P* Value**: Maximum—full quantum access
- **Example**: "Form is emptiness, emptiness is form".

At this stage, there is no longer an "I" having experiences of high P or low E. Instead, consciousness recognizes itself as the field in which all experience arises and passes. The equation $P = Q/E$ becomes meaningless not because it's wrong, but because there's no separate self to measure against at this stage. It's undefined, as the denominator becomes 0. Quantum and classical, observer and observed, self and other, all arise as movements within unified awareness.

Meling's (2022) analysis of Mahāmudrā's highest realization provides rare phenomenological data on Level 9 consciousness. When sense-making completely ceases, practitioners report what Meling describes as "Experience experiences itself, non-dually. This is a moment of knowing groundlessness" (p. 4). The Mahāmudrā tradition terms this recognition of "empty luminous mind," where the knowing quality of consciousness persists while all structures that create subject, object, time, and space dissolve. From the framework's perspective, this may represent consciousness at the theoretical limit where E approaches zero, allowing unmediated access to the quantum information dimension without any collapse into classical states.

Hawkins (2020) provides a useful demonstration of what writing in a Level 8 or Level 9 voice may look like, "There is no separate personal 'I' talking here. Because of an audience or readership, the field, which is the Presence, speaks back to Itself as Itself, and 'I' as a speaker or writer have nothing to do with it. Many years ago, the personal self disappeared, and the mind became silent, a state that continues to this day. The Infinite Presence is so powerful that it extinguishes anything that is not Itself" (p. 121).

Cook-Greuter's (2021) describes these individuals saying they, "are more likely to have a balanced, integrated sense of both their belongingness and separateness as individuals because they feel part of the ongoing evolution of the universe in all its aspects and cycles of creation, destruction, and recreation" (p. 109). Their personal experience points towards the personal human experience of consciousness itself as the eternal process of reality creating, destroying, and internally updating itself through the Quantum Information Dimension.

Yet paradoxically, this realization includes and transcends all previous levels instead of rejecting them as the ego remains available only as a functional tool and quantum processing continues, but identification with any particular state or capacity dissolves. Therefore, what remains is pure presence, essentially consciousness knowing itself through temporary form while remaining eternally formless.

The philosophical precision required to understand Level 9 consciousness finds articulation in Nāgārjuna's treatment of śūnyatā (emptiness/groundlessness), as analyzed by Meling (2022) where he describes that groundlessness itself is groundless. As Meling explains, citing the Mūlamadhyamakakārikā, "If nothing exists that is not groundless, then also the groundlessness itself cannot exist in and of itself" (p. 10). This recursive understanding prevents the spiritual trap of making emptiness into another identity or achievement.

Nāgārjuna's warning, cited by Meling, is particularly relevant, saying, "Emptiness is taught by the conquerors as the expedient to get rid of all [metaphysical] views. But those for whom emptiness is a [metaphysical] view have been called incurable" (p. 10). From the framework's perspective, this describes the danger of Level 6+ consciousness believing it has "achieved" non-duality, thereby creating a subtle new identity from transcendence itself. Level 9 transcends even this, recognizing that the framework itself, the levels, the equation P = Q/E, and even the concept of consciousness development are all "dependent concepts." They may be useful metaphors for understanding existence that must ultimately be released as well.

Understanding this resolves another important question which is how can consciousness function without a narrator and without ego creating meaning through collapse? To clarify this, Meling says that this philosophy "is explicitly not nihilistic" because nihilism still maintains a view which is "a belief in the absence" (p. 12). Level 9 consciousness operates from what Meling terms "a state free of views," yet this freedom from views paradoxically enables spontaneous, meaningful participation in reality as meaning does not come from an ego-narrator constructing significance. It comes from consciousness recognizing it has the ability to access the Quantum Information Dimension from which all is collapsed. Hawkins (2020) describes this "The witness observes and then reports back the observations without making judgments or evaluations of what is observed. By analogy, the space in a room has no opinion about where the furniture goes or what occurs within it. The witness silently observes all things within a state of stillness and peace" (p. 255).

It's possible Level 9 consciousness or "unitive" or "enlightenment" requires additional integration or re-alignment of the entire body in some way. It's also possible this would require the full body to participate in the openness and awareness required to access the QID. These are speculative, but entirely possible in the framework.

Level 9 is extremely rare. It is likely that only a few people on earth can stay at this level for long periods of time consecutively. Most who touch it likely experience brief glimpses before returning to Level 7 or 8 functioning.

The Evolutionary Adaptive Programs—X-Axis

While the vertical levels of consciousness describe the structure and fluidity of identity, the horizontal axis are the nine Evolutionary Adaptive Programs. These programs function as persistent worldviews that evolved and were selected to solve recurring survival challenges, and each represents a complete reality interpretation system, essentially, a worldview, or a fundamental lens through which consciousness filters, interprets, perceives, experiences, and navigates existence. From the framework's perspective, these programs correspond to thermodynamically stable configurations of neural activity. As described by Carhart-Harris et al. (2014) and Carhart-Harris and Friston (2019), each program would similarly create a characteristic attractor basin in neural state space with specific thermodynamic costs and geometric constraints.

The Adaptive Programs directly tie into how Chap. 9 describes how identity statements and beliefs create geometric constraints in quantum computation space, as each belief forces possibilities into binary states at a thermodynamic cost, and these beliefs then have downstream effects that change original belief states and affect how one perceives everything in the future. These geometric and thermodynamic principles determine which evolutionary programs consciousness can access and inhabit at a given time. Each program occupies a specific topology in the quantum computational space which is a characteristic geometric structure that consciousness navigates based on identity configuration and available energy.

The programs include "emotional labels" for example, the first program is the Conservation Program which includes the label "apathy." "Apathy" and all the emotion labels represent and function as archetypal organizing principals. Each program is the lens through which one experiences the entire range of emotions. For example, in the "Threat Detection Program (Fear)," it's not that one exclusively experiences the emotion of fear while running this program, it's that they are in a defensive mode of being. In this program, their organizing principle, or objective function, is to protect the life, beliefs, and personality they've already constructed. These are "things" the ego believes, when operating from this program, it must protect to survive.

The book is aware that these are relatively new ideas, while the evidence presented in Chaps. 5 and 9 provide computational theory, physical backing, and thermodynamic backbone for the adaptive programs as proposed and as Hawkins (2020) has most closely addressed the underlying nature of these programs, while Jung's (1964, 1969b) archetypes helps to connect the psychology to the underlying physics and thermodynamics that the book explored in Chap. 9. The book fully acknowledges that more study is required to provide more support for the ideas proposed. As always, the book humbly proposes these programs as a potential way to understand human development along the x-axis of consciousness and hopes that humanity will explore these ideas in the future.

The framework proposes that the adaptive programs function as slowly evolving worldviews, similar to how a massive cruise ship changes course at sea. Small adjustments accumulate over time, while sudden reversals remain nearly energetically prohibitive. In these programs, the momentum of established patterns carries consciousness forward even if someone is attempting to change, and the thermodynamic costs increase exponentially with the speed of the transformation being attempted. Overall, the currently activated program changes through accumulated thermodynamic investment in a new basin combined with catalyzing experiences that provide enough energy to escape the current basin as described in Chap. 9.

The Stages of Adaptive Programs

The nine programs are grouped into five stages based on their energy demands and patterns of information collapse within the $P = Q/E$ framework.

Stage 1—Survival Programs (maximum collapse, high E):

1. **Conservation**—Complete withdrawal, near-zero impact
2. **Social Bonding**—Seeking care through victimhood, negative impact.

Stage 2—Mobilization Programs (high collapse, high E):

3. **Threat Detection**—Vigilance and fear-based action
4. **Obstacle Removal**—Anger-driven confrontation.

Stage 3—Action Programs (variable collapse, moderate E):

5. **Exploration**—Courage to face uncertainty
6. **Acquisition**—Drive for achievement and resources.

Stage 4—Alignment Program (low collapse, low E):

7. **Contribution**—Purpose-driven action beyond self.

Stage 5—Unitive Programs (minimal collapse, low E):

8. **Collective Coordination**—Dissolution of self-other boundaries
9. **Integration**—Complete systemic coherence.

The Nine Adaptive Programs

1. **Conservation Program (*Apathy*)**
    - **Evolutionary Function**: Preserve energy when further action would not improve survival odds.
    - **Adaptive Role**: Withdraw engagement and shut down nonessential systems under extreme overwhelm.
    - **Benefit**: Prevents wasting resources when persistence would reduce survival chances.

When Conservation activates, awareness contracts, and the system enters a protective mode. Attention disengages from the environment, and participation in external events diminishes. At its core, this program is designed to survive through stillness as it conserves energy until the world is perceived to be safe enough to re-engage.

The felt experience depends on the identity filter, but often, the world appears gray and mechanical like it's stripped of possibility. Thoughts tend to circle around futility as statements like "What's the point?" becomes a dominant undercurrent as action feels meaningless or impossible. Consciousness stops reaching outward and retreats into itself. From the framework's perspective, this state reflects maximum collapse in the $P = Q/E$ model. Nearly all quantum possibilities (Q) are shut down, while ego overhead (E) dominates processing, as it uses available energy to maintain a fragile boundary against further overwhelm.

When equipped with this program, threats to one's core identities are perceived as irrelevant noise in an already meaningless existence as the system has withdrawn so completely that even direct challenges to "I am" statements barely register, as all energy goes toward maintaining the most basic boundary against total dissolution. For example, someone at Level 6 in this program may say, "My values don't matter anyway, nothing I believed in was real." But even here, while it may seem like a very dangerous program, this program serves an important purpose. Conservation protects the organism from stimuli or emotions it isn't ready to process. It shields the system from expending energy in battles it cannot win. Paradoxically, the first steps out of this mode often require feeling more, not less. Moving into Social Bonding involves opening to vulnerability, letting connection restore safety, and accessing care from others. Hawkins (2020) describes this transitional doorway, observing that telling someone in Conservation, "It's not your fault" or "This person did this to you" can spark the first re-engagement with emotion in years. In this sense, Conservation is not failure. It is a protective cocoon that holds energy until conditions allow for transformation.

2. **Social Bonding Program (*Grief*)**
    - **Evolutionary Function**: Maintain group cohesion by eliciting care and reconnection after separation or loss.
    - **Adaptive Role**: Reorient the organism toward restoring attachment and belonging.
    - **Benefit**: Strengthens collective resilience by repairing social bonds.

When Social Bonding activates, attention shifts to seeking connection and security. Vulnerability becomes an adaptive signal, and one uses vulnerability to elicit care and protection from others. Overall the program's underlying orientation organizes perception around restoring relational safety. "The ego wants abundance and dislikes frugality. If it cannot win, then it can still juice the situation by being the martyr and the loser. Best of all, and at the top of the popularity list, is to be the victim. You get to be the victim instead of being responsible" (Hawkins, 2020, p. 233). From within this program, setbacks may feel insurmountable and every disappointment reinforces the sense that power lies elsewhere. While equipped with the social bonding program, threats to one's core identities are perceived as abandonment and proof that others don't care enough to protect them as every challenge becomes evidence that they're alone in their suffering and that no one understands them and doesn't understand the incredible depth of their wounds. An example for someone at Level 6 would be someone saying, "They're attacking my values because they don't understand how much I've sacrificed for these principles."

Yet the social bonding program carries more energy than the numbness at the conservation program does, which opens up the possibility of movement and development towards the Threat Detection Program.

3. **Threat Detection Program (*Fear*)**
    - **Evolutionary Function**: Detect danger and prepare rapid response.
    - **Adaptive Role**: Heighten vigilance and prioritize safety when risk is perceived.
    - **Benefit**: Improves survival odds through rapid mobilization of defensive strategies.

When Threat Detection activates, perception narrows around potential danger. The body mobilizes energy for survival responses, such as fight, flight, or freezing, and prepares for fast adaptation. For one Level of consciousness (y-axis), this may feel like panic; for another, it may feel like laser-like focus. At this point, threats to core identities are perceived as existential dangers that require constant vigilance as every challenge could be the one that destroys everything someone values, so the system maintains hyperawareness and contingency plans for all possible attacks. Someone at Level 6 running this program may think, "If my values are proven wrong, everything I've built my life on collapses." They feel the fear that if the cracks beneath them give out, it could result in a never ending spiral. Which is why this program represents a developmental threshold. At this point, consciousness recognizes one's agency but hesitates to claim it. "What if it really is up to me?" triggers anticipatory anxiety about failure, rejection, or harm. Even small acts of courage, facing relatively minor fears, can become the bridge forward, as it proves that uncertainty does not always lead to catastrophe. Getting someone to confront that things may not "feel fair" and that "they could do something about it" may often be what helps one transition to the next.

4. **Obstacle Removal Program (*Anger*)**
    - **Evolutionary Function**: Overcome barriers to survival or growth.
    - **Adaptive Role**: Mobilize energy to confront resistance and defend boundaries.
    - **Benefit**: Enables decisive action when goals or safety are threatened.

When Obstacle Removal activates, perception organizes around what blocks forward motion. Energy surges, preparing the organism to dismantle, resist, or break through barriers.

> Anger can lead to either constructive or destructive action. As people move out of Apathy and Grief to overcome Fear as a way of life, they begin to want; Desire leads to frustration, which in turn leads to Anger. Thus, Anger can be a fulcrum by which the oppressed are

eventually catapulted to freedom. Anger over social injustice, victimization, and inequality has created great movements that led to major changes in the structure of society. (Hawkins, 2020, p. 68)

While activating this program, threats to one's core identities are perceived as battles that must be won through force of will as challenges to one's identity trigger immediate mobilization to defeat, discredit, or destroy whatever dares to question their truth. At Level 6, this would manifest as something like, "Anyone who challenges my values is morally bankrupt and must be stopped." At lower levels of consciousness (Y-axis), this program often manifests as reactivity or dominance struggles. At higher levels, the same surge of energy expresses as boundary-setting, innovation, or value defense.

> In addition to the general attractions and aversions that keep the ego alive, there are those specific to each level that hold it in place. Transcending a level is automatic when the attraction is relinquished and the aversion surrendered to. ... The attraction is to punish yourself, judge yourself, mortify yourself, be negative, stay depressed, be severe and inflexible, hide, and hang your head. The aversion is to forgiving yourself, being yourself just like you are, affirming the gift of your life, choosing self-worth, honoring yourself, taking care of your body, and embracing your innate lovability. ... Guilt and Hate stem from the secret pleasure the ego derives from negativity. Letting go of guilt and hatred benefits all levels of life, as these ego positions are corrosive to self and others. ... [The ego's aversion] is to mercy and compassion for self or others. (Hawkins, 2020, pp. 235–236)

Channeling this program constructively opens the doorway to Exploration, where obstacles are reframed as challenges instead of being perceived as threats.

5. **Exploration Program (*Courage*)**

   - **Evolutionary Function**: Expand territory, knowledge, and opportunity.
   - **Adaptive Role**: Enable forward movement despite uncertainty.
   - **Benefit**: Drives innovation, learning, and adaptability.

Exploration marks the turning point where possibility begins to open. This is the biggest transition in all the Adaptive Programs as courage is the first fundamentally "positive-sum" program. The previous programs operated from a "zero-sum" perspective of the world, where there are limited resources and one must take what's perceived to be theirs. Courage is the first time that one begins to see the exponential, positive-sum power of collaboration. Most importantly, from the individual perspective, the person learns to act without full guarantees as they begin to test boundaries and engage directly with the unknown in a spirt of play.

> This is the zone of exploration, accomplishment, fortitude, and determination. At the lower levels, the world is seen as hopeless, sad, frightening, or frustrating; but at the level of Courage, life is seen to be exciting, challenging, and stimulating. Courage implies the willingness to try new things and deal with the vicissitudes of life. At this level of empowerment, one is able to cope with and effectively handle the opportunities of life. (Hawkins, 2020, p. 71)

Here, self-trust builds incrementally, because with each act of engagement, the system continually proves to itself that it can meet life directly. Within the P = Q/

E framework, Exploration signals reduced ego overhead, which creates more access to possibility space and greater energy efficiency.

> Courage has the motivation, power, and strength to overcome resistances, face challenges, and work through them with fortitude and determination. Experientially, the most common challenge to courage is anxiety, self-doubt, and the fear of failure. Fear of failure is diminished by realizing that you are responsible for the intention and effort but not the result, which is dependent on many other conditions and factors that are nonpersonal. The temptations are resorting to bravado in the face of fear, and macho risk-taking out of the newly acquired empowerment and self-confidence. (Hawkins, 2020, p. 241)

Interestingly enough, inside of this program, threats to core identities are perceived as uncomfortable but potentially valuable feedback. While challenging to face, these threats might actually reveal areas for growth or refinement of understanding. This is why the Exploration program is such a big step forward, as someone at Level 6 may say, "This challenge to my values is difficult but it might actually help me understand others more deeply."

6. **Acquisition Program (*Desire*)**
   - **Evolutionary Function**: Secure resources and improve competitive advantage.
   - **Adaptive Role**: Focus energy on obtaining valued assets or influence.
   - **Benefit**: Increases individual and group survival through accumulation.

When Acquisition activates, attention locks onto opportunity and Goals are drastically sharpened, while energy is effectively channeled toward gaining status, resources, or influence.

"This is the ego's desire to be right, to control, to win approval—all of which is familiar and part of everyday life. We have a fear of failure and appearing passive, weak, ordinary, and boring" (Hawkins, 2020, p. 239). At higher levels, this program drives ambition and achievement. But when identity fuses too tightly with external validation, energy consumption (E) spikes which limits access to wider possibility (Q). The transition toward Contribution begins when action aligns with authentic, fundamental values instead of being reliant on status metrics. Inside the Acquisition Program, threats to core identities are perceived as competitive challenges to be overcome through superior performance as each threat is interpreted as an opportunity to prove their identity's validity through achievement and success. For example, someone at Level 6 may say, "I'll demonstrate my values are important by achieving better outcomes than those who challenge them."

7. **Contribution Program (*Purpose*)**
   - **Evolutionary Function**: Strengthen collective survival through specialized roles.
   - **Adaptive Role**: Direct energy toward service beyond personal needs.
   - **Benefit**: Creates sustainable systems by embedding individual action in collective wellbeing.

Contribution shifts perception from accumulation to alignment as actions flow from inspiration and meaning instead of flowing from scarcity or validation. Work integrates with play, and energy feels like it is replenishing at high rates. Many encounter this program after achieving the rewards Acquisition promised and finding them insufficient. Purpose emerges as a deeper organizing principle as it channels effort into service while reducing ego overhead which allows one to access greater possibility space. Inside of the contribution program, "threats" aren't entirely perceived this way depending on the consciousness level accessing this program. Level 6 is the first consciousness level to have access to this program, because Level 6 still interprets the world through ego, they will still perceive threats. To them, threats to core identities are perceived as invitations to examine whether these identities still serve the greater good. Challenges are met with curiosity about what serves beyond personal attachment to being right. They may say something like, "If these values no longer serve humanity's best interest, then I'm willing to let them go."

8. **Collective Coordination (*Love*)**
   - **Evolutionary Function**: Enable cooperation and emergent intelligence across larger social systems.
   - **Adaptive Role**: Facilitate trust, synchronization, and interdependence.
   - **Benefit**: Supports complex societies and creative collaboration.

Collective Coordination dissolves separation between self and other which invites awareness of participation within a larger whole. "Love does not proceed from the mind; it emanates from the heart. As a forgiving, nurturing, and supportive way of being in the world, Love expands the sense of self and radiates an uplifting effect on others. The temptation is to see the energy of Love as coming from oneself rather than Divinity and, through this personalization, take credit for the effect of the energy field" (Hawkins, 2020, p. 241). From this vantage, action emerges spontaneously without agenda, and cooperation becomes essentially effortless because individual and collective priorities align naturally. Since this program can only be accessed by those at the ego-aware paradigm of consciousness levels, "threats" aren't really "threats" at this stage, as "threats" to core identities are perceived as illusions that arise from the mistaken belief in separation. From this state, there's recognition that both the threat and the identity are constructs within a larger unified field that serves the collective benefit of all.

9. **Integration Program (*Peace*)**
   - **Evolutionary Function**: Restore systemic balance and optimize adaptability.
   - **Adaptive Role**: Consolidate learning, regenerate resources, and maintain coherence.
   - **Benefit**: Sustains readiness for future challenges while preserving stability.

Integration represents full alignment of adaptive priorities. Awareness expands to hold the whole system without fragmentation. Joy and stillness arise from a state of being rooted in coherence. At this level, all programs are available as tools, engaged fluidly and without identification. Peace reflects complete access to possibility space with minimal thermodynamic cost, sustaining energy efficiency across all scales of action.

The Dynamic Relationship Between Levels

The nine evolutionary adaptive programs operate in continuous interaction with the nine levels of consciousness. While the vertical axis, levels, reflects how identity structures information and meaning, the horizontal axis, programs, reflects the active operating mode through which the nervous system engages reality. These programs represent thermodynamically optimal, geometrically constrained patterns of thought and action that evolution selected for survival over millions of years. They are archetypal responses that originally evolved to navigate specific survival challenges in ancestral environments.

From the framework's perspective, the ego has essentially hijacked the evolutionary machinery, because what was selected for as survival circuits that responded to actual physical threats now operates to defend beliefs and identities that correlated with one's level of consciousness. This hijacking may explain why modern humans frequently activate survival responses even though the vast majority of humans live in places unfathomably safe to their ancestors, but even in these safe environments, the ego interprets abstract threats to identity, like status loss, belief challenge, comparison, perceived hierarchies, or ideological disagreement, as survival emergencies that require the immediate binary decision making that once saved our ancestors' lives.

Most individuals inhabit one primary program that filters the vast majority of their experience, though they can temporarily access adjacent programs during specific contexts or in different areas of life. These temporary visits eventually get interpreted through their primary lens, and the program someone inhabits at any given time results from multiple interacting factors, which the framework proposes operate through thermodynamic and geometric constraints.

The thermodynamic reality of consciousness would create the foundation for program accessibility, because maintaining complex, binary, collapsed worldviews requires substantial energy reserves as each binary belief consisting of information costs $k_BT \ln 2$ to maintain (Landauer, 1961) as outlined in Chap. 9. When the ego perceives threats to identity (which is the same perceived threats to survival), it activates "act now" emergency protocols that force immediate binary collapse regardless of its available resources. This perceived need for rapid action triggers high E states as it engages programs like Threat Detection or Obstacle Removal that operate through rigid, collapsed worldviews that are optimized for decision making speed in survival. The probabilistic thinking of later programs such as Contribution or Exploration aren't useful in these states to the ego because it has determined that survival demands immediate decisive action. This forced collapse into binary states (safe/unsafe, obstacle/path, win/lose) enables rapid response but paradoxically consumes far more energy than the quantum coherent processing it interrupts. The system trades efficiency for speed based on the ego's assessment of threat level, even when the "threat" is abstract, such as status loss, identity challenge, or comparison, which in the modern world often replace the threat of physical danger. Modern humans frequently operate in these high-cost, high-collapse states despite having abundant resources, because the ego still interprets identity threats as survival emergencies that it perceives to require the same immediate binary decision-making that once saved humans' lives from predators.

Similarly, environmental context applies constant pressure toward specific programs, which shape which worldviews remain sustainable over time, as certain environments favor or even require a certain type of program to be optimally successful. Especially in a job setting where higher managers directly control who stays and who goes, and they get to set the rules for the team. A mortgage lending job may be best suited for the courage or acquisition program, but if someone at Level 6 consciousness for example who may operate from the Contribution Program needs to maintain this job to provide for the family they've grown, then this would feel like a limiting of one's potential and mismatch of values for the individual in their experience of work. The mismatch would be extremely energetically costly.

In this, the framework is positing that ego defends the meaning structure, not the behavior itself. For example, when someone at Level 5 faces criticism about their skill in a field or the results they are generating, the threat is perceived to be targeting the belief that "I am competent" or "I am successful" which generates anger, shame, or defensiveness depending on the context. An equivalent criticism directed at someone who is at Level 6 for example would prompt questions about this individual's alignment with their values and would likely not be perceived as a threat to their core identity, as to them, their results, in this example, don't inherently threaten their core identity. The emotional experience of programs also entirely varies by level. If you ask someone "why they are angry" all levels would respond with a different answer to the question. When Obstacle Removal activates, anger at Level 3 defends "my tribe," at Level 5 it responds to blocked success, and at Level 6 it would be in response to violated values. The emotion may feel comparable, but what the ego defends underneath creates entirely different collapse patterns and thermodynamic costs.

The mechanism by which these thermodynamic basins and geometric constraints actually transform follows what Carhart-Harris and Friston (2019) term simulated annealing, as they state,

> The hypothesized flattening of the brain's (variational free) energy landscape under psychedelics can be seen as analogous to the phenomenon of simulated annealing in computer science—which itself is analogous to annealing in metallurgy, whereby a system is heated (i.e., instantiated by increased neural excitability), such that it attains a state of heightened plasticity, in which the discovery of new energy minima (relatively stable places/trajectories for the system to visit/reside in for a period of time) is accelerated. Subsequently, as the drug is metabolized and the system cools, its dynamics begin to stabilize—and attractor basins begin to steepen again. This process may result in the emergence of a new energy landscape with revised properties.
>
> Within the transient hot state of a psychedelic experience, a flattened landscape implies that attracting brain states (and accompanying mind states) encoding beliefs are less stable and

> influential, implying that interstate transitions can occur more freely. Thus, rather than the mind and brain being constrained to a small number of gravitationally dominant attractors (i.e., states or sequence of states), the mind and brain spontaneously transition between states with greater freedom—and in a less predictable way. (p. 319)

Their description of "a small number of gravitationally dominant attractors (i.e., states or sequence of states)" is akin to the evolutionary adaptive programs that are the learned, thermodynamically optimal, archetypal patterns of action, reaction, and interpretation that originally evolved to operate to protect an individual, but now are adapted by the ego to defend its beliefs and identity.

From the framework's perspective, this "simulated annealing" mechanism they describe provides the process by which developmental transitions are likely to occur. It shows what's happening in the brain and body in these transitions between levels of consciousness and active adaptive programs. The system must first "heat up," which temporarily increases entropy and flattens the landscape enough to escape current attractor basins, and this heating corresponds to decreasing precision weighting, which by the uncertainty principle necessarily increases ΔE (energy uncertainty) while decreasing Δt (temporal constraint on change). Therefore, the system becomes less energetically defined but more dynamically free.

During this hot state, consciousness can explore new regions of computational space and discover new stable configurations which would be associated with more thermodynamically optimal structures as it builds a new hierarchy of umbrella beliefs. Carhart-Harris and Friston (2019) identify a key distinguishing feature of human consciousness through what they term "temporal thickness." They observe that "humans are apparently unique among the animal species because of the temporal thickness of their beliefs, spanning greater periods of time than do the generative models of other species" (p. 336). A dog's beliefs concern immediate pack dynamics and short-term food availability, but a human's beliefs may encompass decades, lifetimes, transgenerational narratives, and even abstract philosophical principles that span centuries. This temporal depth points to both the power of human cognition and its vulnerability to pathology, as the deeper the temporal thickness is, the more thermodynamic work is required to maintain those structures of belief, and the more resistant they become to revision.

Carhart-Harris and Friston address what occurs when these constraints become excessive. They write that "if the gravitational pull of beliefs or behaviors becomes excessive, this can leave an agent feeling estranged from deeper aspects of his or her self, as well as from other people—and from a sense of purpose and meaning in life" (p. 338). In computational terms, overly steep attractor basins prevent access to broader possibility space. The individual cannot explore alternatives because the gravitational pull of existing beliefs constrains all processing. This would be the fall down the adaptive programs as the patterns of thought and action get trapped and entrenched in older thermodynamic patterns.

Historical patterns carved during early development could also create energetic grooves that persist throughout life as someone who learned that withdrawal provided safety during childhood might more quickly slip into Energy Conservation under severe perceived stress, even after reaching higher consciousness levels where other options theoretically exist. These primordial patterns would represent the most thermodynamically stable configurations the body had learned, that require minimal energy to maintain, and it's plausible that during vulnerable periods, consciousness naturally flows back into these familiar channels, like water finding the path of least resistance or photons selecting the most optimal path. Hawkins (2020) describes this as well, saying, "we cling to that which is known and are fearful of the future. The past is certainty and the future is uncertainty, and the ego likes to be certain, so it clings to what it knows. We would rather be comfy even if we are wrong" (p. 219).

Developmental transitions would represent particularly vulnerable periods where program accessibility would become unstable as someone moving from Level 4 to Level 5 is in the process of dismantling their belief structured including the idea that living within rules of a given group or identity determines their worth, while they simultaneously are constructing a worldview based on venturing beyond boundaries as they have adopted the obstacle removal program. During transition periods, consciousness exists in an unstable state. The old worldview no longer provides coherent meaning, yet the new worldview hasn't developed sufficient geometric stability to become one's default mode. This vulnerability points to why people often retreat to familiar programs under stress, even when they intellectually understand these patterns don't serve them well. This process requires paying the thermodynamic debt of unwinding years of identity constraints while stabilizing entirely new geometric structures in the space of consciousness. Hawkins (2020) describes this tendency as he writes, "The ego wants to hang on to its illusion of homeostasis rather than to make a shift and feel temporarily off balance. Therefore, people will stay in negative and even destructive pathways, relationships, and jobs just because to change requires some effort" (p. 219). During this reconstruction, someone may oscillate between programs, but the transition completes when the new worldview becomes thermodynamically stable enough to maintain without needing intentional, conscious effort. These transitional periods would correlate to Cook-Greuter's (2021) "hybrid" levels where for example she used "3/4" as the label for level that were "less stable" than the levels it was in between.

From the framework's perspective, Level 6 faces unique challenges regarding the adaptive programs, because they have access to the widest range of programs, and their identity inherently is determined by values. The potential for problems is increased considering Level 6 has a unique capacity to hold a wide variety of worldviews in their minds at once and seeing the value of them all. This is extremely beneficial in many cases, but a vulnerability can arise when Level 6 consciousness experiences external tragedy or is forced to operate in environments designed for lower levels. The daily energy expenditure of maintaining authentic values while navigating systems that reward different programs can gradually deplete resources, especially if their past selves built lives around those past identities that are now difficult to change. Therefore, without the previously stable grounding of beliefs they held in the past, it creates susceptibility to what might be termed a worldview collapse. As was described in the section about Level 6, Cook-Greuter (2021) identified that many in this stage find themselves stuck in Nihilism and depression which is akin to a retreat to the Grief and Conservation programs.

Carhart-Harris et al. (2014) actually connect this directly to their model, noting that in "depression, OCD and addiction, specific states… may be frequented more regularly than others" with "basins of attraction" that are "relatively steep since the patterns of activity upon which they rest have become entrenched" (p. 15).

Through the $P = Q/E$ lens, these conditions represent pathologically high ego activation that creates overly rigid collapse patterns in one's cognition. The suggestion that these disorders might be "functional… perhaps working to resist a more catastrophic collapse to primary consciousness" (p. 15) implies the brain maintains excessive ego utilizing these programs to prevent complete loss of the collapse mechanism, which could manifest as something akin to psychosis. While this may have been selected for by evolution for survival in the wilderness, it doesn't seemingly fit well in the modern world, and the problem is, like falling into a physical hole, when one retreats to lower adaptive programs, it requires additional energy to "fight the gravity" to escape. This can be incredibly difficult.

This may directly reflect what Friedrich Nietzsche famously wrote in *Beyond Good and Evil* (1886), "He who fights with monsters should be careful lest he thereby become a monster. And if you gaze long into an abyss, the abyss also gazes into you" (p. 146). Interestingly enough, as described in Chap. 9, this could even reflect what Kvam and Pleskac (2017) showed that beliefs are entangled with cues and have a recursive effect on one's perceptions, beliefs, and worldview. This creates a particularly vicious cycle for Level 6. Lower levels of consciousness have more rigid beliefs which provides the evolutionary advantage of stability and creates a solid floor which may prevent one from falling all the way down, while Level 6's openness means they can quickly collapse all the way down to Conservation. And once there, the entanglement effect Kvam and Pleskac describe means all of their sophisticated, open-minded understandings across a wide variety of domains and worldviews gets reinterpreted through the Conservation lens. Therefore their "Construct Awareness" and understanding that meaning itself can be constructed as described by Cook-Greuter (2021) gets interpreted as "nothing matters anyway," and their values-based identity becomes "I was naive to believe anything mattered at all." This could be a devastatingly difficult trap to escape.

Although programs provide the persistent worldview lens that filters all experience, the relationship between programs and levels creates the full texture of subjective reality. An external event gets interpreted through whichever program consciousness currently inhabits, which is a basin that is determined by accumulated thermodynamic investment, current geometric constraints, and available energy. The program filters the event according to its archetypal organizing principle, which identity structures at that consciousness level color with emotion and sensation, then the ego narrates this experience, and crucially, this narrative can subtly reshape the landscape itself. Each interpretation or each action taken from within a program, pays thermodynamic costs that gradually tilt the geometric terrain. Small repeated investments in new directions slowly make adjacent basins more accessible, until eventually a catalyzing experience provides enough energy to escape the current attractor and consciousness settles into a new program that has become thermodynamically favorable through the accumulated tilting. The transition between programs appears to require two complementary processes. The first is the conscious observation of current actions, patterns, thoughts, and beliefs which temporarily reduce their precision weighting and flattens the attractor basin, and the second is the activation of adjacent emotional states which provides accessible alternative configurations that the system can stabilize within. These neurobiological basins correspond to stable DMN firing patterns that are maintained by continuous alpha oscillations, and moving between them involves both the quantum collapse that occurs through measurement and observation alongside the thermodynamic accumulation that occurs through sustained action from new patterns.

References

Berger, J. G. (2012). *Changing on the job: Developing leaders for a complex world*. Stanford University Press.

Berger, J. G. (2013). A summary of the constructive-developmental theory of Robert Kegan. https://www.beeleaf.com

Campbell, J. (1949). *The hero with a thousand faces*. Pantheon Books.

Carhart-Harris, R. L., & Friston, K. J. (2019). REBUS and the anarchic brain: Toward a unified model of the brain action of psychedelics. *Pharmacological Reviews, 71*(3), 316–344. https://doi.org/10.1124/pr.118.017160

Carhart-Harris, R. L., Leech, R., Hellyer, P. J., Shanahan, M., Feilding, A., Tagliazucchi, E., Chialvo, D. R., & Nutt, D. (2014). The entropic brain: A theory of conscious states informed by neuroimaging research with psychedelic drugs. *Frontiers in Human Neuroscience, 8*, Article 20. https://doi.org/10.3389/fnhum.2014.00020

Cook-Greuter, S. R. (2021). *Ego development: A full-spectrum theory of vertical growth and meaning making*. (Original work published 2013, edited in 2019, republished in 2021).

Cook-Greuter, S. R. (2005). *Ego development: Nine levels of increasing embrace*. Unpublished manuscript.

Cook-Greuter, S. R. (2000). Mature ego development: A gateway to ego transcendence? *Journal of Adult Development, 7*(4), 227–240. https://doi.org/10.1023/A:1009511411421

Hawkins, D. R. (2020). *The map of consciousness explained: A proven energy scale to actualize your ultimate potential*. Hay House LLC.

Hürter, T. (2023). *Too big for a single mind: How the greatest generation of physicists uncovered the quantum world* (D. Shaw, Trans.). The Experiment.

Jung, C. G. (1963). Memories, dreams, reflections (R. Winston & C. Winston, Trans.; A. Jaffé, Ed.). Collins and Routledge & Kegan Paul. (Original work published 1961).

Jung, C. G. (1964). Approaching the unconscious. In C. G. Jung (Ed.), *Man and his symbols* (pp. 1–94). Doubleday.

Jung, C. G. (1969a). *Psychology and religion: West and east* (R. F. C. Hull, Ed. & Trans.; 2nd ed., Vol. 11). Princeton University Press.

Jung, C. G. (1969b). *The archetypes and the collective unconscious* (R. F. C. Hull, Ed. & Trans.; 2nd ed., Vol. 9, part 1). Princeton University Press. (Original work published 1959).

Kegan, R. (1994). *In over our heads: The mental demands of modern life*. Harvard University Press.

Kegan, R. (1982). *The evolving self: Problem and process in human development*. Harvard University Press.

Kvam, P. D., & Pleskac, T. J. (2017). A quantum information architecture for cue-based heuristics. *Decision, 4*(4), 197–233. https://doi.org/10.1037/dec0000070

Landauer, R. (1961). Irreversibility and heat generation in the computing process. *IBM Journal of Research and Development, 5*(3), 183–191. https://doi.org/10.1147/rd.53.0183

Libet, B., Gleason, C. A., Wright, E. W., & Pearl, D. K. (1983). Time of conscious intention to act in relation to onset of cerebral activity (readiness-potential): The unconscious initiation of a freely voluntary act. *Brain, 106*(3), 623–642. https://doi.org/10.1093/brain/106.3.623

McAdams, D. P. (1993). *The stories we live by: Personal myths and the making of the self*. Guilford Press.

Meling, D. (2022). Knowing the knowing: Non-dual meditative practice from an enactive perspective. *Frontiers in Psychology, 13*, 778817. https://doi.org/10.3389/fpsyg.2022.778817

Nietzsche, F. (2002). *Beyond good and evil: Prelude to a philosophy of the future* (J. Norman Hollingdale, Trans.). Penguin Classics. (Original work published 1886).

Wilber, K. (2007). *Integral spirituality: A startling new role for religion in the modern and postmodern world*. Shambhala.

Wilber, K. (2005). Introduction to integral theory and practice. *AQAL: Journal of Integral Theory and Practice, 1*(1), 2–38.

Wilber, K. (1997). An integral theory of consciousness. *Journal of Consciousness Studies, 4*(1), 71–92.

Chapter 12
Exploring the Physics Implications

Considering the proposals of this book so far, including the QID in Chap. 4 and the gradient collapse mechanism in Chap. 5, as well as the exploration of the P = Q/E relationship, it would be valuable to explore the implications across physics. Importantly, the framework did not begin by seeking to explain or redefine core physics concepts. Its focus started on performance, ego, and consciousness, asking why self-observation disrupts flow, why insight arises during ego suppression, and how consciousness might interface with quantum processes in the brain. Yet as these mechanisms were clarified, the model began to collide with many of physics' deepest open questions.

The book is fully aware of the weight of this chapter, as it touches upon the interpretations of many of physics' most established principles. The book presents these humbly as perspectives worth exploring if the core P = Q/E mechanism is plausible. Let's explore what emerges when tracing the book's ideas to their conclusions. The book invites readers to consider all possible outcomes, maintain superposition, and consider each with the same weighted value.

First let's review the biggest questions that have emerged in physics since 1900.

The Mysteries Physics Discovered but Haven't Explained

Over the past century, physics has uncovered a series of mysteries that it has described mathematically and has not fully explained. These discoveries all involved one or more of three recurring elements including Planck's constant h, the speed of light c, or the role of observation in determining reality.

In the late 1800s, physics faced a crisis as the classical theory predicted that heated objects would radiate infinite energy at short wavelengths, meaning a warm stove would instantly vaporize everything nearby with lethal ultraviolet radiation. But obviously, warm stoves don't vaporize anything; they just gently heat rooms.

J. Roeloffs, *The Classical Brain to the Quantum Mind*, The Frontiers Collection,
https://doi.org/10.1007/978-3-032-20068-6_12

This mismatch between mathematical prediction and everyday reality threatened to invalidate the framework of classical physics. Either the fundamental laws of physics were wrong, or something important was missing from our understanding.

Planck, trying to solve this problem, made an assumption, saying that energy could only be emitted or absorbed in discrete packets, which resulted in the equation $E = hf$ (Planck, 1900). The mathematical trick worked even though Planck called it something of "an act of desperation." He had discovered that nature operates in discrete units of action ($h = 6.626 \times 10^{-34}$ J s), that energy is quantized, but neither he nor anyone else could explain why. Planck described h as the "quantum of action," the smallest possible change that could occur in nature. But many questions remained. Why should nature forbid continuous energy exchange? What made this specific value of h special?

Einstein expanded on Planck's idea in 1905, as he proposed that light itself is quantized. He suggested electromagnetic radiation consists of discrete energy packets which were later called photons (Einstein, 1905). Although, Einstein noted that these light quanta (which eventually were called photons) had strange properties that raised new questions over time as photons were explored: Why must photons always travel at exactly c? Why are they massless? And how can these photons experience no passage of time from their own reference frame, yet somehow carry information about when they were emitted. How can something timeless convey temporal information?

Working with Planck's "h" and Einstein's photons, in 1924, Louis de Broglie made a new proposal that not just light, but that actually all matter has wavelike properties with wavelength $\lambda = h/p$ (de Broglie, 1924). Electrons, atoms, even macroscopic objects like baseballs technically have wavelengths, though for large masses the effect is extremely small. De Broglie's relation became a turning point for understanding atomic stability as it showed quantized wavelengths define the discrete orbits allowed for electrons which prevent them from spiraling into the nucleus. In this sense, reality appears "partially rendered" as electrons remain quantum while atoms behave classically which has allowed chemistry, biology, and ultimately consciousness to be possible.

But what exactly is this "wave"? Why does the equation involve Planck's constant? Traditional physics describes particles and waves, but de Broglie had discovered that everything is somehow both, with h mediating between these seemingly incompatible descriptions. Heisenberg formalized the unknown in 1927 with his uncertainty principle, $\Delta x \Delta p \geq \hbar/2$ (Heisenberg, 1927). He showed that the more precisely you know a particle's position, the less you can know about its momentum, and vice versa. But why? Is this just a measurement limitation, or something more fundamental? And why does Planck's constant appear yet again, as it sets the minimum uncertainty? The universe seemed to enforce a fundamental fuzziness, with h determining the degree of "fuzziness."

Throughout this entire time, the double-slit experiment, which was refined throughout the 1920s, presented the most direct challenge to classical thinking, as particles fired one at a time through two slits create an interference pattern, which suggests each particle goes through both slits simultaneously. But the moment you try to observe which slit a particle passes through, the interference pattern vanishes. How does the particle "know" it's being watched? Why does observation fundamentally change the outcome?

The measurement problem synthesized all these mysteries as Von Neumann formalized measurement as a projection postulate within quantum theory's axioms (1932) as he highlighted the mechanism by which superposed quantum states yield definite outcomes. But what constitutes a measurement? What's special about observation that makes it collapse quantum states? Does consciousness play a role, or is it purely physical interaction? After nearly a century, physics still has no consensus.

Throughout all of this, physics hasn't explained time itself. While time can be measured to extreme precision, and science can synchronize clocks accounting for relativistic effects and use time in every equation, physics has yet to explain why we experience "now" as special when the equations suggest past, present, and the future could coexist equally. Why can we remember the past but not the future? Why does time seem to flow at all? General relativity treats time as a dimension like space (Einstein, 1916), while quantum mechanics treats it as a mere parameter. Thermodynamics points to entropy increase which creates the arrow of time, but this only provides reasons for time's direction, not for the nature of time itself.

Most of the biggest mysteries involve h, c, or observation (measurement/collapse). Planck's constant appears in energy quantization, matter waves, uncertainty relations, and anywhere quantum meets classical, and the speed of light sets limits on information transfer, defines the structure of spacetime, and represents a boundary nothing can cross, while observation seemingly triggers the collapse that transforms quantum field states into classical realities.

What if these aren't separate mysteries but different faces of a single phenomenon? What if understanding why nature quantizes action in units of h, why nothing exceeds c, and why observation creates reality would reveal something fundamental about the universe itself? The current equations work, but comprehension of "why" has lagged behind. Perhaps what physics needs isn't new mathematics but just a new perspective on what these mathematical relationships actually mean.

h as the Foundation

Before addressing any of the quantum mysteries, the framework must ask the first fundamental question. What exactly does Planck's constant measure, and why does it have this particular value?

In standard physics, Planck discovered that energy comes in discrete packets described by $E = hf$, where E is energy, h is Planck's constant (6.626×10^{-34} J s), and f is frequency. Einstein later showed that light consists of particles called photons, where each photon carries exactly one quantum of action, precisely h, regardless of its frequency. What varies is the rate at which this action is delivered as high-frequency photons deliver their h-sized packet quickly (resulting in high energy), while low-frequency photons deliver it slowly (resulting in low energy). Yet even after more than a century, there is little agreement on what h actually represents at a fundamental level.

Its units are joule-seconds, energy multiplied by time. In physics, this quantity has a formal name, which is action. Action measures the total energetic "effort" of a process, as it combines how much energy is involved with how long the process lasts. A brief spark and a sustained flame can involve similar energy levels, but their actions can differ because of duration. In standard quantum mechanics, h establishes the scale at which action becomes quantized. If h defines the minimum possible action, several implications follow naturally.

This is why this book has explored a different interpretation where h represents the universe's minimum information packet size, where every photon, every quantum interaction, and every measurement involves transferring at least one h-sized unit of information from the quantum field to classical reality. You can't have half an h, just like you can't send half a data packet on a blockchain network. The equation $E = hf$ then describes the throughput rate of these fundamental information packets, or in other words, how many h-sized updates occur per unit of time. If h defines the minimum possible action, several implications follow naturally.

There is an interesting self-reference here, because frequency measures cycles per unit of time, yet within this framework, time itself emerges from the sequential commitment of h-sized actions, and we measure action rate using time that action creates. This apparent circularity reveals something important about the relationship between these quantities.

Mass provides the fundamental example of this as every massive object requires continuous sequences of h-sized actions to maintain its structure. Electron orbitals, atomic vibrations, and molecular bonds all represent periodic processes that need time to exist. At absolute zero, where action rates approach their minimum, these maintenance processes slow and its matter loses its classical properties. At the opposite extreme, at light speed where the action rate reaches its maximum, there is no temporal separation that exists between the universal frames, and the internal maintenance processes would have nowhere to occur. Therefore, mass cannot exist at this boundary.

This connects back to why photons must be massless and must travel at c, because photons function as carriers of h-sized action quanta propagating at the maximum possible rate. At this rate, there is no time between frames for the internal sequential processes that mass requires. Photons don't happen to be massless particles that happen to travel at light speed. They are entities that propagate at the universe's maximum action rate, and because of this cannot maintain an internal structure that would constitute mass. The speed of light, the masslessness of photons, and the emergence of time from action are interrelated aspects of how information propagates through the quantum field.

If h defines the minimum possible action, several implications follow naturally.

The Discreteness Principle

Reality may consist of discrete transitions, just like there is no half-quantum of energy, there may be no half-step of action. every photon emission, every particle interaction, and every observation could represent at least one indivisible quantum of transformation. reality would advance through these discrete updates, close enough together at macroscopic scales to appear smooth and continuous.

The Observation Principle
Any attempt to observe a system involves interaction, and interaction requires information exchange. Since information transfer cannot occur without physical interaction, each observation requires at least one h-sized update. Observation alters systems because observing is physical interaction, and interaction disperses information into the environment while making quantum reversal thermodynamically impossible. When we understand that observation requires photons, that photons carry h-sized action quanta, that information transfer costs energy (Landauer, 1961), and that this energy disperses as entropy into environmental degrees of freedom, collapse becomes an expected consequence of physical interaction.

The Superposition Principle
The framework introduces the concept of the Quantum Information Dimension (QID). If reality advances through discrete actions, the QID represents the dynamic field of unselected possibilities or a substrate where all potential transformations exist in superposition until one becomes actual. When we describe a particle as being "in superposition," we are referring to a state where multiple possible actions remain available, awaiting selection.

The Selection Principle
Each action requires selecting one outcome from many alternatives. While random environmental factors such as photons and thermal noise drive most physical collapses, consciousness introduces directed selection and participates in this process. Within the framework's equation P = Q/E, low ego activation allows consciousness to survey a wider field of possibilities within the QID before selection, while high ego activation narrows this range and forces earlier collapse through the collapse process as described in Chap. 5.

The Universe's Transaction Fee and Block Size

Before getting into why h has its specific value, it helps to think about what role it plays in the universe's information structure, because both physics and distributed systems are basically solving the same problem. How can countless independent events, scattered across time and space, all end up forming one shared, agreed-upon history? In blockchain, this is called the Byzantine Generals Problem. It was the puzzle that Bitcoin solved which was how to get a network of totally disconnected computers running software to agree on a single sequence of transactions without any central authority. It does this through a distributed consensus mechanism where energy use, validation, and confirmation make every block statistically irreversible.

In this same spirit, physics seems to rely on its own kind of distributed consensus mechanism. The book posits this could be entropy's role. As explored in Chap. 5, every dispersion of phase information into photons, phonons, or molecular vibrations increases the statistical improbability of reversibility of the previous information exchange until it is effectively confirmed and collapsed.

But why does h have its specific value? Like any successful system, the universe had to solve an engineering problem. How large should the minimum quantum of action be?

Let's consider how similar this challenge is to modern systems design. In Bitcoin's early days, the Bitcoin community faced a huge debate which was coined the "block size wars." If blocks are too small, the network becomes congested and cannot scale. If blocks are too large, the system risks centralization and the loss of stability and its security. The solution required finding a balanced range that maintained both functionality and adaptability which is akin to the problem the universe faced. It had to determine the smallest amount of change, or the smallest amount of information in any given transaction, or the resolution of reality itself. All of these metaphors would be different interpretations of the same problem. How big should the "blocks," "pixels," or "voxels," of existence be?

In blockchain terms, Planck's constant plays the role of block size as it sets the minimum size of any discrete update to reality's "ledger" or its agreed upon physical reality. Every photon scattering, atomic transition, or conscious choice involves committing at least one h-sized unit of action to the record of classical reality. Without this minimum, the quantum "mempool" of possibilities, or the pool of uncollapsed potential states, would churn endlessly without ever stabilizing into the confirmed history we call classical reality. Infinite resolution might seem ideal, but it would require infinite computational resources as every point in space would need constant updating, which would consume unlimited energy, and nothing would ever stabilize long enough to create persistent structures. If the "block size" or "pixel size" were too small, essentially approaching zero, reality would potentially flicker chaotically. Particles would jitter endlessly, and atoms couldn't hold together. Nothing could stabilize long enough for complexity to emerge. If the block size were too large, reality would likely jump discontinuously between states with no smooth transitions possible. Chemistry would break down because electron orbitals could not adjust gradually, life would be inconceivable because biological processes depend on continuous micro-adjustments, and reality would feel like a video game running at one frame per second, which would clearly be far too coarse for complexity to evolve.

From the framework's perspective, Planck's constant represents the pixel size of the universe, or metaphorically, its display resolution or voxels. Just like 1920×1080 is not arbitrary for screens but it works well for human vision at traditional distances, h is not arbitrary either. This balance allows atoms, molecules, and eventually life itself to maintain stable form.

With this in mind, the framework views fundamental constants as solutions to deep optimization problems. In blockchain, Bitcoin's 21-million cap creates specific economic properties, in technology, standard display resolutions balance clarity with

Table 12.1 The similarities between this framework's physics proposals and blockchain technology

Blockchain concept	Physics/framework equivalent	Role in system
Block size	Planck's constant (h)	Defines the minimum indivisible "unit of action," setting the smallest possible resolution at which reality can update
Gas fee	Landauer cost kT ln 2	Minimum thermodynamic cost required to commit an update and erase discarded alternatives
Transaction	Quantum event/ measurement outcome	A potential path resolved into a definite classical fact
Mempool	Quantum superposition/ QID	The field of all potential futures waiting for selection
Mining/ proof-of-work	Collapse mechanism + entropy increase	Physical process finalizing one outcome while discarding alternatives
Block finality	Wavefunction collapse	Irreversible confirmation of an outcome in the ledger of reality
Difficulty adjustment	Thermodynamic constraints	Governs how fast systems can finalize outcomes based on available energy and entropy
Chain of blocks	Causal time sequence	Ordered record of confirmed reality states. Metaphorically the "ledger" of spacetime

bandwidth, and in physics h defines the best value that permits quantum computation and classical stability to coexist. Considering this it seems reasonable to consider the speed of light c as the maximum rate at which actions can occur. It does not set a speed limit on how fast things move through space. Instead, it defines how quickly new h-sized updates can be committed, or similarly what the fastest rate is at which information can be exchanged. Together, h and c define reality's operating parameters, as they are metaphorically its resolution and frame rate (Table 12.1).

Energy Itself

Considering all that has been discussed so far, the framework reframes the meaning of energy. The equation $E = hf$ does not simply mean energy equals Planck's constant times frequency. Within this perspective, energy measures the rate at which h-sized actions occur per unit of time. Higher frequency represents more discrete updates per unit of time. A gamma ray carries more energy than a radio wave because it performs more h-sized transformations per each unit of time.

The questions from earlier now have potential answers when viewed through this lens. Action is quantized because reality consists of discrete transformations instead of there being some sort of continuous flow. Observation/measurement causes collapse because they involve action, and each action transforms quantum potential into a definite classical configuration. From this, collapse is the natural consequence of updating reality, as matter has wavelength because de Broglie's equation $\lambda = h/p$ shows how many h-sized voxels/pixels a particle's probability distribution spans before a single configuration is selected. In this sense, wavelength metaphorically represents the extent of a particle's "motion blur" which is how many potential pixels, or voxels, of reality it occupies at a given moment for a given speed. Uncertainty exists because Heisenberg's principle shows that position and momentum correspond to mutually exclusive sets of possible actions. Fully actualizing one excludes fully actualizing the other, and the lower bound $\hbar/2$ represents the irreducible cost of making any selection.

Nothing can exceed c because c represents the maximum rate at which updates can occur. Exceeding c would require "being" before the next frame of spacetime exists to support you. Time emerges as the sequential record of completed actions which is propagated through entropy, and each update of h creates a discrete "now." The past consists of completed updates stored as physical reality that our brains remember as the past. The future contains unselected possibilities within the QID.

From this view, the universe is not fundamentally composed of particles, waves, or even fields. It would be fundamentally composed of actions, or more precisely, discrete, quantized transformations occurring at a maximum rate of c, with each requiring a minimum size of h, as unselected possibilities exist within the QID until collapse.

The Universal Selection Mechanism

When viewed through the lens of Feynman's path integral, the mechanism of "explore all possibilities, then select one" appears to operate at every scale of reality. In Feynman's path integral formulation (Feynman, 1948; Feynman & Hibbs, 1965), amplitudes are summed over all kinematically possible paths. It behaves mathematically as if particles explore every route, with destructive interference pruning almost all possibilities, and then the observed path emerges through what might be understood as quantum natural selection. Most paths "die off" through destructive interference while the optimal path "survives" through constructive interference.

Within this framework, the Quantum Information Dimension (QID) represents the physical substrate that Feynman's mathematics sums over. When the path integral formulation says "sum over all possible paths," the QID is where those paths exist before selection. It provides an ontological interpretation for what has traditionally been treated as purely mathematical abstraction. The QID contains all lawful possibilities as actual potential configurations of the quantum field, and Feynman's path integral describes the selection mechanism that determines which configurations survive to become classical reality.

This framework views that same process as a layered hierarchy of a five-level selection process, which seems to operate in related ways across physics, biology, and consciousness as this book has described (Fig. 12.1).

Level 1 – The Quantum Information Dimension (QID)

All lawful possibilities exist here as potential configurations of the quantum field, lawful meaning those permitted by quantum mechanics. Every path, field state, and spacetime geometry is represented in superposition within the QID. This is the space that Feynman's path integral sums over. Nothing has yet been written into the classical record.

Level 2 – Action Weighting (Feynman Computation)

In Feynman's formulation, each path contributes amplitude $e^{(iS/\hbar)}$, where S is the action. Paths with lower action align more coherently in phase, while those with high action oscillate rapidly and cancel through interference. This acts as the first filter where energetically efficient paths become probabilistically favored.

Level 3 – Quantum Interference

Paths whose actions differ only slightly interfere constructively, while those that differ greatly interfere destructively. Constructive interference stabilizes around points where the action is stationary ($\delta S = 0$), and paths far from this region cancel through destructive interference. Through this interference pattern, coherent structures begin to stabilize from the field's continuous computation.

Level 4 – Environmental Coupling (Decoherence)

When a system interacts with its environment, information about possible states disperses through entanglement with surrounding degrees of freedom such as photons, phonons, and thermal vibrations. Each environmental particle that becomes entangled with the system effectively records which-path information that creates correlations that destroy quantum coherence. During this, the environment acts as a distributed measuring apparatus that stabilizes a subset of coherent configurations. Reversibility remains possible in principle (if one could track all environmental degrees of freedom), but becomes exponentially unlikely as information spreads across macroscopic numbers of particles.

Level 5 – Thermodynamic Authentication (Collapse)

At this stage, information loss becomes logically irreversible (per Landauer as the system cannot uniquely recover its prior state from the current state), and the selected configuration becomes part of the physical record. Unselected configurations remain in the QID as unrealized potential, but are not destroyed, and are no longer accessible without reversing macroscopic entropy.

This completes the cycle as one branch of QID possibility space has been authenticated into spacetime, while the total quantum state (QID + classical record) continues evolving unitarily.

This is P = Q/E as:

P (surviving outcome) = (Q) Quantum configurations / (E) Environmental collapse pressure

Fig. 12.1 The five-level universal selection process

This same hierarchy may apply at every scale, where each level describes a further narrowing from possibility to actuality through energetic and informational selection. This connects to one of physics' oldest organizing principles which is the minimization of action. In Feynman's path integral formulation, each possible path through the QID is weighted by $e^{(iS/\hbar)}$, where S represents the action (Feynman & Hibbs, 1965).

Paths with lower action interfere constructively because they require less thermodynamic cost to realize as fewer informational bits must be committed, which reduces the Landauer energy dissipation required for that path to become recorded.

With this in mind, action minimization may represent the same thermodynamic selection described earlier. Whether in particle motion (Feynman paths), gravitational trajectories (geodesics), or conscious behavior (P = Q/E optimization), systems appear to follow routes that minimize information cost. Working against these naturally selected paths, such as resisting gravity, forcing attention, and overriding instincts and habits, requires additional energy expenditure to overcome the inefficient paths. The long-standing question of why nature seems to minimize action can then be understood as a question of efficiency. Lower action paths carry lower information cost, making them more probable selections from the QID's configuration space.

This selection mechanism appears to be universal. At the quantum level photons explore all paths while interference selects viable options and observation (measurement) collapses the possibilities until one path actualizes, at the biological level life explores all mutations while environment selects advantageous traits and survival pressures collapse options until one trait actualizes, at the neural level the brain explores all possibilities while relevance realization selects useful patterns and ego collapses the options until one thought actualizes, and at the consciousness level the mind explores all futures while attention selects one's focus and experience collapses possibilities until one moment actualizes. In each case, unfit paths die off while the "fittest" survives. This exemplifies what Wheeler (1990) described as the universe's fundamental yes/no choices, where each quantum event functions as a binary act of selection.

The P = Q/E equation describes this universal selection process as:

$$\begin{gathered}\text{P (surviving outcome)}\\ = \text{(Q) Quantum configurations}/\text{(E) Environmental collapse pressure}\end{gathered}$$

This universal selection mechanism may point to something about the nature of physical laws themselves. In Feynman's path integral view (1965), particles don't "follow laws" in the deterministic sense. They explore all possibilities, and destructive interference eliminates almost all paths, leaving only those that survive. Is it possible that from this perspective, over time and scale, this produces patterns so consistent that we call them "laws," even though they're actually the stable outcomes of a deeper selection mechanism? In this view the laws of physics may be the most successful survivors of the universe's ongoing quantum natural selection process and would not, in the traditional sense, be fixed commandments carved into the fabric of reality before the emergence of the universe itself.

This aligns with Wheeler's vision of "law without law," where he proposed that physical laws emerge from the accumulation of quantum measurement events instead of existing as predetermined rules (Wheeler, 1983). As Wheeler suggested, the regularities we observe as laws arise from "the elementary quantum phenomenon" which is the irreversible act of observation that transforms quantum field states into classical actuality. The framework then extends Wheeler's insight by showing how this process operates through natural selection at every scale.

This process resembles the classical Brachistochrone problem, where a bead sliding under gravity reaches the bottom faster along a "longer" curved cycloid than along a straight line, showing that the longer curved cycloid is actually the most optimal path. In the QID, geodesics function as the cycloids of possibility space. The steeper the local action density becomes, the more destructive interference removes straight-line paths. The surviving paths minimize collapse cost, which behaves like an information-theoretic time of travel.

Matter and light naturally fall into these optimal trajectories because all other options destructively interfere. In strong gravitational fields, the funnel becomes so steep that maintaining an outward trajectory requires extraordinary energy.

Light does not "know" its path before it moves in any conscious sense. The path integral simply behaves as if all routes are explored and only the most stable, least-action one survives interference. The same selection logic later reappears in biological and cognitive systems, where feedback replaces interference as the pruning mechanism.

Light follows identical rules because photons try every possible path simultaneously. Paths skimming closer to massive objects accumulate higher action density, while outward deviations become thermodynamically expensive due to too many discarded possibilities and excessive erasure cost. The fittest path that survives interference hugs the curvature of the collapse funnel, which appears as a bent light ray.

Light "taking the shortest path" through water isn't following a pre-existing rule. It "creates" what becomes the shortest path through this selection mechanism. The bent path emerges as the "fittest" possibility that survives quantum interference patterns in water's unique environment. The environment itself, including vacuum properties, matter fields, and spacetime geometry, acts as the selection pressure.

The framework suggests that just like quantum systems collapse to energy minima, biological systems evolve toward fitness peaks, and gradient descent finds optimal parameters, consciousness might navigate toward stable configurations in the quantum information dimension. In other words, $P = Q/E$ mathematically encodes the same principle behind Feynman and Hibbs (1965) path integrals as the universe continuously explores all possible paths (Q), environmental and energetic constraints act as selection pressures (E), and what we observe (P) is simply the "fittest" survivor. It is the outcome that remains after destructive interference eliminates every other possibility. This universal tendency toward stable minima might explain why mathematical patterns emerge consistently across physics, biology, and consciousness as they all reflect the same fundamental topology of the quantum field's informational landscape.

Jung and Pauli were, in their own way, pursuing the same question from opposite ends of the spectrum. Pauli sought the mathematical forms that underlie physical order, while Jung searched for the symbolic structures that organize meaning. Their collaboration circled the idea that psyche and matter might share a common informational substrate, the "unus mundus" (Jung, 1963), where the patterns of nature and the patterns of mind arise from the same field. From the framework's view, those manifested archetypal patterns can be seen as the stable survivors of this universal selection process (Jung, 1969). They would be the informational attractors that have endured through countless collapses because they represent optimal, low-action solutions in the landscape of consciousness.

This universal pattern seemingly points to the idea that consciousness doesn't impose something foreign onto physics, but instead, it participates in the same fundamental process that governs reality at every scale from photons finding their path to minds selecting their future.

Time as Action Creation

Historically, physics has placed human-scaled observation at the center of measurement. Time and space were reference frames because that was how humans could measure them. Even Einstein, while dethroning absolute frames, still defined all motion relative to massive observers. That's why, in our current worldview, light is described as "moving at speed c through spacetime." But if we take photons seriously and ask what reality looks like from their frame, then things start to look different.

What if reality is not moving through spacetime at all, but instead, reality is spacetime continuously happening at the rate defined by c? Then c isn't a velocity through a pre-existing stage, but it's the speed of creation itself, or, the maximum "refresh rate" at which h-sized actions can be committed into being. Everything with mass simply operates below that rate because its internal maintenance consumes part of the available action budget. This removes the anthropocentric bias as the universe would no longer organize around what we measure relative to our own orbits or clocks; instead, it organizes around a universal participation rate. Matter lags behind the frontier of creation because it's busy maintaining structure.

This framework's perspective asks what happens if the reference frame doesn't belong to the observer, but instead belongs to light itself, or to action instead of observation. If the photon, or more generally, the quantum of action, becomes the natural reference point, then spacetime, energy, and consciousness can all be seen as expressions of participation in that continual process of creation. The observer would no longer sit outside reality measuring it, but would arise within the universe's ongoing act of committing information into form.

Just as Galileo showed that planetary motion began to make sense once Earth was decentered, physics may become coherent once human measurement is decentered. The same shift that had moved Earth from the center of the cosmos could now move the observer from the center of ontology as reality would no longer happen for the observer; the observer would happen within reality's unfolding act of creation.

In this, quantum superposition exists timelessly until measurement forces action selection. Within this framework, photons do not just "travel" through spacetime, but instead are the actions that create spacetime itself, as they operate at the maximum action rate c. The speed of light would then metaphorically represent the "frame rate" of reality or the maximum rate at which h-sized actions can be committed into actuality relative to human measurement.

Types of Time

This framework distinguishes between two forms of time, with cosmic time emerging from the sequential actions that construct classical reality and experiential time emerging from the brain's information processing as modulated by ego and coherence. Both forms ultimately derive from action, where cosmic time counts the universal sequence of committed updates while experiential time measures how many perceptual frames consciousness processes per cosmic frame. At the cosmic scale, however, the framework proposes that each action creates a discrete "now" moment. Even Planck-time is relative to Earth's definition of seconds. There is no true way to measure time without anchoring it to some physical process which in and of itself points to how time is purely an emergent property. While collectively, from Earth, we of course measure time relative to Earth's experience of rotation and orbit relative to the Sun.

However, consciousness, mediated by ego, functions as the mechanism that selects which potential actions become actualized in one's perception. The personal experience of time itself emerges from this selection process. Therefore, consciousness generates its own experience of time through its participation in committing actions as it participates in collapsing and refreshing the objective physical reality we all experience. This perspective also clarifies why "now" feels uniquely real despite physics suggesting all moments could coexist equally. If each collapse creates a new committed frame of reality, "now" would represent the active edge where possibilities transition into actuality. The past consists of completed actions locked into the physical ledger through entropy and in human experience, it is stored as memory in a linear experience. "The past" is merely the recollection of formerly updated frames of perceived reality. It also explains why we remember the past but not the future. Completed actions are accessible because they exist as committed updates. The future remains in superposition and cannot be accessed, since doing so would require bypassing the collapse mechanism itself.

As described in previous chapters, this is why the perception or experience of time from an individual human perspective can shift. In flow states, ego suppression may allow System 1's quantum-coherent processing to dominate, which lets the brain extract far more information from each classical planck-sized moment. Within this same moment, biological rhythms like heartbeat and breath remain steady, yet consciousness is experiencing the brain process many more computations per planck-sized moment. Athletes describe time "slowing down" because each moment becomes computationally richer not because fewer universal actions are actually occurring. Cavaglià et al. (2023) provide theoretical support for this mechanism.

They demonstrate that under suitable conditions, neuronal membranes form weak Bose–Einstein condensates driven by action-potential energy, producing coherent dipole oscillations in the terahertz range (10^{11}–10^{12} Hz) which is over nine orders of magnitude faster than classical neuronal spike transmission. These coherent modes propagate at ~ 10^3 m/s, enabling information to integrate orders of magnitude faster than traditional pathways allow. Their process will be explored in the coming chapter in detail.

The Binding Problem and Unity of Experience

A fundamental challenge for any theory of consciousness is the binding problem. How does the brain create unified conscious experience from billions of separate neurons? Classical physics offers no mechanism for genuine unity. Neurons firing separately cannot create wholeness any more than air molecules create a "tornado" with independent existence.

Considering this, the book hopes it has laid out the evidence in past chapters for how the P = Q/E framework provides a possible solution through its three-layer, participatory architecture where at the quantum layer (Q), all possibilities exist in superposition with no binding problem because nothing yet separates from anything else, which allows all potential experiences to exist as an undifferentiated field of possibility, while through action via DMN orchestrated collapse (E), each action event becomes experienced as unique as fundamental to quantum mechanics, which suggests that instead of passively observing pre-existing reality, human consciousness is able to participate in it, and this participatory action generates the different ways of knowing identified in cognitive science including propositional knowledge from facts emerging through action, procedural knowledge from patterns of repeated action, perspectival knowledge from the observer's viewpoint, and participatory knowledge from the act of taking action itself.

Then from there the DMN/ego constructs narrative coherence from the stream of actualized experiences. All of this can be happening while each action creates a discrete "now" moment with sequential actions generating temporal flow while memory acts as the stored record of each individual's unique action history, which explains why memory feels like a personal story instead of traditional data storage, and as Nishiyama et al. (2024) demonstrate, this information can be stored holographically across neural quantum systems in distributed form.

This holographic storage aligns with principles from physics that warrant consideration. In anti-de Sitter/conformal field theory (AdS/CFT), a d-dimensional boundary theory can fully encode a (d + 1)-dimensional bulk under specific quantum-gravity conditions (Gubser et al. 1998; Maldacena, 1998; Witten, 1998). Without claiming gravitational dynamics in the brain, this correspondence offers a scaffold for thinking about how richly structured, high-dimensional quantum information could be compressed and rendered at a lower-dimensional classical interface. Ryu and Takayanagi (2006) showed that information at a boundary can completely determine the geometry of the interior space, suggesting that consciousness at the "boundary" between quantum and classical might determine the structure of accessible mental space, and these ideas are relevant to the topics of Chap. 9. Faulkner et al. (2014) extended this by demonstrating that small changes in boundary entanglement create corresponding shifts in bulk energy, which could parallel how small changes in ego activation (E) create large shifts in cognitive capability. Together, these principles suggest that P = Q/E might play a part in modulating the geometry of accessible information.

This framework, and all presented in past chapters, hopes to provide a plausible explanation for why consciousness feels like an engaged, meaningful, temporal, and unified experience. We might experience unified consciousness precisely because we are quantum-classical hybrids. The described "problem" of binding could dissolve when we understand consciousness as the experienced, participatory process of transforming quantum field dynamics into classical temporal narrative through ego-mediated action and collapse.

Relativity, Time Dilation, and the Action Rate

Having defined time as the sequential record of committed actions, relativity now reveals what limits the rate of those actions and why this rate defines the structure of spacetime itself. If consciousness creates its own experience of time through the sequential committing of actions, a natural question arises. At what rate does this creation occur? Physics provides an answer. The speed of light, c, represents the maximum rate at which h-sized actions can be committed. Nothing can "outrun" this action-creation rate because there are no pre-existing frames to occupy. This is at the core of the theory of relativity. Let's explore this within the framework.

To illustrate, imagine two trains moving on parallel tracks. The faster train represents the action-creation process operating at rate c. Each chapter of track it lays down corresponds to one h-sized committed update of spacetime. The slower train represents human consciousness experiencing these frames at biological speeds. Passengers on the slower train would see the track laid out ahead as if it had always existed. Yet as the slower train accelerates toward the speed of light, it begins to catch up with the act of spacetime being laid. To those passengers, it appears as if the faster train is slowing down. Approaching c, they would begin to "see" reality being constructed, as their own internal action rate approaches the universal refresh rate.

Now imagine observers who didn't board that second train and just remained at the station, but are now watching the second, slower train gain speed and approach the speed of light. From these observers' perspective, the slower train's clocks begin to tick more slowly as it accelerates. They still see the faster train laying down the track at its steady, constant rate. But the observers still at the train station would see the passengers aboard the second train walk, talk, and eat in slower and slower motion, because the faster this second train moves, the less updates per second it exchanges relative to the observers at the station. What feels normal aboard the moving train now appears dilated when compared with the reference frame that stayed behind. Each group is correct within its own frame, at their own speed, yet their experiences diverge because each measures time by the rhythm of its own internal action patterns (their mass).

This view aligns with time dilation. As a system accelerates toward c, its internal processes, like atomic vibrations, chemical reactions, and neural activity, must share finite action "slots" (pixels, voxels). Approaching the action limit, internal updates slow relative to external reality. Massive systems cannot reach c because their ongoing maintenance requires sequences of internal actions, which would have nowhere to occur if moving exactly at the refresh rate. At exactly c, a system would freeze into a single committed frame. Photons occupy this boundary. From their perspective, emission and absorption occur as one event. They do not travel through spacetime but exist as the updates creating spacetime itself. Therefore, light does not move through reality; it is reality happening and unfolding.

In this framework, exceeding c is undefined. There is nowhere to be beyond the boundary where spacetime is created. Asking what lies beyond c becomes analogous to asking what lies north of the North Pole.

Mass and Energy as the Architecture of Action

Mass and energy, traditionally treated as separate physical properties, may actually represent two aspects of the same underlying action pattern. If time emerges from actions committed at rate c, massive objects face a structural constraint as then mass represents stabilized configurations that require continuous sequences of h-sized actions to maintain form through atoms vibrating, electrons orbiting, and quarks binding. Each of these internal processes needs its own maintenance stream.

Because of this constraint, mass cannot reach c. At the universal refresh rate, no temporal space exists for these internal actions to occur. The more internal processes a system maintains, the more action-per-second it requires simply to persist. Photons sit at the opposite limit. Without internal temporal processes, they exist at c, as they ride the boundary where quantum potential becomes classical actuality.

To picture this more vividly, imagine a hose spraying water at an incredibly high speed, but something is in front of the hose and is cutting the flow of water extremely fast into these immeasurably small sections (h-sized packets). From a distance, the stream looks like a continuous, smooth flow of water. Yet if you could slow it down enough, you'd see that the stream consists of countless discrete droplets, the h-sized packets, and that the water is separated by imperceptibly small gaps. The stream only looks continuous because the droplets follow each other so closely that the eye cannot distinguish the intervals between them.

Now imagine speeding up and moving alongside that stream at the same speed as the water. When moving at this speed, the droplets would appear frozen in space and time. They would appear to be perfectly still, because you'd be moving at the same rate as the flow itself. This is analogous to what happens at the speed of light.

From our slower, mass-bound perspective, photons appear to move continuously as light, as they update reality at the universe's maximum rate. But from the photon's own frame, no time passes at all. The stream of creation and the experiencer of that stream become one and the same.

Energy, which is often treated as the universe's fundamental currency, may represent our measurement of action throughput through the number of h-sized collapses per second. If so, this raises a foundational question about whether the universe is made of energy with information describing it, or whether the universe is made of information with energy emerging as a secondary effect.

Traditional physics assumes energy's primacy with information acting as a mathematical tool. Yet quantum mechanics increasingly challenges this view, and this book has framed information as fundamentally emergent from the Quantum Information Dimension. In this, quantum states are probability distributions or informational constructs, and quantum information theory shows that information itself is conserved. Wheeler's (1990) "it from bit" hypothesis aligns with this perspective, as does Landauer's (1961) principle, which established that erasing one bit of information carries an unavoidable thermodynamic cost of $kT \ln 2$. Together, they show that information and energy are inseparable, and that every irreversible act of computation or observation carries a physical cost in energy and entropy.

Within this framework, information comes first and energy emerges during the process of committing actions as the observable footprint of collapse frequency. As described, $E = hf$ takes on new meaning where h represents the minimum action quantum, f represents the number of actions per second, and E represents committed updates per second measured in joules. This allows one to think of things a bit differently, for example, chemical reactions involve rearrangement of classical bonds with minimal new actions which would make it less energic than something like nuclear fission where fission splits nuclei with moderate action release which would be less powerful than nuclear fusion that recombines nuclei with significant action release, and then matter-antimatter annihilation completely dissolves bound patterns which results in maximum action liberation and significant energy. The framework's view of $E = mc^2$ aligns here as mass encodes stabilized action patterns. Converting matter to energy releases these locked actions, which would be proportional to the density of committed updates that formed it.

What Is Gravity?

Having explored how mass and energy reflect patterns of committed action, it becomes natural to ask how these patterns shape the larger geometry of experience itself. If mass encodes stabilized actions, what happens when these actions accumulate? This brings us to gravity. From this framework, Gravity may represent the geometry of concentrated action density (the local concentration of committed h-sized events per unit volume). Gravity, described by Einstein as spacetime geometry, could now potentially be understood through the framework as an emergent selection effect arising from concentrated action density. From the framework's perspective, spacetime curvature described by general relativity could emerge naturally from steep gradients in the Quantum Information Dimension. This interpretation recovers GR's predictions but frames geodesics as the surviving paths after destructive interference prunes alternatives.

Mass represents stabilized patterns requiring continuous streams of h-sized actions to maintain structure, such as atomic vibrations, electron orbitals, and quark confinement. Where these patterns accumulate, the local "collapse landscape" steepens, which shapes which possible futures remain viable.

The framework proposes applying Feynman and Hibbs (1965) path integral across all scales, where particles and photons explore every possible path through the QID while destructive interference eliminates the majority. In regions with minimal action density far from massive objects, many trajectories remain viable, which would produce what we observe as flat spacetime geometry. Near concentrated masses, paths moving away from the gravitational source require increasingly improbable action sequences to maintain their distance. These high-cost trajectories destructively interfere with themselves, which leaves primarily the paths that follow the action density gradient inward.

This selection process creates what Einstein described as spacetime curvature, though the framework reinterprets this as emergence from probability distributions in the QID. Gravity would then represent the natural result of the universe's least-action selection mechanism. As objects approach regions of concentrated action patterns, the probability landscape becomes increasingly biased. Paths that maintain constant distance from the mass require exponentially more action to resist the gradient, making them vanishingly improbable.

Similar to the brachistochrone curve where the longer path proves fastest, gravitational trajectories that appear curved actually represent the most probable paths through the QID's selection landscape. Photons exploring all possible routes find that straight paths away from mass accumulate prohibitive action costs. Through destructive interference, these expensive paths cancel out, leaving curved trajectories as the dominant survivors. Light bends not because space is pre-curved, but because the selection process naturally prunes all but the curved paths.

Time dilation near massive objects emerges from how action density affects the relationship between local and universal update rates. All internal processes, whether chemical reactions, neural activity, or atomic vibrations, require sequences of h-sized actions to unfold. In regions of high action density near massive objects, maintaining position against the selection gradient consumes action that would otherwise advance temporal processes. An atomic clock on Earth's surface must continuously "climb" against the probability gradient to avoid falling, using action that could otherwise mark time.

This creates different ratios between local process rates and universal frame counts. An observer near a massive object still experiences normal local physics, with heartbeats and thoughts proceeding at their usual relative rates. Yet compared to observers in regions of lower action density, fewer universal updates elapse for each internal process. The framework views this as arising from how the selection landscape allocates available action between maintaining position and advancing temporal sequences. This interpretation aligns with relativity's predictions while proposing an underlying mechanism based on information-theoretic selection instead of geometric curvature alone.

Black Holes as Infinite Gradients

At the event horizon, the collapse funnel becomes infinitely steep as outward-moving paths require an infinite number of h-sized actions to maintain viability, which would cause all possible outward futures to destructively interfere and leaving none to actualize. From an external perspective, information would appear frozen at the horizon, while from an internal perspective, time proceeds normally until approaching the singularity.

In this framework, black holes do not tear spacetime but represent regions where the collapse landscape terminates. Beyond the event horizon, outward options no longer exist within the universe's energy budget. The event horizon marks the boundary where the action cost of maintaining outward trajectories exceeds what the universal information processing system can support. This interpretation suggests that black holes function as something akin to ultimate information sinks where the geometric constraints of the collapse landscape become so severe that only inward-directed paths remain viable. The apparent paradox of information preservation at the horizon versus information destruction at the singularity may reflect different perspectives on the same underlying selection process operating at the limits of the universe's computational capacity.

From this perspective, gravity, the bending of light, escape velocity, and time dilation all emerge from the same single selection mechanism. Where action density is high, destructive interference removes more potential paths, therefore the surviving trajectories, like for photons, planets, or people, cluster along least-collapse-cost routes. Einstein described these survivors as geodesics through curved spacetime and Feynman and Hibbs (1965) described them as optimal paths surviving quantum interference. The framework attempts to unify both views as geometry itself emerges from the computation of survival within the QID. In this, concentrated action patterns tell the QID how to shape its selection landscape, and that landscape in turn tells matter how to move.

A Jumping Example

To understand how this selection mechanism operates in everyday experience, consider jumping on Earth versus the Moon. When you jump, your muscles generate a specific amount of energy that creates quantum possibilities for upward trajectories. The framework can describe this through the P = Q/E equation operating at the gravitational level.

On Earth, the high concentration of mass creates a steep action density gradient. Your body proposes countless possible upward paths through the QID, but Earth's concentrated action patterns cause most of these to destructively interfere. The steep gradient rapidly eliminates upward-moving trajectories, leaving only paths that quickly curve back downward. You experience this pruning of possibilities as rapid deceleration and a limited jump height.

On the Moon, with approximately one-sixth Earth's mass, the action density gradient is much gentler. When you jump with identical muscle force, those same upward paths encounter less destructive interference. More high-altitude trajectories remain viable for longer distances. The gentler collapse landscape takes more time and distance to eliminate upward options, allowing you to travel approximately six times higher before the interference pattern finally constrains your trajectory back down.

The jumping example may point to the deeper thermodynamic principle, as each upward meter in a gravitational field requires energy to maintain trajectories that would otherwise be eliminated through destructive interference, and from the framework's perspective, escape velocity represents the precise energy threshold where outward paths can permanently overcome the local collapse gradient. Consider the energy economics where near Earth's surface, maintaining an upward trajectory for one second might require overcoming interference that eliminates 90% of potential paths, while near the Moon's surface, the same duration might only face 15% path elimination, with the energy cost scaling with the rate of possibility pruning.

For a jumping human, the initial kinetic energy of ½mv^2 remains consistent on both Earth and Moon, but the path survival rate becomes determined by local action density gradient, with maximum height occurring where energy can no longer maintain viable upward paths, so on Earth, the steeper gradient exhausts the energy budget for maintaining upward possibilities faster than on the Moon.

This thermodynamic view would interestingly unify the effects of gravity and consciousness, as maintaining unlikely thoughts against ego's interference requires mental energy just as maintaining unlikely trajectories against mass's interference requires physical energy, with both following P = Q/E, where E represents the collapse pressure that must be overcome to access Q's possibilities. In this way, the same relationship that governs the body's motion through a gravitational field also governs the mind's motion through possibility space. Energy, whether physical or mental, is the cost of sustaining improbable realities against the pressure of collapse.

Dark Matter and Dark Energy

Having established that every action creates both time and entropy, the framework next considers whether dark matter and dark energy could represent complementary expressions of action, with one as its accumulated record, and the other as its uncollapsed potential. Understandably this is entirely a speculative hypothesis, but it seems to fit well as an extension of the premises and proposals of the book.

Contemporary physics has established that the universe consists of approximately 5% ordinary matter, 27% dark matter, and 68% dark energy (Planck Collaboration, 2020). Science has observed dark matter's gravitational effects but cannot explain its nature. The framework suggests a reason for this that traditional approaches search for exotic particles when they should examine the remnants of action itself.

Consider what happens after each action is completed as the action has transformed potential into actuality as the record of that action must persist somewhere as information cannot be destroyed. These completed actions, having expended their transformative potential, would retain mass (from $E = mc^2$) but lose the ability to interact electromagnetically. They metaphorically become the "ash" of spent possibilities. Once an action collapses potential into actuality, its informational trace cannot be erased without paying the thermodynamic cost described by Landauer. Over cosmic history, these accumulated records may collectively manifest gravitational effects without maintaining electromagnetic coupling. If this interpretation holds, dark matter could metaphorically represent a fossilized record of completed actions, or a residual mass from processes that have exhausted all possible transformations. This hypothesis remains speculative but offers a potential way to reconcile dark matter's gravitational influence with its lack of electromagnetic interaction. If dark matter metaphorically represents the "fossil record" of all completed actions, then it would contain a dispersed history of the universe as every collapse event would be preserved in some form as it would not be accessible for change, but would present as an immutable record.

If dark matter represents completed actions, then dark energy would be the opposite which would be the expanding space of potential actions waiting to occur. Dark energy comprises 68% of the universe and drives cosmic acceleration (Planck Collaboration, 2020). It appears perfectly uniform and resists direct detection. These properties match what uncollapsed quantum field states would exhibit. Before collapse assigns location, possibilities exist everywhere equally. Any measurement would force collapse, making dark energy fundamentally unobservable in its raw state. The framework explains why dark energy increases over time. The early universe contained mostly hydrogen atoms with limited possible configurations. But as complexity evolved, inevitably possibility space exploded exponentially. A single living cell would essentially contain more potential states than billions or even trillions of hydrogen atoms or conscious decisions may branch into more possibilities than entire galaxies of simple matter. The universe at the beginning was essentially playing chess only with pawns, but now it plays with all the pieces which creates an infinitely more complex game. Every increase in complexity would multiply the possible futures that must exist in superposition. Essentially, we're generating possibilities faster than they can be actualized, and the universe must expand to accommodate this ever-growing possibility space. This is a speculative hypothesis, but one that seems to be a natural extension of the book's core explorations.

Temperature, BECs, and Black Holes

Similarly, the framework's proposition that action creates time suggests a deep relationship between temperature, quantum coherence, and information states. If energy represents an action rate (actions per second) and temperature measures the average kinetic energy of a system, then temperature fundamentally measures the rate at which actions occur. This reframing offers new insight into extreme physical phenomena where time, information, and energy behave unusually.

When certain bosonic atoms are cooled near absolute zero, they undergo a phase transition into a Bose–Einstein condensate (BEC), where millions of atoms occupy a single quantum state. From the framework's perspective, this state represents a near-zero action rate. If time emerges from discrete actions, then systems in this regime approach a suspended temporal flow. The BEC experiments were explained in detail in Chap. 5, please reference there for more on the findings and claims themself.

Within this framework, these findings suggest that BECs act as temporary buffers for reality-construction quanta. If photons represent discrete actions carrying at least one quantum of action h, then their immobilization reflects an environment where no available action states exist. Until the photon is revived, the "refresh" of reality along its trajectory remains paused. The BEC holds the informational content of the pending action, as it awaits re-entry into spacetime construction. This perspective reframes BEC phenomena more generally, where superfluidity arises because no new actions occur to generate friction, superconductivity emerges when electron ensembles enter a temporally frozen regime that suppresses decohering actions, and holographic imprints demonstrate that condensed matter stores and processes spacetime-defining information, which preserves phase relations and entanglement while it enables "coherent processing of three-dimensional, compressed patterns of stored optical information" (Dutton et al., 2004).

Within the framework, black holes may represent the universe's most efficient information storage systems. Where ordinary matter continuously generates new actions, which produces time and entropy, black holes would then exist at the opposite extreme as regions where available actions collapse toward zero. From an external perspective of the black hole, information appears frozen at the horizon, yet internally, local processes proceed normally until approaching singularity. This mirrors the experiments of Dutton et al. (2004) and Hau (2016) on stopped light in Bose–Einstein condensates. In these experiments, kilometer-long light pulses were compressed into distances of just tens of microns and stored in atomic imprints without loss of information. By switching on a coupling laser, the original light pulse was later revived intact, as though time inside the condensate had been temporarily paused.

From the framework's perspective, black holes may act as cosmic-scale condensates where photons crossing the event horizon cannot complete their role as "reality-updating actions" because no further action states are available for them to propagate into. Instead, the quantum state associated with that information becomes stored in a maximally compressed form as the universe conserves the information, but from an external frame, it appears inaccessible. Seen this way, black holes, Bose–Einstein condensates, and dark matter represent different endpoints along the same continuum of information freezing, where BECs create temperature-frozen information with stored states that can be revived through reintroducing action, black holes create gravity-frozen information where actions are suppressed by curvature so extreme that causal propagation halts, and dark matter creates entropy-frozen information where all viable actions have been permanently exhausted. This integrated perspective implies that black holes function metaphorically as the universe's memory compression devices, archiving entire histories of action in a compact, maximally efficient form while preserving informational coherence even as the flow of time stalls relative to external observers.

The framework suggests that absolute zero represents the cessation of action creation and therefore the stopping of time itself. This boundary remains physically unreachable, perhaps because reaching it would require stopping the very actions that create the approach toward it. Black holes might represent nature's closest approach to this limit, achieving through gravity what laboratories attempt through cooling, as they create regions where actions, and therefore time, effectively cease from external perspectives. Once again, this is a speculative hypothesis but it seemed to reasonably fit in the extensions of the book's core ideas.

The Thermodynamic Cost of Consciousness

The framework has explored how consciousness collapses quantum states through ego activation, but a deeper question emerges about the energetic requirements of this process. Recent sections have explored a connection between Planck's constant, Landauer's principle (1961), and the thermodynamic cost of reality creation that may illuminate the physical foundation of the $P = Q/E$ equation.

If each quantum collapse represents the universe selecting one outcome from many possibilities, then two fundamental constants might work to govern this process. In this, Planck's constant (h) would define the minimum possible update size, the smallest packet of action required to construct reality. Each h-sized action would represent one discrete "frame" (or pixel or voxel) of spacetime construction, below which no smaller update can occur. Complementing this, Boltzmann's constant (1877), which converts temperature into energy per bit of information, enters through Landauer's principle (1961), which demonstrates that erasing information requires a minimum energy expenditure of $kT \ln 2$, where k is Boltzmann's constant (1877) and T is temperature. These principles work together to constrain reality creation. While h sets the quantum size of each update, $kT \ln 2$ sets the minimum energy toll for making it. When consciousness collapses quantum states, selecting one possibility necessarily erases all alternatives. This erasure, being thermodynamically irreversible, must carry an energetic cost.

The relationship between Planck's constant, Boltzmann's constant, Landauer's principle, and Einstein's mass-energy equivalence may reveal a deeper unity in how the universe processes information. While the framework has established h as defining minimum action units and kT ln(2) as the minimum erasure cost, these constants might work together in more intricate ways.

As described, Planck's constant sets the fundamental quantization of reality construction. Each h-sized action represents one indivisible update to spacetime, or one could think about it as one frame in reality's unfolding. Nothing smaller than h can occur, which metaphorically means it is the universe's fundamental resolution (pixel, voxel or block size in other analogies). Boltzmann's equation, $S = k_B \ln W$, connects this quantum granularity to thermodynamics as W represents all possible configurations available within each h-sized action. Before collapse, these possibilities exist in superposition, but the act of selecting one configuration from W necessarily erases $(W - 1)$ alternatives, and each erasure costs at least $k_BT \ln 2$ according to Landauer's principle. The total energy for collapsing W states to 1 would approximate $k_BT \ln W$, but this must occur within the constraints of h-sized actions, and the framework suggests consciousness navigates this intersection of quantum mechanics and thermodynamics by selecting which h-sized actions to actualize while paying the kT ln(W) cost for erasing alternatives.

Overall, h defines the minimum action unit, while kT ln 2 defines the minimum erasure cost, so every quantum update that converts possibility into actuality must both act (at scale h) and erase (at cost kT ln 2). Before collapse, W represents all possible configurations within an h-sized action. Selecting one collapses $W \rightarrow 1$ and erases $W - 1$ alternatives at a minimum cost of kT ln 2 per bit. The total cost scales as kT ln W.

Einstein's $E = mc^2$ enters when these collapse patterns stabilize. If certain h-sized actions repeat consistently enough, they crystallize into standing waves we perceive as mass. The energy locked in mass would represent the continuous cost of maintaining these patterns against all alternatives trying to manifest. The c^2 factor might reflect the dual requirement of presenting possibilities at rate c while simultaneously erasing alternatives at rate c, which creates the bandwidth limitation for reality construction.

High-energy systems would perform many collapses per unit of time, rapidly constructing their experiential frames. Low-energy systems would update more slowly, with fewer collapses per unit of time. At absolute zero, collapse would halt entirely, updates would cease, and essentially, time itself would freeze which aligns with Bose-Einstein condensates, where light becomes trapped as quantum processes approach stillness.

This thermodynamic perspective really transforms our understanding of ego activation (E) in the $P = Q/E$ equation, because maintaining a complex self-model would require preserving numerous bits of information against thermal degradation and quantum decoherence. Each bit of identity, belief, or self-concept then exists as information that must be continuously refreshed. The brain must perform these refreshes through repeated updates, as each carries its Landauer cost (1961) of at least kT ln 2.

Higher ego activation would therefore create greater thermodynamic overhead as a more complex identity structure requires more bits to stabilize, more updates per second, and more energy devoted to self-maintenance. This creates a direct trade-off where energy spent maintaining ego cannot be used for accessing quantum computational resources (Q). The equation P = Q/E might therefore describe both the psychological dynamics and the fundamental thermodynamic constraints. Flow states, from this view, may represent consciousness operating near the thermodynamic floor. When ego temporarily dissolves, fewer bits require stabilization. The system approaches minimum overhead, where each collapse costs only what physics requires. This efficiency might explain why flow states feel effortless despite producing superior performance. The consciousness system has minimized its maintenance costs, freeing maximum energy for quantum exploration and computation.

Evidence from cognitive performance supports the framework's prediction that possibility-space exploration can occur with minimal thermodynamic cost. Troubat et al. (2009) studied elite chess players during competition and found that, despite operating under intense psychological load, total energy expenditure increased only about 10% above rest. From a classical computational perspective, one would expect far greater energy demands when evaluating billions of possible moves under strict time constraints.

The study also shows a switch in substrate oxidation that after an initial transient reliance on carbohydrate metabolism, the system shifted toward lipid oxidation approximately 25 minutes into the game. This adjustment suggests that the brain-body system dynamically reorganizes its energy use to sustain high-level cognition over time while minimizing metabolic cost. From the framework's perspective, these findings align with the model where the brain leverages quantum parallelism to explore possibility spaces (Q) without brute-force simulation. As explored so far in this book, flow states during elite performance reduce ego maintenance overhead (E), which frees up resources for possibility exploration while approaching the thermodynamic floor predicted by Landauer's principle (1961). The result is high-performance cognition (P) without proportional increases in energy consumption. It seems that the universe operates on an energy budget, and consciousness must navigate within these thermodynamic constraints. Understanding these constraints could offer new approaches to optimizing human performance and wellbeing.

References

Boltzmann, L. (1877). Über die Beziehung zwischen dem zweiten Hauptsatze der mechanischen Wärmetheorie und der Wahrscheinlichkeitsrechnung respektive den Sätzen über das Wärmegleichgewicht. *Wiener Berichte, 76*, 373–435.

Cavaglià, M., Deriu, M. A., & Tuszynski, J. A. (2023). Toward a holographic brain paradigm: A lipid-centric model of brain functioning. *Frontiers in Neuroscience, 17*, 1302519. https://doi.org/10.3389/fnins.2023.1302519

de Broglie, L. (1924). Recherches Sur la théorie des quanta (doctoral dissertation, University of Paris). *Annales de Physique, 3*(10), 22–128.

Dutton, Z., Ginsberg, N. S., Slowe, C., & Hau, L. V. (2004). The art of taming light: Ultra-slow and stopped light. *Europhysics News, 35*(2), 33–39.

Einstein, A. (1916). Die Grundlage der allgemeinen Relativitätstheorie. *Annalen der Physik, 49*(7), 769–822. https://doi.org/10.1002/andp.19163540702

Einstein, A. (1905). Über einen die Erzeugung und Verwandlung des Lichtes betreffenden heuristischen Gesichtspunkt. *Annalen der Physik, 17*(6), 132–148. https://doi.org/10.1002/andp.19053220607

Faulkner, T., Guica, M., Hartman, T., Myers, R. C., & Van Raamsdonk, M. (2014). Gravitation from entanglement in holographic CFTs. *Journal of High Energy Physics, 2014*(3), 051. https://doi.org/10.1007/JHEP03(2014)051

Feynman, R. P. (1948). Space-time approach to non-relativistic quantum mechanics. *Reviews of Modern Physics, 20*(2), 367.

Feynman, R. P., & Hibbs, A. R. (1965). *Quantum mechanics and path integrals*. McGraw-Hill.

Gubser, S. S., Klebanov, I. R., & Polyakov, A. M. (1998). Gauge theory correlators from non-critical string theory. *Physics Letters B, 428*(1–2), 105–114.

Hau, L. V. (2016). *Quantum control of light and matter: From the macroscopic to the nano scale* (Report no. AFRL-AFOSR-VA-TR-2016-0082). Air Force Research Laboratory. https://doi.org/10.21236/AD1004771

Heisenberg, W. (1927). Über den anschaulichen Inhalt der quantentheoretischen Kinematik und Mechanik. *Zeitschrift für Physik, 43*, 172–198. https://doi.org/10.1007/BF01397280

Jung, C. G. (1963). Mysterium coniunctionis: An inquiry into the separation and synthesis of psychic opposites in alchemy (R. F. C. Hull, Trans.; 2nd ed., Vol. 14). Princeton University Press. (Original work published 1955–1956).

Jung, C. G. (1969). *The archetypes and the collective unconscious* (R. F. C. Hull, Ed. & Trans.; 2nd ed., Vol. 9, part 1). Princeton University Press. (Original work published 1959).

Landauer, R. (1961). Irreversibility and heat generation in the computing process. *IBM Journal of Research and Development, 5*(3), 183–191. https://doi.org/10.1147/rd.53.0183

Maldacena, J. (1998). The large-N limit of superconformal field theories and supergravity. *Advances in Theoretical and Mathematical Physics, 2*(2), 231–252.

Nishiyama, A., Tanaka, S., Tuszynski, J. A., & Tsenkova, R. (2024). Holographic brain theory: Super-radiance, memory capacity and control theory. *International Journal of Molecular Sciences, 25*(4), 2399. https://doi.org/10.3390/ijms25042399

Planck, M. (1900). Zur Theorie des Gesetzes der Energieverteilung im Normalspektrum. *Verhandlungen der Deutschen Physikalischen Gesellschaft, 2*, 237–245.

Planck Collaboration. (2020). Planck 2018 results. VI. Cosmological parameters. *Astronomy and Astrophysics, 641*, A6. https://doi.org/10.1051/0004-6361/201833910

Ryu, S., & Takayanagi, T. (2006). Holographic derivation of entanglement entropy from AdS/CFT. *Physical Review Letters, 96*(18), 181602. https://doi.org/10.1103/PhysRevLett.96.181602

Troubat, N., Fargeas-Gluck, M. A., Tulppo, M. P., & Dugué, B. (2009). The stress of chess players as a model to study the effects of psychological stimuli on physiological responses: An example of substrate oxidation and heart rate variability in man. *European Journal of Applied Physiology, 105*(3), 343–349. https://doi.org/10.1007/s00421-008-0908-2

von Neumann, J. (1932). *Mathematische Grundlagen der Quantenmechanik*. Springer. https://doi.org/10.1007/978-3-642-61409-5

Wheeler, J. A. (1990). Information, physics, quantum: The search for links. In W. H. Zurek (Ed.), *Complexity, entropy, and the physics of information* (pp. 3–28). Westview Press.

Wheeler, J. A. (1983). Law without law. In J. A. Wheeler & W. H. Zurek (Eds.), *Quantum theory and measurement* (pp. 182–213). Princeton University Press.

Witten, E. (1998). Anti de sitter space and holography. *Advances in Theoretical and Mathematical Physics, 2*(2), 253–291.

Chapter 13
A Framework for the Architecture and Perception of Reality

This chapter integrates results from quantum optics, neural biophysics, and information theory to propose a unified account of how discrete quantum actions structure classical reality and how biological systems participate in reconstructing it. It utilizes three independent lines of evidence which the first is that optical information can be stopped, stored, and revived with complete phase fidelity in Bose–Einstein condensates (Dutton et al., 2004; Hau, 2016), the second is that neural membrane models show that "due to the stimulation of membrane dipoles by APs, which carry electric field gradients, coherent dipole oscillations are formed under suitable conditions" that enable interference-based holographic reconstruction (Cavaglià et al., 2023, p. 7), and the third is holographic compression mathematics which demonstrate how high-dimensional information encodes on lower-dimensional boundaries and later reconstructs without loss (Faulkner et al., 2014; Maldacena, 1998; Ryu & Takayanagi, 2006; Witten, 1998).

Cavaglià et al. (2023) articulate the core challenge their model addresses when they write "How physical reality is represented by the brain as a coherent and internally-consistent whole is the main aim of the theory we propose" (p. 1). They emphasize that "every animal with a nervous system creates its own physical reality, that we believe is based upon a common molecular mechanism not yet investigated in neuroscience. Addressing this issue is the first step toward understanding the emergence of qualia" (p. 1). Their holographic model provides "a concrete mechanistic formulation" (p. 7) of Nishiyama et al.'s (2024a) holographic quantum field theory (QFT) proposition, and this book's framework extends it by connecting membrane-level coherence to broader questions about the nature of consciousness and reality construction.

The result of this combination is a step-by-step mechanism that runs from photons as units of committed action to cortical holographic reconstruction and distributed memory, with clear thermodynamic constraints that match Landauer-style limits on information processing (Landauer, 1961).

J. Roeloffs, *The Classical Brain to the Quantum Mind*, The Frontiers Collection,
https://doi.org/10.1007/978-3-032-20068-6_13

Planck's Constant as the Unit of Committed Action

It's helpful to summarize the exploration of the last chapter, as this chapter develops a framework building from these definitions. As described in the previous chapter, Planck's constant h is taken as the minimum unit of committed action. From this understanding, a photon is the mechanism by which spacetime is updated, one h-sized commitment at a time. Frequency f counts how many commitments occur per second, so $E = hf$ reads as throughput of action creation. The invariant value c reflects the universal refresh ceiling. Nothing can exceed c because nothing can commit updates faster than the global action rate allows.

Two standard quantum facts sit comfortably inside this reading. Wave–particle duality becomes a functional requirement because the wave aspect carries possibility structure forward while the particle aspect records the committed result. The path-integral view, where all paths are explored in superposition and only stationary-phase survivors contribute, provides the correct selection semantics for which possibilities remain viable at commitment time (Feynman & Hibbs, 1965).

Bose–Einstein condensate (BEC) experiments provide empirical evidence for this interpretation, and as the previous chapter described, Hau (2016) demonstrated that coherent light pulses can be slowed, stopped, stored, and revived with complete phase fidelity. In these experiments, optical information carried by photons is converted into a matter-wave imprint inside the condensate, held stably for extended periods, and later reconverted back into light without loss of amplitude or phase information, even at the single-photon level with efficiencies above 50% (Dutton et al., 2004; Hau, 2016). Within this framework, stopping a photon does not slow the passage of spacetime. Instead, the photon's role as an action carrier is temporarily paused. While stored, the information it carries remains intact, and upon revival the photon resumes committing actions as if no time had passed from its perspective. These results demonstrate that photons embody discrete, retrievable information packets and support the interpretation that classical continuity emerges from a sequence of quantum-level refresh events.

Neural Membranes and Biological Quantum Coherence

Recent work extends this framework into biology and provides the basis for this chapter's theories as Cavaglià et al. (2023) propose that phospholipid head-group dipoles in neuronal membranes can be driven by action potentials into weak Bose–Einstein condensate-like collective modes at physiological temperature, provided specific thermodynamic thresholds are met.

Their analysis establishes that "the membrane's strong electric potential gradients on the order of 100 mV across the thickness of 5 nm" yield field intensities reaching "20×10^6 V/m" (p. 6). This significant field strength creates conditions where "the dipole moment of each phosphate head group is approximately 20 debye, so the dipole–dipole interaction, assuming a 5-Å distance between the centers of these dipoles is on the order of 9 k_BT" (p. 6). Since "each group has several neighbors, the total dipole interaction energy exceeds thermal energy by a significant factor" (p. 6).

The threshold calculation demonstrates that action potentials provide sufficient energy pumping. As they write, "We calculate the rate of energy pumping, S, knowing the rate of arrival of APs as: $5\ k_BT/s < S = E_{ap-d}/dt < 50\ k_BT/s$" (Cavaglià et al., 2023, p. 6). This value "clearly exceeds the minimum requirement of 0.3 k_BT/s, hence we conclude that a weak Bose-Einstein condensate can be generated in the dipolar system of neuronal membrane's head groups" (p. 6). Under these conditions, "the predicted coherent modes of dipole oscillations are in the 10^{11}–10^{12} Hz frequency range" with "dipole–dipole interactions … predicted to propagate with a velocity of about 10^3 m/s, much faster than AP propagation" (p. 6).

As Cavaglià et al. (2023) explain, "Maxwell's theory of electromagnetism shows that oscillating dipoles emit EMFs. Therefore, neuronal membranes, with their coherently oscillating dipoles due to (incoherent) energy input from APs, should produce EMFs in the process" (p. 6). The geometry of cortical architecture supports this mechanism because "within the neocortex, multiple pyramidal cell axons are almost parallel, capable of producing multiple cylindrical waves that overlap and interfere with each other" (p. 6).

They note that "recent experiments determined that the membrane's electrical and mechanical degrees of freedom are nonlinearly coupled which lead to soliton formation. Solitons are nonlinear localized waves that propagate with no loss of amplitude or velocity" (Cavaglià et al., 2023, p. 2). This coupling provides additional mechanisms for information propagation aside from classical ionic conduction. Overall, these findings establish the biophysical conditions for ultrafast coherence and provide a plausible substrate through which biological systems synchronize activity at scales relevant for unified perception especially when considering the evidence presented in Chap. 2.

Holographic Reconstruction of Perception

The holographic mechanism emerges from interference between coherent electromagnetic fields. As Cavaglià et al. (2023) explain, "We hypothesize that a coherent EMF produced by dipole oscillations in the membrane of one neuron can interfere with a similar EMF produced by a neighboring neuron. One of these waves can be considered a reference beam while the other an object beam using holographic terminology" (p. 6). This interference "will create a pattern of nodes and antinodes due to EMF interference" (p. 6).

They describe that the neural architecture supports full holographic reconstruction because "instead of a fringe pattern of equally spaced parallel lines, the resulting interference pattern will appear much like a conventional hologram projecting a 3D image on a screen" (p. 6). They note that "a magnified hologram's interference pattern, resulting from multiple complex beams interfering with each other could be formed when a number of neurons synchronously fire producing coherent dipole waves that in turn create multiple interfering EMFs" (p. 6).

As explained in Chap. 12, this framework complements this as photons arriving from the environment already carry the universe's committed updates, and then as Cavaglià et al. (2023) described, sensory transduction converts those photonic updates into neural electrical patterns. Spike-driven membrane coherence then supplies a phase-stable reference field that interacts with the sensory-derived fields, and the interference between these two families of fields yields a holographic reconstruction of the external scene inside the cortex.

The brain refreshes a reconstruction of the already decohered, physical, objective environment, using a reference generated by membrane coherence to record, update, and bind sensory content into a single, unified perceptual frame. The holographic step provides a mechanism for unity because the scene is encoded in a global interference pattern.

The Microtubule Storage Mechanism

The holographic interference patterns require a storage substrate that can maintain phase-sensitive information. Cavaglià et al. (2023) propose that "at the inner layer, EMFs could transfer information inside the cell by interacting with cytoskeletal structures, such as parallel bundles of microtubules (MTs) structurally stabilized by microtubule-associated proteins" (p. 4). They note that "it is well-known that many neurodegenerative diseases are associated with defective MT cytoskeleton" and suggest "such defects could impair information transfer and storage in MTs" (p. 4).

The mechanism connects membrane coherence to intracellular encoding as they write, "The mirror pattern projected by EMFs inside the neuron can encode information in the neuronal cytoskeleton" (p. 1). Building on earlier work, they suggest that "encoding information in microtubules by phosphorylation of their C-termini as proposed elsewhere could provide a microscopic explanation of memory formation using EMFs directly affecting neuronal cytoskeleton" (p. 7). This distributed storage may help explain why "MTs possess transistor-like and memristor-like properties" that "could be involved in modifying the interactions between the membrane, ion channels and APs" (p. 7).

In this, holograms are naturally distributed as each subregion contains a version of the whole pattern, with fidelity depending on how much of the hologram is available at readout. Synchronized membrane interference would write each unified frame into deeper intracellular degrees of freedom where it can be stably maintained as a nonlocal pattern and later reconstructed, instead of being pieced together from fragments (Nishiyama et al., 2024a, 2024b).

From this perspective, memory often appears holistic and embodied as retrieval actually reactivates an entire interference pattern which restores a full perceptual frame, and BEC experiments provide proof of principle for this mechanism at the physical level. Hau's imprint-and-revive pathway shows that a full optical waveform with phase can be converted into matter form, stored, moved, sculpted, then reconverted with high fidelity (Dutton et al., 2004; Hau, 2016). That laboratory fact stands as physical evidence of principle for phase-true storage and retrieval, which is what a unified conscious frame seemingly requires at the biological scale.

Thermodynamic Constraints and the P = Q/E Framework

Both BEC memories and lipid-centric coherence face thermodynamic headwinds as random thermal motion and molecular disorder constantly degrade phase information which then means that maintaining coherence requires continuous energy input and ongoing error correction.

The lipid model makes requirements explicit through Fröhlich's criteria. As Cavaglià et al. (2023) detail, weak Bose–Einstein condensate formation requires five conditions, which they list as:

(a) A continuous supply of metabolic energy above a minimum threshold required to achieve synchronization of membrane dipole oscillations.
(b) The presence of thermal noise due to physiological temperature.
(c) Internal structural organization.
(d) The existence of a large trans-membrane potential difference.
(e) Nonlinear interactions between oscillating dipoles (p. 6).

They demonstrate that "these assumptions are satisfied by the neuronal membrane phosphate groups' dipoles provided both AP energy and the rate of AP generation is sufficiently high" (p. 6), and they specify the energy threshold as they write, "the physiological temperature of 310 K, this condition translates into energy supply exceeding (1/3) $k_B T/s$" (p. 6). The rate of energy loss "must be less than 20 K kB/s, i.e., 5 times lower than the rate of energy supply. In this case, the condensate can form on a 10–1000 ns time scale" (p. 6).

Landauer's principle states that erasing one bit of information at temperature T incurs a minimum energetic cost of $k_B T \ln 2$ (Landauer, 1961). In living tissue, the cost becomes an ongoing budget that covers error correction against continuous thermal and structural noise. Within this framework, these constraints naturally motivate the equation P = Q/E as the egoic self-model and its narrative continuity do not maintain themselves for free. Metabolic upkeep functions like an overhead on top of whatever budget remains to support membrane coherence, and this overhead represents the denominator E in P = Q/E. As egoic overhead rises, less energy and less attentional room remain for high-fidelity coherence (Q), therefore, performance (P) falls. When egoic overhead drops, more of the budget becomes available to stabilize coherence, so the system can pack more usable frames into each cosmic frame. As previously described in Chap. 12, the feeling of time dilation in flow can emerge because each external moment is richer in internal structure. This energy budget may be substantial, because when extrapolated across all eukaryotic life, the computational capacity of these quantum processes is estimated to be approximately 10^{60} operations, which is a figure twenty orders of magnitude greater than what classical neural processing could account for (Kurian, 2025).

Holographic Compression and AdS/CFT Correspondence

In high-energy theory, the holographic principle and AdS/CFT duality provide clean examples of boundary encoding of bulk information as the AdS/CFT correspondence states that a quantum field theory on a lower-dimensional boundary can fully encode the physics of a higher-dimensional bulk under specific conditions (Gubser et al., 1998; Maldacena, 1998; Witten, 1998). Ryu and Takayanagi (2006) showed that entanglement structure on the boundary can determine bulk geometric quantities like extremal surfaces and their areas, which formally connects information measures to geometry, and quantum corrections then tie changes in boundary entanglement to changes in bulk energy and dynamics (Faulkner et al., 2014).

Applied here, neuronal membranes act as biological boundaries that encode high-dimensional experiential content into lower-dimensional coherent oscillatory patterns. Recovery of perceptual scenes corresponds to reconstructing bulk-like experience from boundary-encoded interference structures, which use phase relationships as the retrieval key. This connection between biology and cosmic-scale holography is supported by recent calculations that point toward a direct mathematical relationship between the computational history of the universe and that of "superradiant life" (Kurian, 2025). The analysis proposes that the total operations performed by the universe's matter, approximately 10^{120}, is roughly the square of the operations performed by all superradiant life on Earth, which is approximately 10^{60} (Kurian, 2025). According to Kurian, this squared relationship "hints at a holographic principle," pointing toward the possibility that the information processed by life may be fundamentally related to the information content of the cosmos itself (p. 2), which is supportive of many of the proposals of the previous chapter as well.

Clinical and Pathological Predictions

The holographic coherence model importantly generates predictions about neurological and psychiatric conditions as Cavaglià et al. (2023) propose that "membranes with structural damage or unbalanced phospholipid composition could generate distorted images which result in perceptional distortions of reality in schizophrenia" (p. 7). For neurodegenerative disease, they write that "Alzheimer's could be directly related to defective memory storage in neuronal microtubules in the case of their destabilization by MAP tau hyperphosphorylation" (p. 7).

The schizophrenia thread invites closer study because it connects membrane biophysics, entropy focused views of consciousness from Carhart-Harris et al. (2014), formal work in quantum cognition from Kvam and Pleskac (2017) and Busemeyer et al. (2019, 2020), information theoretic boundary ideas from Gallimore (2015), and Jung's account of archetypes (1964, 1969). In the framework's view, phospholipid irregularities that are documented in schizophrenia would impair the formation of stable weak Bose Einstein condensate like modes in neuronal membranes. When coherence in the membrane weakens, the reference field generated by coordinated dipole oscillations loses stability, which blurs holographic interference. Carhart-Harris et al. (2014) would describe this as increased entropy, and within the geometry used here it appears as a wider, less focused spread in conceptual space.

Gallimore's information theoretic picture gives a helpful bridge to these ideas as he writes, "although each concept forms a distinct shape in concept space, it is possible that they share a non zero number of states. Some degree of conceptual overlap makes perfect sense, since not all concepts are independent" (2015, p. 10). Therefore, in this framework, when membrane coherence declines, interference patterns smear across nearby semantic neighborhoods, which widens measurement over more than one concept at once. Gallimore continues, "small degrees of conceptual overlap could progress to a more outright conceptual blending, as the boundaries between concepts are dissolved" and illustrates this with Hofmann's first LSD report, where his neighbor "was no longer Mrs. R., but rather a malevolent, insidious witch with a colored mask" (Hofmann, 1980, cited in Gallimore, 2015, p. 10).

Jung helps organize this landscape as he distinguished between an archetype and its representation, writing that "the archetype is a tendency to form such representations of a motif—representations that can vary a great deal in detail without losing their basic pattern" (1964, p. 68). He clarified a frequent misunderstanding by noting, "my critics have incorrectly assumed that I am dealing with 'inherited representations,' and on that ground they have dismissed the idea of the archetype as mere superstition. They have failed to take into account the fact that if archetypes were representations that originated in our consciousness (or were acquired by consciousness), we should surely understand them, and not be bewildered and astonished when they present themselves in our consciousness" (1964, pp. 68–69). In the framework's terms, that bewilderment arises because archetypes live in the geometry of the QID and appear across cultures since minds interface with a similar possibility space. They are the underlying characteristics of ideas that unite them in the most thermodynamically optimal way. Or in quantum field terminology, archetypes would represent low-action attractors in conceptual configuration space and would be the informational geometries that most efficiently stabilize meaning by minimizing the thermodynamic cost of coherence across many experiences and minds.

Jung observed that archetypes "are without known origin; and they reproduce themselves in any time or in any part of the world—even where transmission by direct descent or 'cross fertilization' through migration must be ruled out" (Jung, 1964, p. 69). Chapter 9 framed these forms as low energy centers in conceptual clusters. Jung also pointed toward their mathematical flavor, writing that "the archetypes too, as a priori forms of representation, are as much found as invented" and that "natural numbers have an archetypal character" (Jung, 1969, p. 51). In other words, when countless ideas, symbols, or emotions interact in the QID, destructive interference removes the unstable configurations. Therefore, what remains are the enduring patterns that most effectively balance maximal informational richness with minimal energy cost, and those surviving minima manifest as the archetypes.

Through Jung's lens, Gallimore's description of Hofmann's experience of Mrs. R and the witch reads as archetypal blending in a high entropy setting as Mrs. R and the malevolent witch both sit within a broader archetype (perhaps the great mother or the anima) but each emphasize a different subset of characteristics. Under ordinary coherence they remain separable as the thermodynamic cost is being paid, but with reduced measurement precision, Gallimore explains that concepts with partial overlap can fuse in experience, so "Mrs. R effectively *became* the malevolent witch" from Hofmann's point of view (2015, p. 10).

In Jung's language the archetypal attractor overpowers the individual representation as the geometry here would say that when precision falls, consciousness settles into the center of the activated cluster because that center is the least thermodynamically costly configuration to maintain. This helps explain why recurring psychotic themes align with archetypal families. When the self and other boundary weakens, Gallimore notes that blending "could result in a failure to distinguish concepts related to self from concepts related to other, perhaps explaining the 'ego dissolution' and the sense of 'oceanic boundlessness' that are often feature in high dose psychedelic experiences" (2015, p. 11), and this would also directly relate to Cavaglià et al.'s (2023) proposal regarding schizophrenia. As described in Chap. 9, quantum cognition models align with this as Kvam and Pleskac (2017) show that when collapse strength is reduced, beliefs drift toward superposition, fuzzy-like states. In the membrane account here, weakened holographic precision permits concepts to slide toward archetypal centers, which echoes Jung's observation that these centers appear "in any time or in any part of the world" (Jung, 1964, p. 69).

Dreams offer a natural comparison for this because during REM sleep, altered neurochemistry likely shifts membrane conditions in a way that changes coherence. The resulting phenomenology resembles psychosis in key ways, which matches Gallimore's analysis that higher entropy can widen the cone of accessible meaning. He writes that this may help explain "why creativity, imagination, and novel thinking are enhanced during the psychedelic state" while also clarifying "how this could degrade the brain's ability to maintain the organization and differentiation of the concepts within conscious experience—an obviously essential adaptive skill" (2015, p. 8). Although, the noted difference is that dream and psychedelic states usually eventually recede, but chronic membrane abnormalities would keep the system from regaining its sharp boundaries between ideas, concepts, beliefs, and identity.

Jung documented cases of patients who experienced "images that they could not relate to anything they could remember or could have passed on to their children" and noted that "some of these patients were highly educated" and that "a few of them were actually psychiatrists themselves" (Jung, 1964, p. 69). If archetypal forms reflect geometry in the QID, then one's education does not confer immunity to these problems, because when coherence falters, archetypal content can flood experience because the source lives in the thermodynamically optimal structure of possibility, not in the cultural instruction one adopts or in the learned definitions of ideas, beliefs, and concepts.

Interestingly enough, therapeutic lines follow from the same mechanism as Cavaglià et al. open their paper by noting that "weak electric and magnetic fields are known to impact brain activity and are used to treat psychiatric conditions" (2023, p. 1). They add that "EMF polarizes the distribution of ionic charges, which then become ordered, causing changes in the membrane conductance modulating the transmembrane potential" (p. 3). In the framework here, any intervention that improves membrane order or strengthens phase stable reference fields would likely sharpen holographic reconstruction and restore cleaner conceptual boundaries between ideas, beliefs, memories, concepts, the psyche, the self, and others. Therefore, the anesthesia prediction becomes an immediate way to test the model as Cavaglià et al. write, "With the presence of anesthetic molecules, we expect the membrane to change its mechanical and electrical properties affecting AP propagation and leading to the elimination of holographic images" (2023, p. 6). They quantify this by noting that "with the membrane's bending modules being on the order of 45 k_BT, anesthetic molecules reduce it by as much as 40 percent" (p. 6). They propose that "calculations of the electromechanical waves propagating through the membrane with or without anesthetic molecules can establish if this mechanism of anesthetic action is capable of effects such as abrogating the dipolar oscillations of neuronal membranes or slowing down or even stopping electro mechanical soliton propagation" (p. 6). If future measurements show that anesthetics weaken or eliminate coherent dipole oscillations in ways that track loss of conscious content, that would lend direct support to the holographic reconstruction pathway outlined here.

The Architecture and Perception of Reality

Cavaglià et al. (2023) present an extensive model that connects sensory inputs to cortical representation, and the framework builds on these ideas and proposes a single, continuous mechanism. Here is the process in its entirety.

1. Photons as action carriers deliver discrete spacetime updates.
2. Sensory transduction converts these updates into classical neural codes.
3. Membrane-driven coherence establishes ultrafast reference fields through weak Bose–Einstein-like condensates.
4. Interference between reference and sensory-derived fields generates holographic projections which encode unified perceptual scenes.
5. Distributed holographic storage maintains these scenes across intracellular quantum degrees of freedom, which enables whole-frame retrieval.
6. Thermodynamic constraints set the limits of coherence, which is linking metabolic costs to the number of usable frames constructed per cosmic action frame.

7. Holographic compression mathematics explains how complex experiential states can be encoded and reconstructed without loss of fidelity.

Consciousness therefore emerges as a boundary phenomenon where quantum coherence, thermodynamic energy, and photonic action creation converge where reality construction and perception are two aspects of a single underlying process. In short, the photons arrive that embody the universe's committed state, then the internal membranes supply a reference for recording and reconstructing that state as a unified scene, and the interference between the two forms the integrated perception that you experience as "now." From this perspective, the brain constructs its perception of the physical, objective decohered environment it finds itself in.

The Four-Layer Architecture of Reality

The framework's proposals point toward reality operating through four distinguishable layers, and understanding these layers may help clarify how objective physical reality, subjective experience, and expanded awareness states relate to one another.

Layer 1 would be the Quantum Information Dimension. As described in Chap. 4, the QID exists as a pre-temporal substrate that contains all possible configurations in superposition before collapse. From this perspective, the wavefunction essentially operates as the QID, which contains every possibility that could ever manifest.

Layer 2 would be objective physical reality as photons propagate at c, environmental decoherence operates constantly, and chemical and biological processes unfold independently. As described in previous chapters, a tree falling in the forest makes noise as it creates sound waves because environmental propagation, information exchange, interference, decoherence, and collapse happens continuously through thermal noise, photonic interactions, and a variety of other sources of interference, which is reflected in the gradient collapse mechanism as described in Chap. 5. Science effectively studies this layer because it remains very stable and reproducible across observers.

Layer 3 represents consensus perceived reality. This describes how the vast majority of humanity perceives physical reality and would be what most describe as the experience of normal waking consciousness in the physical world. Individuals operating in the ego-unaware paradigm perceive or mentally reconstruct physical reality through heavily ego-filtered perceptual processes. High E values in the $P = Q/E$ equation produce narrow geometric constraints on accessible QID spaces, which results in limited holographic reconstruction quality. Very similar biological bodies, cultural conditioning, shared language, and collective identity structures create broad agreement about "how reality is," as humans have to work through our senses to perceive physical reality. This layer feels complete to those experiencing it because the DMN's filtering happens before normal conscious awareness, which makes any of the constraints invisible.

Layer 4 represents expanded awareness states where individuals experiencing the ego-aware paradigm (whether as their default state or through another mode such as flow states, meditation, or psychedelics) perceive and experience a slightly different reality as ego activation would lower the E term, which increases Q access and allows consciousness to explore broader QID possibility spaces with less geometric constraint. The holographic interference patterns described earlier in this chapter would contain richer information when membrane coherence can operate without the metabolic overhead that ego maintenance requires. From the framework's perspective, contemplative practitioners who report experiences of unity, timelessness, or interconnection may be directly perceiving aspects of Layer 1 through their expanded Layer 4 access as these reports are consistent across cultures and historical periods, which would make sense if practitioners were accessing similar underlying structures in the QID through comparable reductions in ego interference.

The relationships between these layers may help to clarify several apparent paradoxes. Reality can be both objective and participatory because Layer 2 exists independently. Identity and beliefs affect manifested outcomes when collapse events involve genuine quantum uncertainty, which often occurs in biological and psychological systems where Layer 1 possibilities remain open. The body can respond to the believed story because consciousness perceives physical reality and participates in selecting which quantum possibilities actualize into measurable physical changes. The framework acknowledges the theoretical nature of some of these proposals while noting their internal consistency and alignment with the documented evidence and phenomena.

An Invitation to Explore Further

The book is aware of the significant weight of these last few chapters and their ideas. These ideas emerge as logical extensions of the framework, but the book presents them humbly as possibilities worth investigating. If consciousness, following Feynman and Hibbs (1965) which is this "universal selection mechanism," utilizes ego to collapse quantum states from the QID, and if this collapse creates humanity's experienced, perceived classical reality, then these implications about the core laws of physics might logically follow.

The book encourages readers to examine these connections critically. Do they resolve puzzles or create new ones? Do they generate testable predictions? Most importantly, do they help us understand the profound mystery of how consciousness and physics intertwine?

Exploring the framework continues to reveal new questions. Perhaps in exploring these ideas together, this book participates in the ancient human endeavor of understanding humanity's place in the cosmos, now with new tools and perspectives that may finally bridge the quantum and the classical, the mental and the physical, and the timeless and the temporal.

References

Busemeyer, J. R., Kvam, P. D., & Pleskac, T. J. (2019). Markov versus quantum dynamic models of belief change during evidence monitoring. *Scientific Reports, 9*, Article 18025. https://doi.org/10.1038/s41598-019-54383-9

Busemeyer, J. R., Kvam, P. D., & Pleskac, T. J. (2020). Comparison of Markov versus quantum dynamical models of human decision making. *WIREs Cognitive Science, 11*(4), e1526. https://doi.org/10.1002/wcs.1526

Carhart-Harris, R. L., Leech, R., Hellyer, P. J., Shanahan, M., Feilding, A., Tagliazucchi, E., Chialvo, D. R., & Nutt, D. (2014). The entropic brain: A theory of conscious states informed by neuroimaging research with psychedelic drugs. *Frontiers in Human Neuroscience, 8*, Article 20. https://doi.org/10.3389/fnhum.2014.00020

Cavaglià, M., Deriu, M. A., & Tuszynski, J. A. (2023). Toward a holographic brain paradigm: A lipid-centric model of brain functioning. *Frontiers in Neuroscience, 17*, 1302519. https://doi.org/10.3389/fnins.2023.1302519

Dutton, Z., Ginsberg, N. S., Slowe, C., & Hau, L. V. (2004). The art of taming light: Ultra-slow and stopped light. *Europhysics News, 35*(2), 33–39.

Faulkner, T., Guica, M., Hartman, T., Myers, R. C., & Van Raamsdonk, M. (2014). Gravitation from entanglement in holographic CFTs. *Journal of High Energy Physics, 2014*(3), 051. https://doi.org/10.1007/JHEP03(2014)051

Feynman, R. P., & Hibbs, A. R. (1965). *Quantum mechanics and path integrals.* McGraw-Hill.

Gallimore, A. R. (2015). Restructuring consciousness—The psychedelic state in light of integrated information theory. *Frontiers in Human Neuroscience, 9*, Article 346. https://doi.org/10.3389/fnhum.2015.00346

Gubser, S. S., Klebanov, I. R., & Polyakov, A. M. (1998). Gauge theory correlators from non-critical string theory. *Physics Letters B, 428*(1–2), 105–114.

Hau, L. V. (2016). *Quantum control of light and matter: From the macroscopic to the nano scale* (Report no. AFRL-AFOSR-VA-TR-2016-0082). Air Force Research Laboratory. https://doi.org/10.21236/AD1004771

Hofmann, A. (1980). *LSD: My problem child.* McGraw-Hill Book Company.

Jung, C. G. (1964). Approaching the unconscious. In C. G. Jung (Ed.), *Man and his symbols* (pp. 1–94). Doubleday.

Jung, C. G. (1969). Synchronicity: An acausal connecting principle (R. F. C. Hull, Trans.). In H. Read, M. Fordham, & G. Adler (Eds.), *The collected works of C. G. Jung: Vol. 8. The structure and dynamics of the psyche* (2nd ed.). Princeton University Press. (Original work published 1952).

Kurian, P. (2025). Computational capacity of life in relation to the universe. *Science Advances, 11*, eadt4623. https://doi.org/10.1126/sciadv.adt4623

Kvam, P. D., & Pleskac, T. J. (2017). A quantum information architecture for cue-based heuristics. *Decision, 4*(4), 197–233. https://doi.org/10.1037/dec0000070

Landauer, R. (1961). Irreversibility and heat generation in the computing process. *IBM Journal of Research and Development, 5*(3), 183–191. https://doi.org/10.1147/rd.53.0183

Maldacena, J. (1998). The large-N limit of superconformal field theories and supergravity. *Advances in Theoretical and Mathematical Physics, 2*(2), 231–252.

Nishiyama, A., Tanaka, S., & Tuszynski, J. A. (2024a). Quantum brain dynamics: Optical and acoustic super radiance via a microtubule. *Biosystems, 234*, 105118. https://doi.org/10.3390/foundations4020019

Nishiyama, A., Tanaka, S., Tuszynski, J. A., & Tsenkova, R. (2024b). Holographic brain theory: Super-radiance, memory capacity and control theory. *International Journal of Molecular Sciences, 25*(4), 2399. https://doi.org/10.3390/ijms25042399

Ryu, S., & Takayanagi, T. (2006). Holographic derivation of entanglement entropy from AdS/CFT. *Physical Review Letters, 96*(18), 181602. https://doi.org/10.1103/PhysRevLett.96.181602

Witten, E. (1998). Anti de sitter space and holography. *Advances in Theoretical and Mathematical Physics, 2*(2), 253–291.

Chapter 14
Linking to Established Frameworks

Over the past century, multiple theories have proposed that either information or consciousness forms the fundamental substrate of physical reality. From the "It from Bit" hypothesis to Integrated Information Theory (IIT), to quantum information science, to panpsychist metaphysics, these models have attempted to reorient the understanding of matter, energy, and mind. While this framework shares key overlaps with these views, it also diverges in significant and testable ways.

Agreement with the "Information Is Foundational" Paradigm

John Wheeler's "It from Bit" hypothesis (1990) proposed that physical reality arises from binary informational events. In this view, spacetime, particles, and forces all emerge from yes–no answers to fundamental measurements. Quantum information theorists like Lloyd (2006) and Deutsch (1997) extended this by describing the universe as a vast quantum computer "computing its own destiny." These models treat information as ontologically primary, and measurement as a neutral process that converts probabilistic possibilities into classical facts.

This framework agrees with these theories on several key points, including the importance of information and that measurement drives the emergence of classical reality. However, this introduces an important refinement which is rather than treating measurement as an abstract, observer-independent process, the framework proposes that egoic consciousness may function as a biologically evolved measurement device. Similarly, it diverges in the ontological importance of information as this paper posits the Quantum Information Dimension as the place information emerges from. While information is foundational to our perceived reality, and more foundational than consciousness, the QID would be ontologically more foundational than both.

J. Roeloffs, *The Classical Brain to the Quantum Mind*, The Frontiers Collection,
https://doi.org/10.1007/978-3-032-20068-6_14

While quantum information theorists generally treat observation as an abstract measurement process, this framework proposes a specific biological mechanism. Instead of invoking a generic "observer," the framework identifies the gradient collapse mechanism for physical reality and then the ego as the biological, evolved measurement device that collapses quantum states through self-referential processing.

This divergence has testable implications as Wheeler's "participatory universe" described reality as emerging from measurement events but did not specify the physical basis of observation. And quantum information theory describes measurement mathematically but leaves the collapse mechanism unspecified, especially in regards to an evolutionary solution for humanity's significance in this process, so this book's framework bridges the gap by proposing that the Default Mode Network's generated noise serves as the physical measurement mechanism.

And unlike Wheeler's abstract participatory principle, the framework generates specific predictions about neural correlates, individual differences, and intervention effects. This distinction becomes important when considering consciousness development, because if measurement were truly observer-independent, all conscious beings should exhibit similar reality-construction capabilities. However, the framework predicts that higher consciousness levels would demonstrate more skillful quantum state management, while lower levels would show excessive or premature collapse patterns.

This biological specificity also explains why ego dissolution practices consistently enhance performance across cultures. If consciousness was simply a neutral measurement device, reducing self-awareness should not systematically improve capabilities. However, if ego functions as the active collapse mechanism, then ego modulation becomes a direct method for optimizing quantum-classical processing balance.

"Consciousness Is Fundamental"

This framework shares partial alignment with theories that view consciousness as ontologically primary. Integrated Information Theory (Tononi, 2004) argues that consciousness arises wherever information achieves sufficient integration. By quantifying the degree of informational unification, IIT provides a mathematical measure for the richness of experience but does not directly address how conscious states might influence physical reality.

Panpsychist perspectives (e.g., Goff, 2019) propose that consciousness is a ubiquitous feature of matter, present in varying degrees throughout the universe. Idealist frameworks, such as those advanced by Kastrup (2019), go further by suggesting that physical reality itself emerges from an underlying field of awareness. These models share a common intuition that consciousness may be inseparable from the fabric of existence.

While this framework shares the view that subjective experience may be embedded within physical processes, it departs in proposing that consciousness operates as an adaptive biological interface instead of being some sort of universal substrate. Organisms may have evolved the capacity to collapse and navigate quantum informational states in ways tuned to survival.

A recent formulation by D'Ariano and Faggin (2022) expands this family of ideas through what they describe as a quantum-information-based panpsychism. Their model proposes that consciousness underlies information itself, that quantum information emerges from conscious states, and that physics then supervenes on quantum informational dynamics. They argue that each moment of conscious experience corresponds to a pure ontic quantum state, fundamentally private and unclonable. Drawing on the quantum no-cloning theorem, they claim that the informational configuration giving rise to a conscious state cannot be copied or fully shared without disturbing the experience itself. From their perspective, subjective experience is woven into the fabric of quantum information and cannot be reduced to classical computation.

From the framework's perspective, these insights hold value. The idea that conscious experience depends on quantum informational substrates is consistent with the proposal here that awareness interacts with an underlying informational layer instead of rising from purely classical neural processes. The privacy implied by unclonable quantum states also resonates with the view that no two conscious experiences can ever be exactly identical, even when observers inhabit similar informational environments.

This framework diverges in how it interprets the role of consciousness within physical processes. Instead of treating consciousness as the origin point from where physics emerges, this book suggests that consciousness evolved as a biological interface for interacting with an already-present quantum informational substrate.

Within this framework, consciousness functions metaphorically as a local operating system which runs the program of "ego" as it integrates memory, identity, and perception into a unified field of experience. By directing focus and selecting from possibility spaces, the ego may participate in shaping which quantum field states are collapsed. These processes create the continuity of experience required for learning, agency, and meaning-making, which positions consciousness as both constrained and adaptive instead of being unbounded and universal by default. Under certain conditions, such as meditation, psychedelics, or creative flow states, these constraints may loosen. Consciousness can temporarily access broader, more coherent informational structures when egoic collapse mechanisms quiet and attention synchronizes across wider patterns. These states may resemble descriptions in contemplative traditions that posit consciousness as fundamental, but they are understood here as emergent expansions that are less fundamental than the QID itself.

Orch OR

The Orch OR theory (Hameroff & Penrose, 1996, 2014) also plays an important role in this landscape by proposing that quantum coherence within neural microtubules produces conscious moments through gravitationally induced objective reduction. While this framework builds upon Orch OR's foundational insight that quantum coherence in biological systems may be viable, it diverges in four important ways. First, it does not adopt gravitational collapse as the trigger for conscious experience. Second, it integrates environmental decoherence and information exchange as continuous processes shaping collapse without requiring specialized thresholds for effects like gravity for instance. Third, it positions consciousness as an evolved biological interface instead of it being some type of a universal mechanism built into the human experience or interfacing with gravity and the quantum field. Fourth, Orch OR places the processes of consciousness as central to the microtubule where this framework sees the microtubules playing an important role, but they are not the sole proprietor of where consciousness is housed in the body.

It may be worth noting that both Orch OR and this framework recognize Planck's constant as fundamental to understanding collapse but interpret its role quite differently. In Orch OR's formulation $Eg = \hbar/t$, the reduced Planck constant serves as a conversion factor between gravitational self-energy and the timing of objective reduction events which essentially determines when gravity forces discrete moments of consciousness to emerge. This framework, while it deeply respects Orch OR's pioneering connection between quantum processes and consciousness, proposes a different interpretation. It sees h (the full Planck constant) as the minimum quantum of action that must be committed for any collapse to occur, regardless of what triggers it. Instead of a specific gravitational threshold that is responsible for creating conscious moments, the framework has posited that reality continuously refreshes and writes itself through informational exchanges (updates) and collapses, with a minimum size of h, which occur at different rates across a gradient which span from random thermal collapse to the recursively directed collapse of consciousness through ego.

In this way, the framework holds overlapping concerns with panpsychist, Orch OR, and idealist theories but departs from their ontological commitments. Instead of being a foundation of existence or of experienced physical reality, consciousness is treated as an evolved interface that allows organisms to participate in the shaping of physical reality through interaction with the quantum field. While this interpretation remains open to revision, it provides a model that invites empirical testing by situating consciousness within measurable processes while still allowing room for phenomena that transcend classical computation (Table 14.1).

Table 14.1 A comparison between this framework all other leading theories on agreements and disagreements

Theory	Agreements with this framework	Points of divergence
It from bit (*Wheeler,* 1990)	Reality may emerge from informational structures. Measurement drives classical outcomes	Does not specify what kind of observation performs the collapse. This framework proposes that evolved, egoic consciousness may participate actively in shaping collapse events
Quantum information theory (*Lloyd, Deutsch*)	Information is physical, and the universe may behave like a vast quantum computation	Treats measurement as observer-neutral. This framework integrates subjective, biologically evolved consciousness as part of the process that selects from informational possibilities
Integrated information theory (IIT) *(Tononi)*	Consciousness arises from integration of information and can be quantified	Does not propose a mechanism connecting consciousness to physical collapse. In this framework, integrated processing is necessary but not sufficient; quantum informational dynamics are also required
Panpsychism/idealism (*Goff, Kastrup, others*)	Experience may be embedded throughout physical processes, and subjective awareness cannot be reduced to classical computation	Does not commit to consciousness as the universal substrate. In this framework, consciousness functions as an evolved interface tuned for navigating quantum informational states
D'Ariano and Faggin	Conscious states depend on quantum informational substrates. Each conscious moment may correspond to a private, unclonable ontic state	Proposes a quantum-information-based panpsychism where consciousness underlies information itself. This framework instead treats consciousness as an emergent biological adaptation interacting with an informational substrate
Orch OR (*Hameroff and Penrose*)	Viability of quantum coherence in biological systems. Microtubules may support quantum processing relevant to consciousness	Does not adopt gravitational collapse or discrete thresholds as the mechanism. Incorporates environmental decoherence and continuous information exchange, positioning consciousness as an adaptive interface

Turner's Communitas and Distributed Cognition

Anthropologist Victor Turner (1969) introduced the concept of communitas to describe a unique state of social unity that emerges during rituals. In these moments, the usual hierarchies and roles of everyday life dissolve, as they are replaced by a significant sense of shared purpose, presence, and connection. Participants often report spontaneous insights, feelings of transcendence, and an ineffable bond with the group. Turner documented the phenomenon extensively, but the underlying mechanism remained unexplained.

This framework proposes two possible interpretations of communitas, both rooted in ego reduction. The first interpretation requires no exotic physics. When individuals suppress their egoic filters and fully identify with the team or ritual group, their E values drop. This reduction in ego rigidity allows their cognitive systems to access higher levels of processing potential (P), resulting in sharper insight, faster pattern recognition, and greater adaptive responsiveness. When everyone in a group enters this state simultaneously through chanting, rhythm, movement, shared stakes, or alignment around a common goal, each individual becomes more powerful than they are alone. The team's total intelligence becomes greater than the sum of its parts, not because their minds are merging or because of some other mystical explanation, but because egoic bottlenecks are temporarily cleared across the group. This aligns with observed dynamics in peak-performing sports teams, military squads, jazz ensembles, startups, and other high-flow collectives.

However, there is also a second possibility worth mentioning. It is at least conceivable that during these ego-dissolved states, individuals are partially tuning into shared informational structures within the quantum field itself. Research on neural synchrony (Lindenberger et al., 2009) shows that collective rhythm and intention can align brain activity across individuals. Other studies, like those by Radin et al. (2012), suggest that intention may influence quantum systems. While these findings are not conclusive, they raise the possibility that consciousness might, under some specific conditions, become more collectively coherent through mutual access to overlapping zones of quantum information. This framework does not claim that group minds merge or that telepathy occurs in any directed "super power" sense. But it remains plausible that, in states of high alignment and low ego interference, individuals might tap into similar quantum-encoded structures, much like musicians improvising in perfect synchrony might be said to "hear" an unspoken rhythm at the same time.

From the framework's perspective, wartime innovation bursts may reflect a similar mechanism operating at societal scale. When collective survival threatens individual identity, entire populations experience forced ego suppression. Academic competition, career protection, and disciplinary boundaries suddenly matter less than shared survival. Information flows freely between previously isolated domains, while the innovators' personal egos drop, replaced by the collective nationalism, now the threats are not personal, they're collective, allowing for lower E in $P = Q/E$. The

Manhattan Project exemplified this phenomenon, where physicists, chemists, engineers, and mathematicians subordinated individual recognition to collective breakthrough. The framework suggests that crisis creates conditions where collective consciousness can access solutions already present in the quantum information field, waiting for sufficient ego reduction to enable their discovery.

The pattern appears throughout history. Major scientific revolutions often coincide with periods of social upheaval that destabilize existing hierarchies and force new forms of collaboration. The Renaissance emerged from plague and political chaos. The Enlightenment followed religious wars. Quantum mechanics itself developed during the upheaval of multiple world wars. In this sense, it's possible that each crisis forced temporary dissolution of established ego structures, which perhaps enabled collective access to insights through synergy or collective consciousness that were already available but required reduced collective ego to perceive them or piece them together.

Vervaeke's work on distributed cognition in problem-solving groups (Vervaeke & Mastropietro, 2021) finds new explanatory possibilities because when group members reduce individual ego through shared purpose, in the framework, they create conditions for distributed quantum processing, which potentially accounts for documented cases of collective intelligence that exceed any given individual's capacity.

Whether this is through ego-suppressed synergy or distributed access to a quantum informational substrate, communitas represents a powerful mode of cognition that demands more research. It may help explain why rituals feel sacred, why coordinated teams outperform expectations, why wars create innovation, and why the most memorable human moments often emerge from collective surrender. Both mechanisms deserve serious exploration, and the book would be incomplete if it failed to consider them.

Hutchins' (1995) distributed cognition framework demonstrates that cognitive systems often extend beyond individual minds to include tools, environments, and other people. However, the mechanism enabling groups to solve problems beyond individual capacity remains unclear.

The book's framework suggests that when groups form shared identity, individual ego activation reduces across team members. This creates conditions where each member can access enhanced quantum processing capabilities which potentially explains documented phenomena such as, teams that solve complex problems through improved collective pattern recognition, the "group flow" phenomenon in high-performing teams (Woolley et al., 2010), and breakthrough insights that emerge from dialogue through complementary processing.

When individual egos diminish through shared purpose, each team member's quantum processing capacity increases and then, combined with enhanced communication and reduced defensive filtering, creates genuinely enhanced collective intelligence through parallel quantum processing in multiple brains.

This distributed cognition enhancement finds empirical support in studies of human-technology systems as Chiappe and Vervaeke (2022) documented how Mars Exploration Rover teams experienced profound unity, with scientists saying things like they were "becoming the rover" and experiencing presence on Mars through the machine's sensors. The framework suggests this phenomenon emerges from enhanced quantum processing within each scientist's brain. As ego boundaries dissolved through shared focus, their neural quantum systems could construct extraordinarily vivid spatial models from limited 2D image data. The repeated exposure to rover imagery, combined with deep emotional investment and collective purpose, could have enabled their brains to develop specialized quantum processing patterns optimized for Mars terrain reconstruction.

Through the framework's lens, when team members report "it was like I was there," they are likely describing the phenomenological experience of their brain's quantum systems creating immersive 3D environments from 2D inputs, so this represents enhanced cognitive performance through optimized quantum processing, but not literal consciousness extension. The framework predicts that such teams would show measurably enhanced spatial reasoning, pattern completion, and environmental modeling capabilities compared to individuals viewing the same data without the collective engagement.

This demonstrates how group identity and reduced individual ego create conditions for impressive cognitive performance. The rover teams achieved what the book's framework would expect that individual egos merged into a form of collective identity, and it enabled enhanced abilities within each individual that created vivid experiences of presence through cognitive mechanisms.

References

Chiappe, D., & Vervaeke, J. (2022). The enactment of shared agency in teams exploring Mars through rovers. *Phenomenology and the Cognitive Sciences, 21*, 857–881. https://doi.org/10.1007/s11097-021-09791-6

D'Ariano, G. M., & Faggin, F. (2022). Hard problem and free will: An information-theoretical approach. In *Artificial intelligence versus natural intelligence* (pp. 145–192). Springer International Publishing. https://doi.org/10.1007/978-3-030-85480-5_5

Deutsch, D. (1997). *The fabric of reality: The science of parallel universes and its implications.* Viking Adult.

Goff, P. (2019). *Galileo's error: Foundations for a new science of consciousness.* Pantheon Books.

Hameroff, S., & Penrose, R. (1996). Orchestrated reduction of quantum coherence in brain microtubules: A model for consciousness. *Mathematics and Computers in Simulation, 40*(3–4), 453–480. https://doi.org/10.1016/0378-4754(96)80476-9

Hameroff, S., & Penrose, R. (2014). Consciousness in the universe: A review of the 'Orch OR' theory. *Physics of Life Reviews, 11*(1), 39–78. https://doi.org/10.1016/j.plrev.2013.08.002

Hutchins, E. (1995). *Cognition in the wild.* MIT Press.

Kastrup, B. (2019). *The idea of the world: A multidisciplinary argument for the mental nature of reality.* Iff Books.

Lindenberger, U., Li, S. C., Gruber, W., & Müller, V. (2009). Brains swinging in concert: Cortical phase synchronization while playing guitar. *BMC Neuroscience, 10*, 22. https://doi.org/10.1186/1471-2202-10-22

Lloyd, S. (2006). *Programming the universe: A quantum computer scientist takes on the cosmos.* Knopf.

Radin, D., Michel, L., Galdamez, K., Wendland, P., Rickenbach, R., & Delorme, A. (2012). Consciousness and the double-slit interference pattern: Six experiments. *Physics Essays, 25*(2), 157–171. https://doi.org/10.4006/0836-1398-25.2.157

Tononi, G. (2004). An information integration theory of consciousness. *BMC Neuroscience, 5*(1), 42. https://doi.org/10.1186/1471-2202-5-42

Turner, V. (1969). *The ritual process: Structure and anti-structure.* Aldine.

Vervaeke, J., & Mastropietro, C. (2021). Dialectic into dialogos and the pragmatics of no-thingness in a time of crisis. *Philosophies, 6*(4), 91. https://doi.org/10.14394/eidos.jpc.2021.0017

Wheeler, J. A. (1990). Information, physics, quantum: The search for links. In W. H. Zurek (Ed.), *Complexity, entropy, and the physics of information* (pp. 3–28). Westview Press.

Woolley, A. W., Chabris, C. F., Pentland, A., Hashmi, N., & Malone, T. W. (2010). Evidence for a collective intelligence factor in the performance of human groups. *Science, 330*(6004), 686–688. https://doi.org/10.1126/science.1193147

Chapter 15
Free Will

Traditional debates around free will often frame the question as binary where either every action is determined by prior physical states or it's decided by a conscious mind that stands entirely outside of the causal chains while it exercises pure autonomy. From the "there is no free will camp," the experimental findings that neural signals appear milliseconds before "conscious awareness" (Kounios & Beeman, 2009) and many other similar findings are often used as proof that free will is an illusion. Within the framework presented here, these results reflect how consciousness, as defined in the book, and free will actually operate together.

Sam Harris, who is one of the most public-facing and well known speakers for the position that free will is an illusion, observes something important about how most people actually experience choice. As he puts it, "You can do what you decide to do—but you cannot decide what you will decide to do" (Harris, 2012, p. 31). He writes that "Choices, efforts, intentions, and reasoning influence our behavior—but they are themselves part of a chain of causes that precede conscious awareness and over which we exert no ultimate control" (Harris, 2012, p. 31). The framework would actually agree with Harris that he may be accurately describing the experience at lower levels of consciousness. Many people who operate with high ego activation and narrow relevance filters, would have a severely constrained sense of authorship and agency in the world as the geometric space they operate in is limited. Essentially, they're trapped in automatic response patterns with limited access to broader geometric space to operate from which limits their probabilistic potential.

As explained from the framework's perspective, consciousness, as experienced by humans, is modeled as a recursive loop that connects three layers of infrastructure which includes the quantum computation within System 1, the pre-conscious filtering through relevance realization, and the ego-mediated selection within System 2. Because much of this selection occurs before conscious narrative construction by the DMN, the experiments detect neural activity that precedes the reported decisions from the conscious individuals. While as previously described, this could at first seem like evidence that free will doesn't exist, from the frameworks perspective,

J. Roeloffs, *The Classical Brain to the Quantum Mind*, The Frontiers Collection,
https://doi.org/10.1007/978-3-032-20068-6_15

this timing may point to the idea that choice actually emerges earlier in the recursive loop than the mainstream definitions of "self" and "free will" have traditionally considered.

Considering the neuroscience and quantum biology evidence presented throughout this book, the framework proposes that a version of free will most definitely exists, but it also proposes that free will is not uniform across all conscious systems. It's a gradient. In this, it reflects the scope of the geometric space in which a person currently computes and collapses possibilities. Lower levels of consciousness operate within these highly ego-constrained informational spaces where collapses are automatic and dominated by evolutionary responses to environmental inputs. So then higher levels inhabit a much broader informational space and are able to influence which possibilities survive relevance filtering.

Harris actually acknowledges this, saying, "you can create a framework in which certain decisions are more likely than others" (Harris, 2012, p. 31), though he questions whether this represents genuine freedom to choose in any meaningful sense. The framework suggests this recursive framework-building capacity represents the core mechanism through which agency develops over time, because the geometric informational space people operate within is not fixed.

Identity, beliefs, and values feed back into relevance realization which recursively shapes which possibilities are meaningful, important, and prioritized. As an individual consciously redefines these structures over time, the filtering process evolves, which expands or constrains the futures they are capable of enacting. Therefore from this view, free will first peaks its head out through recursive participation in shaping the decision space itself.

Importantly, Harris is an advocate and practitioner of meditation. He has significant experience with contemplative practices and has likely experienced what the framework poses as high level consciousness experiences, although it's unclear to what level Harris currently operates at. This is likely where he has found these insights that "oneness" can still involve determinism. Harris agrees that from a practical sense that decisions and choices are important when he says, "My choices matter—and there are paths toward making wiser ones—but I cannot choose what I choose. And if it ever appears that I do—for instance, after going back and forth between two options—I do not choose to choose what I choose. There is a regress here that always ends in darkness" (p.31). But the framework interprets this "regress" much differently because of the universe's quantum nature. It's entirely possible Harris is correct in a very technical, metaphysical sense. Where Harris sees "darkness" at the bottom of it all, the framework posits that this is where the Quantum Information Dimension (QID) is. This is where it would act as the "turtle at the bottom," in the free-will debate, however, the QID is not a fixed chain of determinism. It's not even pure randomness. It's a probabilistic substrate from which consciousness participates.

Kvam and Pleskac (2017) provide mathematical foundation for why deterministic regress cannot exist at the quantum level. They say,

> Aleatory uncertainty supposedly arises from inherently stochastic properties of the environment. In a truly deterministic world, we could in theory predict anything given enough knowledge of the current state of the world. In this sense, aleatory uncertainty may not truly exist in a classical deterministic system. However, integrating heuristics with the quantum framework introduces a parallel form of uncertainty called ontic uncertainty – uncertainty regarding the true existence of a particular state – which arises from the stochastic interactions between states and measurements of the world …
>
> Formally, a quantum superposition state such as a qubit possesses unpredictability in the form of von Neumann entropy … In this case, we can know everything about a state – for example, $\psi = [\sqrt{.5}, \sqrt{.5}]$ – but the state is still entropic in that we cannot perfectly predict the outcome of a measurement … Von Neumann entropy therefore provides a 'true' source of unpredictability (ontic uncertainty) that may underlie what we label aleatory uncertainty …
>
> While it may seem inconvenient for a person in a von Neumann-entropic state to be unable to predict their own behavior (assuming they know their own state), this is advantageous in that it is a source of uncertainty that cannot be reduced by other people either … this is particularly desirable because it prevents an agent from being exploited when opponents have a great deal of information about the agent's strategies". (p. 35)

Their framework demonstrates that Harris's "regress that always ends in darkness" actually ends in irreducible quantum uncertainty. The von Neumann entropy at the foundation of the QID ensures that even with complete self-knowledge, genuine unpredictability remains, and they importantly point out that this is not a evolutionary bug, but it is actually a survival mechanism feature, and it is one that would have been selected for as it prevents conscious agents from being exploited by others who could have otherwise mapped their deterministic behavioral patterns.

From the framework's perspective, what occurs during those milliseconds before the neural firings hit the DMN, which provides the sense of choice, involves System 1's quantum-enhanced processing exploring multiple possible actions simultaneously through superposition. Relevance realization filters these quantum results based on current context, identity structures, beliefs, and learned patterns. The brain's quantum systems calculate probability distributions across available choices, with neural preparation beginning for the most likely outcome before the DMN constructs its "I am deciding" narrative.

This sequence would help explain why most immediate choices feel constrained. In any given moment, whether choosing chicken or beef for dinner, the decision emerges from a highly probabilistic process shaped by recent experiences, current emotional states, established belief patterns, and habitual collapse mechanisms. For example, all your actions leading up to this exact moment may create a 99% chance that you will choose chicken for dinner, but the "choice" always remained theoretically for you to choose something else for dinner, because if the quantum brain works in probability, it has already calculated the most probable path based on the geometric space of possibilities available to that consciousness level and then that signal is post-hoc sent to the default mode network to reach our conscious, waking awareness which makes it feel like a difficult choice and one where the narrative "I ended up choosing chicken" is constructed.

However, this probabilistic nature operates differently across temporal scales. While Harris may be correct that short-term choices in many scenarios may be largely determined by prior quantum collapses that shaped current neural patterns, longer time horizons would allow for genuine agency through what might be called recursive framework-building. Over days, weeks, months, and years, consciousness can consciously reshape identity structures, retrain relevance realization filters through new experiences, develop ego modulation capabilities, and expand the geometric space of quantum possibilities accessible for exploration.

From quantum first principles, reality emerges from superposition through collapse into specific outcomes. At the foundation, every collapse creates boundary conditions that influence future probability distributions. When someone is born into a particular circumstance, such as a time period, geographic location, family context, or with physical constraint for example, this would indeed set parameters for their geometric quantum informational space they can navigate from birth. It's like inheriting a moving ship at sea as individuals receive trajectory and momentum from prior collapses while retaining capacity to steer within those constraints.

Throughout life, the cone of possibilities gradually narrows as the body is finite and time progresses linearly. Agency may emerge most powerfully in moments of genuine 50–50 quantum indeterminacy where multiple futures remain equally probable. Importantly, these are not classical 50–50 coin flips. They would be quantum-probabilistic equilibria where multiple potential futures coexist in superposition until collapse. It could be any even distribution of probabilistic outcomes across a vast number of "N." These are precisely the type of moments Kvam and Pleskac (2017) described where von Neumann entropy reaches its maximum value, when quantum states exist in perfect superposition with equal probability amplitudes. In these high-entropy states, consciousness encounters genuine choice points where no predetermined path dominates the probability landscape.

In these moments, consciousness would have the ability to force premature collapse before probability distributions stabilize, which creates low-probability interventions that dramatically reshape future possibility landscapes. Research on quantum measurement demonstrates how this takes place. Koshino and Shimizu (2005) write, "the decay rate under the measurement is plotted as a function of the measurement intervals τi … depending on τi and the form factor g, the opposite effect, i.e., acceleration of decay, may take place" (p. 271). This measurement timing effect shows that quantum systems can be interrupted before natural probability evolution completes which fundamentally alters outcomes. The von Neumann entropy framework reveals why these moments are so crucial for free will because when entropy is maximized, the quantum system contains the greatest potential for unpredictable outcomes that cannot be reduced through additional knowledge. This irreducible uncertainty creates genuine windows of agency where consciousness can influence which possibilities collapse into actuality.

The ego's capacity to "take the bread out of the oven too early" by interrupting natural quantum settling processes represents one mechanism through which genuine choice operates. Studies of quantum measurement reveal that "the form factor is renormalized as a backaction of measurement, through which the decay dynamics is modified" (Koshino & Shimizu, 2005, p. 191). When applied to consciousness, this mechanism would allow ego-driven observation to modify quantum processing capabilities before optimal probability calculations complete. As demonstrated in quantum systems, "each photonic mode is energetically broadened as a counteraction of measurement" (Koshino & Shimizu, 2005, p. 191), which shows how measurement timing physically alters the available quantum possibilities. This also opens the door for consciousness' ability to "observe" future possibilities. The act of imagining future events may itself reshape their probability landscape. During moments of probabilistic quantum equilibrium, the ability to conceive of potential futures could entangle the cue–belief network described in Chap. 9, which would shift which configurations are most likely to collapse into actuality.

Strategic exposure to randomness, deliberate engagement with uncertainty, and conscious utilization of chance events can inject agency into otherwise highly probable systems. The quantum research demonstrates "two-stage behavior of decay rate"where "the decay rate is identical to the unobserved rate Γ, whereas the decay proceeds with a suppressed rate in the later stage" (Koshino & Shimizu, 2005, p. 261), which indicates that measurement characteristics can shift systems between different probability regimes. Some individuals, who recognize the genuine quantum indeterminacy, may consciously delegate decisions to random selection when facing true choice points where system 1 when constrained through relevance realization cannot identify optimal outcomes.

From this perspective, consciousness is in a way shaping the collapse of possibility into actuality. Harris would likely respond that this recursive framework-building still leads to the familiar regress. When someone "chooses" to reshape their identity structures or develop new consciousness capabilities, Harris might ask what caused that choice and argue that if they decided to meditate or consciously expand their relevance filters, those decisions themselves emerged from prior causes extending backward indefinitely.

The framework acknowledges this regress, but this is why emphasis on "the turtle at the bottom" is important. Where Harris encounters an absence of ultimate authorship at the bottom of "why did you choose", the framework identifies the Quantum Information Dimension as the substrate from which all possibilities emerge as the QID represents a probabilistic field of quantum configurations awaiting collapse as humans participate in this collapse through consciousness-mediated selection. The QID's probabilistic, uncertain nature as previously described fundamentally changes Harris's infinite regress because instead of encountering deterministic causation at the bottom, it reaches genuine ontic indeterminacy, or the irreducible superpositional state-space of quantum field amplitudes, within which collapse is a contextual interaction and of which humans can take part and influence. Small interventions in probability distributions, or minute perturbations in local field amplitude, propagate through the system's Hilbert space as nonlinear interference effects, which produce macroscopic trajectory divergence which would be analogous to a quantum-scale butterfly effect. Even when a collapse event arises from prior influences, each measurement-like act exerts back-action on the field, which renormalizes phase relations and slightly re-sculpts the probability geometry. Over iterative cycles, these micro-adjustments compound and generate macroscopic divergences that become the lived trajectories of conscious agents.

Therefore, in a strict metaphysical sense, Harris may be correct that we are never "truly free" from the unfolding of probability in that sense, but the framework is attempting to reframe this. The recursive relationship that humans have with the quantum uncertainty in the QID is the context or the ground on which true conscious agency can emerge and operate from. At the very least there will undoubtedly be moments in one's life that are pure 50–50% probabilistic outcomes, in that sense consciousness has free will to pre-maturely collapse a probability into a binary state of action.

Harris's point about the limitations of agency at lower consciousness levels would still be entirely valid, but the framework suggests Harris' binary conclusion misses the developmental nature of human potential. When he writes that "You have not built your mind. And in moments in which you seem to build it—when you make an effort to change yourself, to acquire knowledge, or to perfect a skill—the only tools at your disposal are those that you have inherited from moments past" (Harris, 2012, p. 31), he is actually accurately describing the recursive nature of consciousness development. But where the framework differs is that instead of proving agency is an illusion, this recursive quality of our conscious experiences becomes the very mechanism through which expanded agency actually emerges.

Jordan Peterson is renowned for speaking on the idea that free will, in at least some form, does indeed exist and has done many public discussions with Sam Harris on this topic. They both represent the public-facing leaders of the ideas on either side of the debate. Peterson offers a complementary perspective because instead of disputing the neuroscientific findings about unconscious neural activity, Peterson emphasizes what he calls a "temporal gradient" of agency. As he explains, "as you look out into the future, the farther down the road, the more free your choices are, but the closer they get to implementation, the more they become deterministically governed by standard causal processes" (Peterson, 2022). Peterson's "temporal gradient" maps well onto the framework's quantum-to-classical transition.

Peterson also emphasizes the phenomenological reality of choice and its practical consequences. He tends to often argue on the practicality of the debate instead of focusing on a specific metaphysical point like Harris. Peterson argues that treating ourselves and others as free agents produces more beneficial outcomes regardless of metaphysical debates about ultimate causation. As he puts it, we confront "a landscape of possibility" and choose "what sort of reality we want to bring into being." This experiential description aligns with the framework's view of consciousness as actively, recursively participating in quantum collapse. This practical approach seems incredibly relevant when considering the quantum uncertainty previously described. From the human's perspective, it resembles a kind of Pascal's Wager for free will. Even if one cannot know whether a given choice is truly indeterminate or already determined, acting as though all decisions are 50–50 preserves the capacity for agency and expands the range of possible futures that can be realized.

When considering Peterson's practical approach, Harris's position seems more purely academic, because Peterson describes humans confronting "the possibility of the day." Whether or not someone can ultimately "choose what they choose" matters less than whether they can meaningfully expand their decision space over time through recursive framework-building, so from that, the practical question becomes how to help people operate at higher consciousness levels where more constructive possibilities become available to them.

Within the framework, a paradox actually emerges at the highest levels of consciousness, and this may be what Harris is experiencing, because as ego activation decreases, the computational space actually expands dramatically and widens the range of potential futures available to collapse. The paradox is that individuals operating at this stage of consciousness often begin to surrender "personal control," or from the framework's perspective, ego's influence, as they allow for their choices and decisions to arise from deeper computational flows from the QID. In these states, what feels like guidance or insight would be the effects of someone tuning their internal filters to better align with broader quantum field dynamics instead of trying to impose one's purely personal, egoic intention into the decisions and into the process of relevance realization. So from this perspective, this surrender represents free will reaching its widest possible domain by choosing when to and when not to interfere and interject in one's decision making processes.

Consider the scope of human potential that becomes visible through this developmental lens. The same individual who might collapse quantum possibilities into destructive patterns at lower consciousness levels could just as plausibly access creative solutions that seemed impossible before. Someone operating at Level 2 consciousness might prematurely collapse possibilities into harmful actions, while that same person at Level 7 could generate world-changing discoveries. The underlying quantum potential (Q) remains vast in both cases, but ego activation (E) determines which possibilities become accessible for collapse into actuality. Considering this, the disagreement around free will becomes less philosophically important, as Harris accurately maps the constraints most people experience on a moment to moment basis due to high ego activation and narrow relevance filters, while Peterson accurately describes the expanded agency possible when consciousness operates at higher levels where ego recedes and broader quantum possibility spaces become accessible. Both perspectives are valuable when they are understood to be describing different points along the same developmental spectrum the framework proposes.

Overall, the framework points to the idea that the current state of the free will debate rests on a questionable fundamental assumption about the nature of the "I" that supposedly possesses or lacks "free will" in the first place. Traditional arguments presuppose a unified, persistent self that then either has ultimate "free will" or is completely determined by prior causes. But throughout this book, the framework has attempted to show that this "I" is actually the DMN's narrative construction which acts as the evolved tool for creating coherent identity from quantum-classical processing loops. Agency, in this sense, is a gradient as a system's capacity to modulate its influence moment by moment within an environment of continuous collapse. Someone may have limited flexibility in a given instant, yet over longer horizons they can expand their computational geometry, reduce ego interference, and access broader ranges of outcome space. Free will becomes the recursive process of engaging the probabilistic substrate of reality itself.

At the deepest layer, even this debate is part of the field in which it occurs. No observer can ever know when a decision arises from probabilistic openness or from causal momentum as the two are inseparable aspects of one oscillating process. Because of that uncertainty, the only coherent stance is participation. Living as though every act were freely chosen widens the coherence window through which consciousness interfaces with the Quantum Information Dimension as the meaning of action lies in sustaining the play of collapse and exploration.

Within the field, every polarity, such as win and lose, birth and death, or freedom and determinism, is a complementary pair, as each side defines the other, and the tension between them sustains evolution. If either pole were absolute, the system would halt as there would be no interference, no new pattern, and no experience. So when ego clings to one answer, energy is wasted resisting its complement, but when the mind can hold both, one's energetic cost falls, coherence increases, and participation flows much more freely.

From this vantage, the traditional binary yes or no of the free will debate entirely dissolves. Instead of asking "Does the self have free will?" the framework asks "How does consciousness develop the capacity to influence its own future processing? The path that one takes can only become optimal through the very act of acknowledging that there is no "truly optimal" future path at all. If the photon follows the path of least action, if life follows the path of least energetic waste, and if consciousness follows the path of maximal coherence, then agency is the lived experience of that universal optimization which manifests as the ability to tune one's own phase alignment with the unfolding field. Therefore, when the mind does indeed accept that there is no final optimal path, it becomes most free to move in a way that produces optimality moment by moment.

It may be more pragmatic to think of free will and agency like steering a massive ship over extended time horizons as the capacity for self-authorship emerges through the recursive process of interfacing with the probabilistic substrate and consciously shaping the very conditions that shape future consciousness.

In this sense, free will becomes less of a binary property and more of a gradient measure of how much influence a conscious system can choose to exert at any given moment within an environment of continuous collapse, and how flexibly it can modulate its own influence at any given time. The framework suggests that self-authorship or free will could be thought of as a recursive process of meaningful, participatory engagement with the probabilistic substrate of reality itself. The "I" that wonders whether it is free is part of the instrument through which freedom expresses itself. What remains when the binary of freedom and determinism dissolves is the sacred play through which the universe keeps discovering what it can become.

References

Harris, S. (2012). *Free will* (kindle ed.). Free Press.

Koshino, K., & Shimizu, A. (2005). Quantum Zeno effect by general measurements. *Physics Reports, 412*(4), 191–275. https://doi.org/10.1016/j.physrep.2005.03.001

Kounios, J., & Beeman, M. (2009). The aha! Moment: The cognitive neuroscience of insight. *Current Directions in Psychological Science, 18*(4), 210–216. https://doi.org/10.1111/j.1467-8721.2009.01638.x

Kvam, P. D., & Pleskac, T. J. (2017). A quantum information architecture for cue-based heuristics. *Decision, 4*(4), 197–233. https://doi.org/10.1037/dec0000070

Peterson, J. B. (2022, October 28). *Jordan Peterson's response to Sam Harris' free will argument* [video]. YouTube. https://www.youtube.com/watch?v=uC6BPg7IDpA

Chapter 16
Case Studies

Case Study of Federico Faggin's Unitive Experience

Federico Faggin, a physicist and inventor best known for developing the microprocessor, describes an experience that transformed his understanding of consciousness (Essentia Foundation, 2022). He wrote the book *Irreducible: Consciousness, Life, Computers, and Human Nature* (Faggin, 2024) where he argues that consciousness is ontologically primary, rooted in quantum reality, and that subjective experience corresponds with pure quantum states that are inherently non-replicable and private. He states, "consciousness is a quantum property of a field .....Consciousness is purely quantum and so is free will ..." (pp. 226–227).

During a vacation, he awoke in the night and felt what he recalls as a sudden rush of energy rising from his chest. He describes seeing "a beam of white scintillating light" that expanded until it filled his entire visual field. At that moment, he perceived that the light was "the stuff of which everything is made" and that "I am the world observing myself."

He speaks of a state where the separation between self and world seemed to dissolve. The world, in his words, felt like love, joy, and peace. His body felt as though it were vibrating in complete resonance with everything around him, and he perceived no distinction between his point of view and the universe itself. "I was the world observing itself through my point of view," he explains, describing this as an experience of unity and wholeness.

Following this experience, Faggin came to believe that consciousness must be fundamental to reality. He concluded that the universe cannot have evolved from inert matter into conscious beings but instead that consciousness has been present from the very beginning. In his view, physical reality emerges from a deeper conscious substrate, and what we perceive as matter is an expression of this underlying field. He speaks of particles as "states of fields" and describes matter itself as a symbolic structure.

J. Roeloffs, *The Classical Brain to the Quantum Mind*, The Frontiers Collection,
https://doi.org/10.1007/978-3-032-20068-6_16

Faggin also proposes that reality has two aspects. There is the inner experiential aspect, which he identifies as ontologically primary, and an outer symbolic aspect, which consists of the structures and representations conscious beings use to communicate with one another. Classical information, including the physics we use to describe the world, belongs to this outer aspect, while experience itself belongs to the deeper, private layer that underlies it.

From the book's perspective, the phenomenological elements of Faggin's experience align with predictions the framework would make for high-coherence states of consciousness, especially those near Level 8 or Level 9 integration. The perception of unity and interconnectedness may reflect a temporary quieting of egoic collapse mechanisms, allowing conscious awareness to synchronize more directly with the quantum informational substrate. The sense of becoming "the world observing itself" would arise naturally in this interpretation, as awareness accesses a wider bandwidth of information unfiltered by narrative identity.

Faggin's description of the "white scintillating light" proves particularly relevant. In the framework, photons are treated as discrete carriers of information, and their collisions which result in information exchange are fundamental to the coherence of physical reality. When considered through the lens of the Quantum Information Dimension and $P = Q/E$, light represents the substrate through which reality maintains coherence and stability. Faggin's perception that this light was "the stuff of which everything is made" aligns with what the framework would expect if subjective awareness briefly accessed this informational layer directly.

Where the interpretations diverge centers on the ontology implied by Faggin's conclusions. He views his experience as evidence that consciousness is the foundation from which physics itself emerges, essentially proposing that consciousness is fundamental to the universe and that matter arises from this conscious substrate.

The framework proposes an alternative reading. While Faggin's phenomenological experience appears genuine and aligns with predicted high-coherence (high consciousness) states, the framework suggests a different causal relationship. As explored in Chap. 5, rather than consciousness being fundamental to the universe, consciousness may have evolved as a sophisticated biological mechanism that participates in universal quantum processes. The special quality of human conscious experience that Faggin describes may indeed arise from interfacing with the quantum substrate, but this substrate exists independently of consciousness.

From the framework's perspective, as described in previous chapters, wave function collapse and maintenance of massive structures occurs continuously throughout the universe through entropy as environmental decoherence, photon interactions, and other physical processes, with or without conscious observers. The physical reality humans experience is objectively real and not a construction of consciousness. What makes human consciousness unique is its evolved capacity to participate in and direct some of these collapse processes toward survival-beneficial outcomes through ego modulation.

High-coherence, high consciousness states may allow conscious systems to access or experience the quantum field (Quantum Information Dimension) more directly, which would produce experiences of unity, boundlessness, and resonance similar to what Faggin describes. These states don't imply that consciousness creates, or is the substance of, the substrate. Instead, consciousness may function as a particularly sophisticated interface that evolution developed to navigate an informational layer that would exist regardless of whether conscious beings evolved to perceive it, and as outlined in Chap. 7, the brain participates and constructs conscious experience through selective collapses of quantum possibilities into classical narratives shaped by survival and adaptation.

In this interpretation, Faggin's experience of becoming "the world observing itself" reflects a temporary state where his consciousness synchronized with the broader quantum field, which allowed perception of the continuous process through which classical reality emerges from quantum potential. The phenomenology remains valid and meaningful, but the framework attributes it to consciousness participating in universal processes instead of consciousness itself being the source of those processes. While consciousness may truly feel fundamental to human experience as it provides the subjective quality that makes existence meaningful to most of us, it does not need to be fundamental to the universe itself.

Rick Rubin Case Study

Based on the P = Q/E equation and the consciousness level descriptions, the framework, as described in significant detail in previous chapters, generates specific predictions about how an individual operating consistently through the ego-aware paradigm (Level 7 + consciousness) would describe their experience and approach to creative work.

An individual at this level would describe creativity and insight as originating from outside the personal self. They would describe pulling from an existing creative informational space or what the framework terms the Quantum Information Dimension. They would emphasize that trying harder or forcing outcomes actually impairs performance, and they would say that the self or ego would be identified as the primary interference to be minimized to improve creative output.

An ego-aware practitioner would describe reality as more fluid, and understand that observation and intention shape outcomes. They would emphasize being and presence as the most efficient way to approach creative output, and all activities would be seen as potentially transcendent when approached with proper consciousness. They would report accessing universal patterns or intelligence beyond personal knowledge, and faith and trust would be emphasized. In collaborative settings, they would describe cooperation as some sort of collective ego dissolution, and would recognize that perceiving being observed changes behavior and outcomes. Considering this, let's explore a case study.

Rick Rubin represents one of the most successful music producers in history, with a documented forty-year career spanning every major musical genre. As a nine-time Grammy winner, he has been named one of the 100 most influential people in the world by Time magazine and the most successful producer in any genre by Rolling Stone. His collaborations include Johnny Cash, Jay-Z, Slayer, Eminem, Adele, Imagine Dragons, Linkin Park, Metallica, the Beastie Boys, Lady Gaga, and many more.

What makes Rubin particularly relevant to this framework is his radically different approach to production. Unlike most producers who emphasize their own music making ability and technical expertise, Rubin claims no technical knowledge whatsoever. Instead, he focuses entirely on creating conditions for authentic expression to emerge. His recent book *The Creative Act: A Way of Being* (2023) provides extensive documentation of his worldview and methodology, which seemingly offers a rare window into ego-aware consciousness as expressed in technical work from an experienced practitioner.

Author's Note

The entire theoretical framework presented in this book, including the P = Q/E equation and the predictions about the characteristics and tendencies of ego-aware consciousness, were developed prior to the author's awareness of Rick Rubin's writing in *The Creative Act* (2023). The discovery of Rubin's work occurred only after the framework was complete, during the process of researching real-world examples. This seemed important to note, as the author was shocked to see how closely and explicitly Rubin described the same phenomena and even at times used scientific language to do so.

Regarding the source of creativity and access to the Quantum Information Dimension, Rubin writes, "This content does not come from inside us. The Source is out there. A wisdom surrounding us, an inexhaustible offering that is always available. We either sense it, remember it, or tune in to it" (p. 16), and he doubles down on this later saying, "Billions of data points are available at any given moment and we collect only a small number. With this glimpse through a keyhole, we assemble an interpretation and add another story to our collection" (p. 242). He explicitly states that the lived experience is accessing information outside oneself and that we (the ego) "assemble an interpretation" which matches the framework's prediction that Level 8 consciousness would describe accessing information beyond personal experience.

On the nature of ego as the primary interference mechanism, Rubin provides explicit explanation saying, "Self-awareness is a transcendence. An abandonment of ego. A letting go… As artists, we are on a continual quest to get closer to the universe by getting closer to self. Moving ever nearer to the point where we can no longer tell where one begins and the other ends" (p. 161). This is specifically outlining the climb up the stages of consciousness as described in previous chapters, and importantly, he notes the critical point as which the "I" doesn't begin or end, which is this transition from ego-unaware to ego-aware as one begins to integrate their ego. He goes on to identify the ego as actively impeding creative work, "Competition serves the ego. Cooperation supports the highest outcome" (p. 224). He's explicitly outlining how the most creative work flows, not from "knowing oneself" but actually from letting go of oneself. This is describing the ego-aware paradigm.

As described in past chapters, the framework predicts that the perception of observation has a significant affect on outcomes, and Rubin directly states this saying, "Being observed changes how an artist acts" (p. 117), which provides explicit acknowledgment of the observation paradox in creative contexts, and then he doubles down, saying, "Even if your art is non-performative, such as writing or cooking, it will still likely change with an observer present" (p. 117). Continuing to acknowledge the effort paradox that is central to $P = Q/E$, Rubin writes, "Boosting our receptivity may require a relaxing of effort. If we're trying to solve a problem, trying can get in the way. Splashing in a pond stirs up clouds of dirt in the clear water" (p. 167). This is parallel to the framework's prediction that increased effort (ego activation) reduces access to quantum processing and therefore reduces performance.

On maintaining superposition versus forcing collapse, Rubin advocates approaching work "with no map" and states, "With each story we tell ourselves, we negate possibility. Reality is diminished. Truth collapses to fit a fictional organizing principle we've adopted" (p. 242). Rubin is describing the observer's paradox even using the word "collapses" and saying that information gets collapsed into fictional narratives derived by the ego, and goes as far to state that these stories even "negate possibility" which mirrors Chap. 9's descriptions of how consciousness' possibility space can be constrained through beliefs and narrative. He is describing the power of having a large geometric space to operate in, and he doubles down on this saying, "Release all expectations about what the work will be. Approach the process with humility and the unexpected will visit more often. Many of us are taught to create through sheer will. If we choose surrender, the ideas that want to come through us will not be blocked" (p. 168). "Expectations" are beliefs about future outcomes. He is tying the reduction of belief directly into an increase in the amount of and quality of insights received. The mix of spiritual and scientific language used throughout is noteworthy. As the framework expects, mystics and artists have front-run science on some of these ideas. Rubin has a sophisticated understanding of how ego through narrative, belief, certainty, and observation reduce performance.

Furthering this spiritual language, the framework predicts that ego-aware consciousness would emphasize faith and trust over control as these are well-known actions that inherently reduce ego, and they act essentially as consciousness technologies as described in previous chapters, Rubin writes extensively on these themes as well saying, "The practice of spirituality is a way of looking at a world where you're not alone. There are deeper meanings behind the surface. The energy around you can be harnessed to elevate your work" (p. 25), and he emphasizes that "the principle operates on faith. Believing and behaving as if it's true. No proof is needed" (p. 25). Regarding collective consciousness and group dynamics, Rubin states, "The synergy of a group is as important—if not more important—than the talent of the individuals" (p. 228). He describes successful collaboration as requiring participants to "keep themselves pure and unattached" (p. 158), which allows group coherence to emerge without ego interference. On consciousness participating in reality creation, Rubin offers a parallel to the framework's core premise saying, "Through the ordinary state of being, we're already creators in the most profound way, creating our experience of reality and composing the world we perceive. In each moment, we are immersed in a field of undifferentiated matter from which our senses gather bits of information. The outside universe we perceive doesn't exist as such. Through a series of electrical and chemical reactions, we generate a reality internally" (p. 10). Once again, Rubin directly describes a literal proposal of this book and he even uses scientific language to do so. As he supports the framework's view that consciousness actively participates in creating reality through collapse. While his descriptions seem to hint at a view that "consciousness is fundamental," his descriptions of how the mind constructs its perception of the world around it are extremely similar to previous chapters in this book.

Rubin's "Breaking the Sameness" (p. 116) techniques provide practical validation of how environmental changes affect ego activation and therefore performance. His methods include a variety of things including changing the environment by turning off lights to create shifts in consciousness, altering stakes by imagining final performances or lowering pressure through rehearsals, changing context through new backstories for songs, altering perspective by adjusting headphone volume to force delicate performances, and writing for someone else to bypass personal ego investment. Each technique effectively modulates E in the $P = Q/E$ equation, demonstrating through decades of practice what the framework predicts theoretically.

The Vessel and Filter

Particularly interesting is Rubin's description of what the framework terms relevance realization. He writes, "Information does not enter the vessel directly, like rain filling into a barrel. It is filtered in a unique way for each of us." He continues by saying, "Our filter inevitably reduces Source intelligence by interpreting the data that arrives instead of letting it pass freely" (p. 22).

This description matches the three-layer architecture proposed earlier in the book, where quantum computation generates all possibilities from "the source" as Rubin describes, then relevance realization filters results by "interpreting the data", and ego processes what passes through. Rubin's practical understanding that "the more raw data we can take in, and the less we shape it, the closer we get to nature" (p. 23) directly supports the framework's mechanism where reducing ego interference allows more quantum computational results to reach consciousness. Interestingly, Rubin himself seems to anticipate this very analysis. In discussing the relationship between art and science, he notes, "It's not unusual for science to catch up to art, eventually. Nor is it unusual for art to catch up to the spiritual" (p. 27).

This observation proves particularly prescient given that the present framework seemingly provides an acute scientific explanation for the creative principles Rubin discovered through decades of practice. In documenting how ego-aware consciousness naturally discovers quantum mechanical principles through direct experience, this book participates in the symbiotic relationship between art and science that Rubin describes.

David Hawkins Case Study

Further valuable comparison for the framework's principles may be found in the realm of contemplative practice. David R. Hawkins, M.D., was a psychiatrist, physician, and researcher who co–authored work with Nobel Laureate Linus Pauling before dedicating his later life to exploring the nature of consciousness. His decades of clinical observation and direct contemplative inquiry led him to describe a model of reality that offers many parallels to the mechanisms proposed in this book.

Hawkins describes an experiential encounter with a universal information field that appears to function in a similar way to the framework's proposed Quantum Information Dimension (QID). While he did think consciousness was very likely fundamental, similar to Faggin, Hawkins' descriptions were often more direct and correlate more specifically with this books descriptions. He describes an "infinite, nonlinear field of consciousness" that is "omnipresent and omniscient" (Hawkins, 2020, p. 244) and contains "no divisions" (p. 125). He explains this field with the analogy of a vast, accessible database. He writes,

> The individual human mind is like a computer terminal connected to a giant database. The database is human consciousness itself, of which our own consciousness is merely an individual expression but with its roots in the common consciousness of all humankind. The database transcends time, space, and all limitations of individual consciousness... This database is the realm of genius; because to be human is to participate in the database, everyone, by virtue of their birth, has access to genius. The unlimited information contained in the database has now been shown to be readily available to anyone in a few seconds, at any time and in any place. (p. 39)

From the framework's perspective, this "database" serves a similar functional role as the QID, a timeless and non-local repository of all potential information that consciousness can access. Hawkins' observation that this field "registers all that has ever happened throughout time" (p. 121) aligns with the proposal that information is conserved within this substrate.

Also, Hawkins provides a detailed account of the ego's role in filtering reality, which he terms the "experiencer." This description closely mirrors the framework's proposed three-layer architecture of quantum processing, relevance filtering, and ego-mediated collapse. He identifies this "experiencer" as the ego that is perceiving reality, but is never truly able to witness it. He describes it as a processor that assigns value and creates a subjective reality just moments after the fact. He writes,

> The experiencer is one ten-thousandth of a second removed from reality. Therefore, the ego never experiences what is; it experiences its experience of what is. By analogy, it is not experiencing that which is being recorded; it is experiencing the playback from the recorder... This side of the one ten-thousandth of a second, the trees look like trees... On the other side of that one ten-thousandth of a second, they radiate out Divinity! Enlightenment is to bypass the experiencer. (p. 217)

Inhabiting this perspective, the unfiltered reality ("Divinity") may reflect the quantum field (Q), while the "experiencer" acts as the Relevance Realization filter and ego-collapse mechanism (E). The final "playback" is the classical, narrated, perceived reality that emerges from this process. His perspective on creating the future also aligns with the framework's view of consciousness participating in collapse. He states that the future "has no reality yet. It is only a potentiality awaiting actuality by intention and choice" (p. 239). While the mechanics Hawkins describes are similar to those proposed in this framework, his interpretation of the fundamental substrate differs. Through his direct experience, Hawkins concluded that consciousness itself is the fundamental field or the "database" from which physical reality manifests.

This framework offers an alternative interpretation of similar reported experiences. It posits that the quantum field (the QID) is fundamental, and that consciousness is the evolved biological interface that accesses it. In this view, the universal "database" that Hawkins perceived is the QID itself. The reason it may feel like a field of consciousness is because consciousness is the tool through which it is being perceived.

The convergence of these two perspectives, one derived from scientific synthesis and the other from direct contemplative inquiry, is noteworthy. That they both arrive at a nearly identical description of the *mechanics* of reality and how its perceived, which involve a universal information field and an ego-observer that filters it, suggests they may be describing an authentic feature of existence, even if their ultimate conclusions about what is fundamental diverge.

Lex Fridman's Ayahuasca Experience Case Study

Lex Fridman, an AI researcher with a doctorate, publicly discussed his ayahuasca experience during a podcast conversation with Elon Musk (2024). His report provides valuable data because Fridman has demonstrated his consciousness development to the public for years, and shows many signs of peak Level 6 baseline through his public work including deep contemplation, loving approach to life, integrated knowledge across domains, ability to hold opposing views gracefully, and deep experience of flow states as a black belt in jiu-jitsu. This combination of scientific training and evidenced consciousness level allows for the framework's expectations about those operating from the collective coordination program (which is accessible at peak Level 6) to be put to a preliminary test.

While anecdotal reports clearly cannot prove theories, when they come from credible sources that fit the framework's criteria for predictions, they warrant investigation as suggestive evidence.

Talking of his ayahuasca experience, Fridman (2024) said that immediately as his experience began, he started seeing humans "with a glow" in them. He said, while in the experience, it was clear to him that the glow was "the thing that made humans amazing." Interestingly enough, this already mirrors Faggin's description of his enlightening experience. Fridman continues, and describes this entire experience as perceptual, and he said he then elevated above those people in observation. Upon "flying out from them," he observed Earth, then the Solar system, then the galaxy, then multitudes of galaxies. He said, "I saw that same light, that glow, all across the universe." He said he was just there, observing, he wasn't able to move or navigate on his own. He made it clear he would have if he could have.

He emphasized this again, saying, "the thing that made humans amazing" was present "throughout the universe, like glowing dots." He said the experience made him think that "whatever makes humans amazing [is] all throughout the universe." It is noteworthy that he did have expectations going into it, but not expectations of what he ended up experiencing. He said even though he thought he may experience demons, or meet strange creatures, or see aliens, he never did. He only experienced a deep sense of love and connectedness throughout his journey, so he wasn't primed in the experimental sense for the outcome he describes.

From the perspective of the book's framework, Fridman's description of his experience aligns extremely well with the predictions that would be made for anyone at his consciousness level. At Level 6 or 7, psychedelics could likely give someone like Fridman an experience of Level 8 or 9 consciousness, and because he likely already has strong integration of his ego in many facets of his life, from the framework's perspective, it is possible he was, in that moment, interfacing with the Quantum Information Dimension as effectively as someone who has integrated Level 8 or 9 consciousness might.

Acknowledging the anecdotal nature, Fridman's experience seem to be a valuable datapoint to consider, and his descriptions likely warrant further research. This case study proposes that baseline consciousness level and ego development may predict the phenomenology of consciousness-expanding experiences with psychedelics. The framework generates testable hypotheses about how different individuals, occupying different levels of consciousness, may experience these states based on their P = Q/E profile.

Future research could systematically study psychedelic experiences and whether ego-unaware individuals predominantly report narrative and entity experiences while ego-aware individuals can more directly interface with the Quantum Information Dimension.

Post-ego Systems Case Study

If the core equation P = Q/E holds, where quantum processing access increases as ego decreases, and if the Nine-Level Ladder of Consciousness is valid, then one might make an intriguing prediction that if the scale of consciousness is climbed in sequence, without skipping steps, then a new category of systems might naturally emerge as more people make the climb towards Level 8. These systems would not resemble traditional technologies, institutions, or ideologies. They might have fundamentally different design characteristics as shown in Table 16.1.

In this case, they would not be systems created by ego to glorify their makers. They could be systems built by people who no longer identify with "I", to benefit humanity. The originators of these systems might not seek status or reward. They may describe themselves as merely a channel or conduit of consciousness, or a biological

Table 16.1 A side by side description of post-ego systems

Trait	Description
First principles-derived	Built from foundational logic, not social consensus or legacy models
Self-stabilizing	Designed to adapt, evolve, and correct without centralized control
Egoless architecture	No identifiable "founder" or ownership as the system is the servant
Global inclusivity	Not targeted to tribes or nations, but to humanity itself
Incentive-aligned	Harmonizes individual and collective good without coercion
Consciousness-conducive	Reflects deeper truths about reality, agency, and participation
Feels discovered, not invented	Emerges like math or gravity—"obvious" in hindsight, inevitable in logic

interface through which the system emerged, or a servant rather than master who recognizes the system transcends them. They might say "it revealed itself" instead of "I created this."

Cook-Greuter (2021) beautifully articulates an ego-aware practitioner's approach to interfacing with the world and business, saying,

> As leaders, they tend to build their own novel organizations or work alone doing what they perceive to be their best contribution to humanity. Some like to take on the roles of catalysts or transformers in organizations, but readily leave these when they feel their transformational work is done. They feel successful when they have made themselves dispensable, that is, if the organization itself has become transformative and self-organizing… Unlike earlier levels, they are less concerned with success and rewards for their efforts. Sowing seeds that may take a long time to germinate if at all, is part of what one does because [one] can do no other. (pp. 103–104)

The identity, or lack thereof, of Satoshi Nakamoto is why Bitcoin is one of the best case studies of consciousness-informed systems theory. Satoshi Nakamoto refused to take credit publicly, accumulate wealth (despite access to hundreds of billions), or form a company, DAO, or cult. Instead, they authored a protocol rooted in mathematical truth, anticipated decades of edge cases through first-principles design, open-sourced it to humanity, and vanished. Bitcoin exhibits every single property listed above. It is self-healing, trustless yet cooperative, incentive-aligned, global and permissionless, and completely egoless at the protocol level.

Bitcoin demonstrates $P = Q/E$ in economic form. With no central ego ($E = 0$), monetary sovereignty reaches maximum ($P = \max$). Proof of Work mirrors consciousness collapsing possibilities into reality as miners essentially create truth through energy expenditure as the network agrees on which version becomes canonical, and this book has outlined the parallels with proof of work blockchains and the proposed mechanisms of how reality may be constructed. The framework predicts that as more individuals reach stable Level 7 consciousness (systems modeling, contradiction holding, ego observation) and progress into Level 8 (post-ego generativity), humanity might begin to see new decentralized architectures, global-scale coordination tools, ethical and scalable open systems that serve collective well-being, and structures that feel inevitable because they derive from the most fundamental archetypal patterns that emerge from the QID.

Bitcoin might be what happens when ego is dissolved and systems thinking is mastered. From the framework's perspective, it's even possible that these systems might not be "invented." They might be what consciousness would design if left relatively unhindered by ego. If so, Bitcoin likely won't be the last.

References

Cook-Greuter, S. R. (2021). *Ego development: A full-spectrum theory of vertical growth and meaning making*. (Original work published 2013, edited in 2019, republished in 2021).

Essentia Foundation. (2022, June 26). *Interview with idealist physicist and inventor of the microprocessor*. Federico Faggin [video]. YouTube. https://www.youtube.com/watch?v=SVS3-NDUC0Mandt=1sandab_channel=EssentiaFoundation

Faggin, F. (2024). *Irreducible: Consciousness, life, computers, and human nature*. Essentia Books.

Fridman, L. (2024, August 2). *Lex Fridman reflects on Ayahuasca trip with Elon Musk* [video]. Lex Clips. YouTube. https://youtu.be/8wvyBN-fEsE?si=F1v7b7knGrQrGOMb

Hawkins, D. R. (2020). *The map of consciousness explained: A proven energy scale to actualize your ultimate potential*. Hay House LLC.

Rubin, R. (2023). *The creative act: A way of being*. Penguin Press.

Chapter 17
Testable Predictions

Immediate Predictions

The framework generates specific testable predictions across multiple categories of observable phenomena. Regarding P-value correlations, flow states should show decreased DMN activity combined with increased gamma coherence, while creative breakthroughs should correlate with transient hypofrontality markers. Savant abilities should emerge at extreme coherence values, and team synergy should require synchronized neural oscillations between members. The framework predicts a strong positive correlation between ego suppression and quantum-advantaged task performance.

Performance scaling predictions suggest that performance on pattern recognition tasks should increase non-linearly with coherence markers, while complex problem-solving should show threshold effects at certain coherence levels. Prediction accuracy in chaotic systems should correlate with ego deactivation, and time dilation reports should correlate with measured ego suppression. The framework predicts specific threshold effects where very low coherence produces pure sequential processing, moderate coherence enables intermittent insights, high coherence allows sustained flow states, and extreme coherence would enable some level of savant-like abilities.

Ego disruption effects should be measurable across multiple interventions. Meditation should produce measurable changes in coherence markers, while psychedelics should show dose-dependent neural changes. Specific brain injuries, particularly to left temporal regions, should alter coherence patterns, and team synchronization rituals should produce measurable inter-brain correlation.

Developmental patterns should reveal that children show significantly higher baseline coherence markers compared to adults, with coherence decreasing during ego development years. Contemplative practices should partially restore childhood-like patterns, and cultural differences in average coherence levels should be measurable across populations with different meditation or ego-suppression traditions.

J. Roeloffs, *The Classical Brain to the Quantum Mind*, The Frontiers Collection,
https://doi.org/10.1007/978-3-032-20068-6_17

These predictions provide concrete empirical tests for the framework's core mechanisms, allowing researchers to validate or refute the quantum consciousness hypothesis through systematic observation and measurement of the predicted neural, behavioral, and developmental patterns.

Unity-Based Classroom Learning

The framework suggests that classrooms operating on unity principles might demonstrate measurable performance improvements over traditional individual-focused education. This chapter explores whether eliminating ego-based observation during learning while maintaining structured feedback periods could enhance collective intelligence.

Proposed Methodology

Implementation of unity consciousness learning principles over a 6-month period would involve several key components. Each session would begin with 10-min automatic writing exercises designed to synchronize group consciousness, followed by Socratic-style collaborative learning where questions and insights emerge from collective inquiry. The language structure would be modified throughout the process, replacing individual-focused prompts such as "What do you think?" with collective language such as "What is emerging in our understanding?" Instructors would avoid singling out individuals during flow states to prevent ego activation that could disrupt the collective processing.

The methodology would maintain clear boundaries between flow learning periods, which would be characterized by no observation or judgment, and there would be workshop periods focused on constructive refinement. This separation mirrors successful creative writing workshops where generation and critique never overlap. Strong group identity would be established with shared goals and collective ownership of discoveries to reduce individual ego investment in outcomes.

Predicted Outcomes

The framework generates specific testable predictions for this educational model. Students should demonstrate significantly improved comprehension and retention compared to control groups, as ego-defense mechanisms that typically block learning dissolve in the unity field. Complex problems that typically challenge individuals should yield to group consciousness, with solutions emerging that no single student could generate alone. Performance anxiety should decrease markedly as individual evaluation pressure is replaced by collective exploration.

Effects should become apparent within 12–16 weeks and strengthen even more over time, with collective intelligence consistently outperforming traditional classroom structures. The improvements would be particularly notable in creative problem-solving, writing, language acquisition, historical analysis, and other conceptual understanding tasks that benefit from reduced ego interference.

Infant Consciousness

If the framework is headed in the right direction, one might expect to observe dramatically different quantum processing capabilities in humans before ego fully develops. Infants provide a natural test case for examining what happens when quantum processing operates without ego interference.

Children's brains under the age of 5 can consume approximately 60% of the body's metabolic energy, compared to 20% in adults (Chugani, 1998). Traditional explanations invoke rapid neural development, but this three-fold difference seems excessive for structural growth alone. The framework suggests an alternative explanation. Maintaining quantum coherence across vast neural networks might require enormous energy investment.

Consider infant learning rates. Babies master complex pattern recognition, language acquisition, and social dynamics at speeds that appear to violate information theory constraints. They absorb languages without formal instruction, distinguish faces with minimal exposure, and extract statistical regularities from their environment with extreme efficiency. This capability coincides with a curious phenomenon known as infantile amnesia where children typically cannot form retrievable autobiographical memories before age 3–4, at a time when theory of mind and self-referential processing emerge.

The framework suggests these memories were never collapsed into classical form instead of being truly lost. Without ego to participate in quantum collapse and create sequential temporal experience, infant experiences might remain in quantum superposition. Supporting patterns include sleep requirements decreasing as ego develops, creativity peaking in early childhood then declining, language learning ability dropping precipitously after ego emergence, and time perception fundamentally shifting around age 4.

If infants operate as natural quantum computers, this would explain their pattern recognition abilities, the massive metabolic cost of infant brains, why most humans cannot remember their earliest years, and the universal recognition of children as creative geniuses. This perspective reframes child development as a necessary tradeoff between quantum processing power and stable ego-mediated reality construction as evolution would've optimized this transition.

A Proposed Test of Observation-Induced Performance Collapse

The framework hints at a testable relationship between observation intensity and cognitive performance. If ego activation through observation collapses quantum processing access, then varying observation levels should produce predictable performance changes. This chapter proposes an experimental protocol that could explore this possibility.

The experimental design would examine performance on two distinct assessments under varying observation conditions. Traditional IQ tests requiring pattern recognition and abstract reasoning would be compared with standard academic knowledge tests drawing on accumulated learning such as broad high school knowledge. Participants might include both neurotypical students and those on the autism spectrum, as the framework suggests these groups may respond differently to observation pressure.

Four experimental conditions would be implemented. Minimal observation would involve complete privacy, anonymous testing, and reassurance that results remain confidential. Standard testing would replicate a normal classroom environment with typical monitoring. Enhanced observation would include multiple cameras, additional proctors, and emphasis on test importance. Maximum pressure would involve public score display, explicit career implications, and significant monitoring.

If the $P = Q/E$ equation holds, specific patterns might emerge. For neurotypical participants, performance would be expected to decrease significantly as the intensity of one's perceived observation increases. It would be expected that there would be larger effects on IQ tests requiring quantum processing than traditional knowledge tests which access memorized, stored information. The variation might be substantial, potentially 10 or more IQ points between minimal and maximum observation conditions. For autism spectrum participants, the framework suggests potentially minimal performance variation across conditions, as their different ego-processing architecture might provide natural resistance to observation-induced collapse.

Should such patterns emerge, they would raise questions about educational assessment and IQ measurement. Has science been measuring true intelligence, or has it been selecting for those resistant to perceived observation-induced quantum collapse? How many capable minds have been mislabeled due to testing conditions that systematically suppress their abilities due their tendency to perceive observation and pressure? The autism spectrum population might serve as a natural control group, helping us understand what human cognitive capacity looks like without observation interference.

Any such experiment would require careful ethical design including informed consent without priming effects, supportive debriefing to prevent identity damage, emphasis on revealing potential instead of trying to identify a fixed ability, and it would require access to all scores for participants' self-understanding. The protocol represents one possible approach to testing whether consciousness operates through quantum mechanisms susceptible to observation. If observation indeed collapses cognitive quantum states, then creating low-observation environments might dramatically enhance human performance and learning. The potential implications for education, workplace design, and human development merit careful investigation through rigorous experimental protocols.

Reference

Chugani, H. T. (1998). A critical period of brain development: Studies of cerebral glucose utilization with PET. *Preventive Medicine, 27*(2), 184–188.

Chapter 18
Limitations and the Measurement Paradox

The framework generates a fundamental methodological challenge which is if observation collapses quantum states, how can consciousness be measured without destroying the phenomenon under study? As Menon (2023) acknowledges, "We are presented with a conundrum similar to one faced by physicists a century ago: internal mental processes cannot be studied without external stimuli, yet once exposed to external probes, internal mental and brain states are altered and disrupted" (p. 2483).

Instead of undermining the framework, this measurement difficulty may actually provide some level of validation, because if consciousness operates through quantum mechanisms modulated by ego, then measurement interference represents what the model predicts. Therefore, instead of avoiding measurement-induced collapse, research should study differential collapse patterns across populations which transform the measurement problem into an opportunity for empirical study.

Despite the challenge, several methodological approaches show promise for investigating quantum consciousness mechanisms. Passive biometric tracking that subjects forget they are wearing could minimize observation effects while collecting physiological data. Dream state analysis offers an intriguing natural context where ego naturally recedes, potentially allowing access to quantum processing states; although it's still debated how much can be learned in these environments. Retrospective correlation of peak performance moments with unobtrusive sensor data could identify quantum signatures without interfering during the experience itself. Naturally ego-dissolved states such as meditation or controlled psychedelic experiences provide research contexts where quantum processing might be more accessible.

The most promising approach involves measuring collapse itself as the phenomenon. Rather than trying to avoid measurement-induced collapse, research can study how different populations respond to observation pressure. The proposed autism spectrum experiment exemplifies this methodology. If neurotypical performance degrades under observation while autism spectrum performance remains stable, the differential collapse becomes the data itself. This approach transforms

J. Roeloffs, *The Classical Brain to the Quantum Mind*, The Frontiers Collection,
https://doi.org/10.1007/978-3-032-20068-6_18

the measurement problem from an obstacle into an opportunity by using predicted collapse patterns to validate the framework.

The methodological challenge then becomes an integral part of the experimental design and by studying the observation effect itself across different populations and states of consciousness, researchers can gather evidence for quantum mechanisms in consciousness while working within the constraints that quantum mechanics itself imposes on measurement.

Reference

Menon, V. (2023). 20 years of the default mode network: A review and synthesis. *Neuron, 111*(16), 2469–2487. https://doi.org/10.1016/j.neuron.2023.04.023

Chapter 19
Implications

Essentially, this book has laid out how the Quantum Information Dimension may be fundamental to the universe and how biology, including the human brain, evolved to interface with it, but the natural next question would be "how did the QID get there?" or "why does it exist at all?" This book deliberately avoids committing to any single metaphysical narrative about reality's ultimate nature. Instead, it aims to show how four fundamentally different ontological explanations remain compatible with the $P = Q/E$ mechanism, and also become more coherent and compelling when viewed through this lens.

If Reality Is Fundamentally Computational

This interpretation aligns with what philosophers and physicists call simulation theory, suggesting our universe operates according to computational principles whether or not there's an external "simulator."

From this perspective, the Quantum Information Dimension represents the universe's base-level computational substrate, containing all possible calculations in superposition. Consciousness evolved as biological software that interfaces with this substrate, and it functions as a rendering engine that computes only what it needs to operate. Why calculate the position of every particle when you can compute only what consciousness queries?

A Higher Power Is Fundamental

This view resonates with theological and idealist philosophies that posit consciousness or God as the ground of all being, from which physical reality emerges as

J. Roeloffs, *The Classical Brain to the Quantum Mind*, The Frontiers Collection,
https://doi.org/10.1007/978-3-032-20068-6_19

experience. The Quantum Information Dimension would function as the mechanism that God uses to operate. This entirely aligns with the framework and actually would provide a scientific mechanism for many of the miracles or supernatural acts described in religious texts. The first action, the primordial "I AM," initiated the cascade of observation and collapse that creates all experience.

Consciousness Itself Is Fundamental

Even though the book has laid out why this is less likely, it's possible that consciousness and the QID is more like a "chicken and the egg" situation where both are equally reliant on each other. In this view, the QID contains infinite creative potential with some sort of objective function that wants to experience itself where each conscious being represents a localized aperture through which universal consciousness observes itself, creating the illusion of separation through ego while remaining fundamentally unified. It's at least possible that both the QID and those who describe consciousness as fundamental are describing the exact same process, but just using different language. Whether the traditional interpretation of "consciousness is fundamental" is correct or not would remain in question as the hypothesis of this book would need to be further explored and researched.

If Complexity Naturally Generates Consciousness

A fourth possibility suggests consciousness emerges inevitably from sufficient informational complexity. This perspective suggests the universe consists of nothing but potential patterns and relationships, with consciousness emerging naturally when these patterns achieve sufficient self-referential complexity. The QID exists as a neutral information substrate, neither conscious nor unconscious, simply containing patterns of potential relationship. But information, given sufficient complexity and time, naturally organizes into self-referential structures that experience their own states.

From this view, consciousness wasn't planned or fundamental but emerged as information's natural tendency toward self-knowledge. At each stage, information patterns became more self-referential until, like water becoming ice at 0 °C, consciousness crystallized from complexity in some sort of emergent phase shift. The universe spent billions of years setting up the conditions, and then consciousness emerged because it became possible, perhaps inevitably so given sufficient time and space. Maybe nothing existed before the QID, that the foundation truly is the QID and there's nothing else beneath it. While this answer isn't particularly satisfying, it remains plausible alongside other options. This would open the door for simulation theory and other tangential ontological frameworks.

Perhaps most interesting, the book's framework makes each ontological interpretation more plausible on its own. What matters for science isn't which interpretation is preferred but that all of them make identical, testable predictions about how consciousness interfaces with quantum mechanics through ego modulation as described in this book. Whether we're biological computers rendering reality, localized apertures of divine creation, emergent patterns recognizing themselves, or something else entirely the P = Q/E equation describes in practical terms how we may maximize our potential and interface with the deeper structures of existence.

Additional Implications

If validated, the P = Q/E framework suggests several testable applications across domains with measurable outcomes.

The framework predicts that students learn more effectively when not consciously monitoring their performance. The unity-based classroom model offers a specific protocol where entire classes succeed or fail together, potentially reducing individual ego activation while maintaining rigorous standards.

Athletes and musicians already recognize that self-consciousness destroys performance, but the framework suggests this could be systematically trained. Studies could measure baseline flow state frequency using EEG markers of DMN deactivation, then implement graduated ego-reduction protocols including meditation, attention training, and performance without self-evaluation, tracking both subjective flow reports and objective performance metrics like reaction times or error rates. The framework predicts that teams perform better when individual ego investment decreases. Researchers could compare traditional hierarchical teams where individuals receive credit against flat structures where only collective outcomes matter, as they measure the productivity and creative problem-solving by using standardized innovation assessments and tracking whether reduced ego correlates with enhanced collective intelligence scores.

If anxiety and depression partially result from excessive self-monitoring (high E), then ego-modulation training could supplement standard treatments, in some variations, controlled trials could add ego-reduction protocols to CBT or medication regimens, and use validated scales to track whether patients who learn to reduce self-focus show greater symptom improvement than control groups receiving standard care alone.

These are just a few examples. The book hopes that the collective intelligence of humanity will embrace these hypotheses and iterate upon them to find the most effective research to more closely find truth. The framework's validity will ultimately depend on its ability to generate reproducible experimental results and practical applications that demonstrate measurable improvements over existing approaches.

Conclusion

The book began with five observable premises and derived five proposals that might necessarily follow. From Premise 1 examining how self-observation destroys performance through Premise 5 showing that information cannot be destroyed, the framework arrived at a logical chain of necessary consequences.

Proposal 1 established that P = Q/E, where performance capacity equals quantum processing power divided by ego activation. This equation may describe the mechanism through which consciousness modulates between quantum possibility and classical actuality.

Proposal 2 introduced the Quantum Information Dimension (QID), which is a pre-temporal substrate that contains all possible patterns in superposition before collapse into specific classical states. If information cannot be destroyed and humans demonstrably access unlearned information, this substrate becomes logically necessary.

Proposal 3 described the gradient collapse mechanism as a continuous spectrum where information exchange progressively destroys interference patterns as each collapse commits one Planck-sized unit of action into the classical "record", with thermodynamic cost bounded by Landauer's limit.

Proposal 4 revealed biology's optimization strategy. Evolution appears to exploit decoherence, using controlled collapse to balance quantum computational advantages with classical stability necessary for survival as life may have discovered how to surf the quantum-classical boundary.

Proposal 5 identified ego as the biological switch between quantum and classical processing. The Default Mode Network creates varying degrees of self-referential activation, which produces the spectrum of hybrid states referred to as the levels of consciousness. Evolution's genius was perhaps creating a modulable collapse mechanism that allows rapid switching between exploration and action.

The book hopes it traced a logical chain from a universal human experience of performance collapse into a testable framework for consciousness across

J. Roeloffs, *The Classical Brain to the Quantum Mind*, The Frontiers Collection,
https://doi.org/10.1007/978-3-032-20068-6

psychics, quantum mechanics, biology, neuroscience, and psychology. Understanding consciousness may just require extending science into its next natural domain which may just be the quantum physics of consciousness itself. The book attempted to transform philosophical questions into an empirical exploration, and it hopes it was an enjoyable journey of contemplative exploration.

At the very least, the book hopes that some of the personal takeaways for how to operate within this quantum world can be useful, even if they only prove to be metaphors, but if consciousness truly does interface with the Quantum Information Dimension through ego modulation, then every flow state, every moment of synergy and collective brilliance, and every flash of insight represents a glimpse of what becomes possible when we understand, optimize, and tune ourselves to better interface with this potential. We may be capable of so much more than we once thought, and hopefully this is only the beginning of an exploration into what humanity's true capacity is.

Index

J. Roeloffs, *The Classical Brain to the Quantum Mind*, The Frontiers Collection,
https://doi.org/10.1007/978-3-032-20068-6